Frontiers in Biotransformation
Vol. 4

Frontiers in Biotransformation Volume 4

Microbial and Plant Cytochromes P-450: Biochemical Characteristics, Genetic Engineering and Practical Implications

Edited by
Klaus Ruckpaul and Horst Rein

Taylor & Francis
London, New York and Philadelphia 1991

UK Taylor & Francis Ltd., 4 John St., London WC 1N 2ET

USA Taylor & Francis Inc., 1900 Frost Road. Suite 101, Bristol, PA 19007

British Library Cataloguing in Publication Data

Microbial and Plant Cytochromes P-450
1. Organisms Cytochrome P-450
I. Ruckpaul, Klaus II. Rein, Horst III. Series
574.19258

ISBN 0-7484-0029-X

Library of Congress Cataloging-in-Publication Data is available

Printed in Germany

Contents

Introduction

H. Rein and K. Ruckpaul

Volume 4 is focussed on microbial and plant cytochromes P-450. Disregarding the microbial cytochrome P-450_{cam}, the **mammalian** cytochromes P-450 have long been and still are the main subject of research in the cytochrome P-450 field, most likely due to their medical implications in biotransformation of endogenous substrates and drugs and in carcinogenesis as well. It was molecular biology and genetic engineering which opened up a new research area with exciting possibilities for application in biotechnology, medicine, pharmaceutical industries and ecotoxicology, thus stimulating scientific interest in the study of cytochromes P-450 of **microbial** and **vegetable** origin.

This process has just begun to develop but first results have become obvious such as cytochrome P-450 catalyzed single cell protein production by yeasts, use of highly specific cytochrome P-450 inhibitors as antimycotics, usage of transformed microbial cells for bioconversional processes. Within the next decade remarkable progress is to be expected in that very field. Moreover, genetic engineering techniques will provide appropriate means to understand deeply the structure-function relationship, molecular recognition processes and the molecular basis of structural dynamics and stability.

Since the first communication on the occurrence of cytochrome P-450 in bacteria — 10 years after its detection in mammalians — our knowledge about microbial cytochrome P-450 has grown enormously. Indeed the essential processes of the reaction mechanism, electron transfer and oxygen activation of cytochrome P-450 catalyzed reactions but also the regulation of the hydroxylase gene have become obvious from the microbial cytochrome P-450_{cam} from *Pseudomonas putida*. Particularly the elegant studies by the Gunsalus group provided results which made this bacterial cytochrome P-450 an excellent model for all other cytochromes P-450. The soluble state of this monooxygenase was advantageous for these investigations which culminated in the elaboration of its three-dimensional structure in 1985, up to now this being the first known structure of a cytochrome P-450. The results presented in the contribution of Martinez, Ropp, Sligar and Gunsalus prove the value of site-directed mutagenesis of cytochrome P-450_{cam} and its redox partners as a valuable tool in the exploration of the various mechanisms involved in camphor hydroxylation by cytochrome P-450_{cam} and thus reflect the remarkably high level in the understanding of biochemical and biophysical properties of this monooxygenase.

This volume commences with a substantial overview on the occurrence of cytochrome P-450 in bacterial species, considering historical aspects and especially describes the properties, distribution and diversity of bacterial cytochromes P-450. Remarkable is the very different level of our knowledge concerning several bacterial cytochromes P-450 such as specific catalytic activities and spectral evidence only, on the one hand, and detailed data on molecular structure and function of other species, on the other. At length the authors of the first chapter, ASPERGER and KLEBER, describe the hydrocarbon inducible cytochrome P-450 from *Acinetobacter calcoaceticus*. This microorganism has been intensively studied in the past as a model for the assimilation of long chain n-alkanes by bacteria.

In the meantime it is well known that in alkane utilizing yeast cytochrome P-450 catalyzes the first step of long chain n-alkane catabolism. The molecular organization of the alkane-hydroxylating cytochrome P-450 system is similar to the cytochrome P-450 systems of mammalian cell endoplasmic reticulum. However, the catalytic activity of yeast cytochrome P-450 exceeds those of mammalian cytochromes P-450 by 2–3 orders of magnitude. Up to now the reasons for this large difference are not known. But detailed knowledge of the alkane hydroxylating cytochrome P-450 as it is summarized in chapter 3 by MÜLLER, SCHUNCK and KÄRGEL could be useful for solving this question, which is important for the use of cytochromes P-450 in biotechnology.

Although a lot of papers have been presented concerning the molecular mechanism of the cytochrome P-450 action, details of the catalytic mechanism are still not completely understood. More than one oxygenation step is obviously necessary for the C-C cleavage as already shown for the side chain cleavage of cholesterol by cytochrome $P\text{-}450_{scc}$. Also, the 14α-demethylation of lanosterol, being an obligatory step in the biosynthesis of ergosterol — the principal sterol of yeast and fungi, is connected with three successive monooxygenations without release of any intermediate. In chapter 4, YOSHIDA and AOYAMA deal with details of the molecular mechanism of cytochrome $P\text{-}450_{14DM}$, which was isolated and purified from *Saccharomyces cerevisiae* and identified as lanosterol 14α-demethylase. Moreover, by reconstitution experiments the authors could evidence that the fungicidal activity of azole antifungal agents originates from its interaction with cytochrome $P\text{-}450_{14DM}$. This form is the only cytochrome P-450 which is widely distributed in eukaryotes with essentially the same function, i. e. 14α-demethylation of lanosterol and cholesterol. Therefore, knowledge about the properties of this monooxygenase is not only of practical interest because of its target function for antifungal agents but it may be an important object for studying the evolution of cytochrome P-450, too.

The review by HUDNIK-PLEVNIK and BRESKVAR contains interesting data about cytochrome P-450 in *Rhizopus nigricans*, one of the biotechnologically important microorganisms. Because the regio- and stereospecific introduction of hydroxyl groups into therapeutically important steroid molecules by chemical methods is complex and difficult, the use of microorganisms as chemical tools represents a great economic improvement because it substantially reduces the cost of production, e.g. the synthesis of glucocorticoids from naturally occurring sterols is reduced from 37 to only 11 steps. The involvement of cytochrome

P-450 in hydroxylations performed by the fungus *Rhizopus nigricans* was discovered in 1977. Although cytochrome P-450 with steroid 11α activity localized in the microsomal membranes of *Rhizopus nigricans* is a very labile enzymatic system, it was possible to isolate and reconstitute the system. The review of HUDNIK-PLEVNIK and BRESKVAR recounts the purification story, which represents the first successful of a steroid hydroxylating cytochrome P-450 from a filamentous fungus.

From the review of YABUSAKI and OHKAWA we learn the usefulness of the expression of cytochrome P-450 in heterologous cells including bacterial, yeast and mammalian cells for identification, characterization and functional analysis of soluble and membrane bound cytochrome P-450. Modern genetic engineering technology made it possible to produce cytochrome P-450 in large quantities in *Escherichia coli* and yeast and to improve the enzymes in their reactivity and stability for practical purposes. Remarkably, from the expression of truncated cytochrome P-450c in yeast it emerges that the amino terminal hydrophic region of this isozyme is not only important for the correct orientation for the newly synthesized protein into endoplasmic membranes but also for sufficient enzyme activity. Moreover, a fused cytochrome P-450/reductase enzyme consisting of the whole cytochrome P-450c and the soluble reductase moities (lacking the amino-terminal anchor region) is functional in yeast cells. The highest activity in the conversion of acetanilide to acetaminophene was observed for such recombinant yeast cells carrying both cytochrome P-450c and overproduced but not fused yeast reductase. All these results are the basis for the use of cytochrome P-450 for practical bioconversion processes.

Obviously the isolation, purification and characterization of cytochrome P-450 of higher plants is much more difficult than those of animals. This results from different reasons such as the very low titer of cytochrome P-450 in most plants, the high concentrations of proteases and lipases in plants destroying the cytochrome P-450 enzyme system and, moreover, the difficulties in isolation resulting from the rigidity of the plant cell wall, on the one hand, and the instability of the enzyme, on the other. Despite these facts, during the last few years the study of plant cytochrome P-450 has been expanded rapidly. In the last chapter DURST presents an overview about recent data of biochemistry and physiology of plant cytochromes P-450 demonstrating the physiological significance of cytochrome P-450 in higher plants because of its catalytical activity in the biosynthesis of plant hormones and of numerous secondary plant products. Moreover, it is recognized that genes coding for certain cytochromes P-450 are factors of resistance to pesticides, e. g. herbicide detoxification. Therefore interest exists from the practical point of view to study plant cytochrome P-450 at the molecular level. An important potential of plant cytochrome P-450 for biotechnology results from regio- and stereospecific hydroxylation of secondary plant metabolites which may display biological and pharmacological activity.

Looking ahead in the field of microbial and plant cytochromes P-450 we can expect in the near future the detection of further cytochromes P-450 with specific properties in bacteria, yeast and fungi and higher plants, too. The resulting multifariousness of different substrate specific catalytic action and regio- and sterospecific hydroxylations opens up high potentialities for cytochrome P-450 application in the biotechnological industry. Using the exciting developments in genetics and protein engineering including commercial activities such as the bank of expression vectors with unique cytochromes P-450 as well as stable expression systems we are not far from constructing cytochromes P-450 with improved properties for application in different fields.

Chapter 1

Distribution and Diversity of Bacterial Cytochromes P-450

O. ASPERGER and H.-P. KLEBER

1. Introduction

Bacteria are able to degrade or to convert numerous natural and xenobiotic compounds by a variety of oxidative reactions. These reactions, mainly used for the biodegradation of waste products and other pollutants or for the production of biomass, are of increasing interest with respect to the production of pharmaceuticals, diagnostics or other chemicals by microbial or enzymic biotransformation. Many types of bacteria-mediated biotransformations are also recovered in the broad spectrum of reactions catalyzed by cytochromes P-450 of higher organisms, above all by those of liver microsomes. This poses the question, of whether P-450-dependent enzymes may be responsible for various of the microbial monooxygenatic activities. Rich fundamental knowledge of the structure and function of P-450-dependent monooxygenases and the facilitated analytical accessibility, which bases on the unique Soret maximum of their CO compounds, provide worthwhile advantages for the study and the application of this kind of enzymes and make the exploration of this question highly attractive.

Microbial P-450s have been described in the past in several reviews (Gunsalus et al., 1974; Yoshida, 1978; Müller et al., 1984; Käppeli 1986; Sligar and Murray, 1986; Unger et al., 1986), but most of them had been focused mainly on the P-450 of *Pseudomonas* and yeasts. The reader is kindly referred to these reviews or to the corresponding chapters of this issue.

It is the intention of the present review to give a comprehensive overview of the P-450s detected in various other bacterial strains. For some of them only their occurrence is known, whereas for others, such as for the fatty acid monooxygenase of *Bacillus megaterium*, a lot of data on structure, function and regulation had been accumulated in recent years. For the sake of completeness, also the longer known bacterial P-450s for which new information has not been reported recently, have been included in this review.

In the first part of this article, the distribution of P-450 among diverse bacterial taxons is discussed. The second part describes the individual P-450 systems and in the third part main features of the various bacterial P-450 systems are summarized. For fundamental information on P-450 proteins, volume I of this series may be consulted.

2. Overview on the occurrence of cytochrome P-450 in bacterial species

It took about ten years from the detection of P-450 in mammalian liver microsomes (Garfinkel, 1958; Klingenberg, 1958) to the first communication on the occurrence of P-450 in bacteria (Appleby, 1967). About the time Appleby, (1967), more occasionally, detected P-450 in *Bradyrhizobium* bacteroids from

able 1. Occurrence of cytochromes P-450 in bacteria

Bacterium	Putative or established predominating function	Reference
Pseudomonadaceae		
Pseudomonas putida ATCC 17453	camphor 5-exo-hydroxylase	KATAGIRI et al., 1968
Pseudomonas putida PpG 777	linalool 8-hydroxylase	BHATTACHARYYA et al., 1984
Pseudomonas aeruginosa PaG 158	p-cymene 7-hydroxylase	GUNSALUS et al., 1985
Rhizobiaceae		
Bradyrhizobium japonicum (P-450a; P-450b; P-450b_1; P-450c)	unknown	APPLEBY, 1967
Bradyrhizobium sp. (*Lupinus*)	unknown	KRETOVICH et al., 1972
Neisseriaceae		
Moraxella sp. GU2	guaiacol demethylation	DARDAS et al., 1985
Acinetobacter calcoaceticus	alkane oxidation	MÜLLER et al., 1978
strains: EB 102; 103; 104; 113; 10C; 10D; 11	alkane oxidation	ASPERGER et al., 1981
strain: A6E	alkane oxidation	WYNDHAM, 1987
Vibrionaceae		
Vibrio fisheri MJ-1	alkanal oxidation	ISMAILOV et al., 1979
Vibrio harveyi	alkanal oxidation	BARANOVA et al., 1982
Photobacterium phosphoreum	alkanal oxidation	BARANOVA et al., 1982
Different gram-positive bacteria		
Bacillus megaterium ATCC 13368	steroid 15β-hydroxylase	BERG et al., 1975
Bacillus megaterium ATCC 14581 (P-450_{BM-1}; P-450_{BM-2}; P-450_{BM-3})	fatty acid monooxygenase	HARE and FULCO, 1975
Bacillus megaterium strains: ATCC 13368; 21181; 19213; 25833; 33164; 33165; 33166; 33167; 33168; 33169	fatty acid monooxygenase	FULCO and RUETTINGER, 1987
Corynebacterium sp.	cholesterol oxidation	YAMAMOTO et al., 1974
Nocardioforms		
Nocardia sp. NH1	p-alkoxybenzoate dealkylation	CARTWRIGHT et al., 1971
Rhodococcus rhodochrous	alkane oxidation	CARDINI and JURTSHUK, 1968
Saccharopolyspora erythrea (P-450_{I}; P-450_{II})	macrolide hydroxylation	CORCORAN and VYGANTAS, 1982

Table 1 (continued).

Bacterium	Putative or established predominating function	Reference
Streptomycetaceae		
Streptomyces setonii	veratrole dealkylation	SUTHERLAND, 1986
Streptomyces griseus strains: ATCC 13273; 10137 and NRRL B8090	unknown	SARIASLANI and KUNZ, 1986
Streptomyces griseolus ($P\text{-}450_{con}$; $P\text{-}450_{SU1}$; $P\text{-}450_{SU2}$)	sulfonylurea transformation	ROMESSER and O'KEEFE, 1986

soybean nodules during the spectral investigation of their cytochrome composition, P-450 was identified as the terminal oxidase of the camphor 5-exo-hydroxylase in consequence of a systematic research by the group of GUNSALUS (KATAGIRI et al., 1968) on the biodegradation of camphor by *Pseudomonas putida*. These early discoveries were followed by a search for further P-450s presumed to be involved in degradative metabolic pathways. Consequently, shortly after P-450 was detected in alkane-oxidizing *Rhodococcus rhodochrous* and in an alkylaryl ether-degrading *Nocardia* species (Tab. 1). But further P-450s, apparently involved in initial oxidations of hydrophobic compounds making them more accessible to biodegradation, have been reported only recently (Tab. 1). Compounds initially attacked by P-450-dependent hydroxylation or dealkylation comprise terpenes, cholesterol, alkanes and alkylaryl ethers. Other reactions brought about by bacterial P-450s are the selective hydroxylation of steroids, subterminal hydroxylation of fatty acids, hydroxylation of macrolides, and de-esterification, dealkylation or hydroxylation of ring substituents in sulfonylurea herbicides. It remains to mention the quite hypothetical function of P-450 as an alkanal-oxidizing component of bacterial luciferase and the bacterial P-450s of, until now, unknown function that occur in *Bradyrhizobia* and *Streptomyces griseus* or those present as multiple forms in *Bacillus megaterium*, *Streptomyces griseolus* or *Saccharopolyspora erythrea*.

But, considering the P-450 dependence of an oxygenatic biotransformation it has to be kept in mind that reactions resembling with respect to the type and substrate P-450-dependent conversions are brought about also by monooxygenases having other kinds of catalytic centers. Thus, the only examples of the occurrence of P-450 in alkane-assimilating bacteria are the P-450s of *Rhodococcus rhodochrous* and *Acinetobacter* (Tab. 1), whereas other alkane-assimilating bacteria such as *Pseudomonas oleovorans* (GUNSALUS et al., 1974), several strains of *Acinetobacter* (ASPERGER et al., 1978; 1981; ENSLEY and

FINNERTY, 1980) or the methane-oxidizing bacteria (ANTHONY, 1986) definitely do not contain P-450. Otherwise, only P-450-dependent n-alkane monooxygenases have been found in yeasts (MÜLLER et al., 1984; KÄPPELI, 1986). Another example of the existence of different types of enzymes is given by the demethylation of p-methoxybenzoate for which P-450 dependence is postulated in the case of *Nocardia* sp. NH1 (Tab. 1) but an iron-sulfur protein has been shown to function as monooxygenase in a strain of *Pseudomonas putida* (BERNHARDT et al., 1978). Furthermore, it has been described that unique non-P-450 monooxygenases are also versatile enzymes catalyzing several different, reactions and transforming rather different substrates. Thus, methane monooxygenases do not only catalyze the hydroxylation of methane, but also that of higher n-alkanes, or aromatic compounds and furthermore the epoxidation of alkenes, the dealkylation of ether, dehalogenations and N-oxidations (STIRLING et al., 1979). Epoxidation of alkenes, O- and S-dealkylations, sulfoxidation and the oxidation of alcohols to aldehydes have also been reported for the alkane monooxygenase of *Pseudomonas oleovorans* (MAY and KATOPODIS, 1986).

Therefore, to characterize a biotransformation as a P-450-dependent reaction, proof for P-450 by means of CO and substrate binding difference spectra, inhibition of enzymic activity by CO, photoactivation spectra and above all the reconstitution of activity with the highly purified P-450 should be provided as evidence. With regard to these aspects, only those bacteria are reviewed in this article for which P-450 has been demonstrated at least by the typical absorbance maximum at 450 nm in optical spectra of its CO complex.

As shown in Table 1, P-450 has been detected in 38 bacterial strains, distributed among 16 species that comprise 12 genera. Compared to the myriads of bacterial species and strains, this information is not sufficient to answer the question as to the ubiquity of P-450 among bacteria. In order to facilitate the elucidation of this problem, the P-450s from a taxonomical point of view are listed in Table 1. The bacteria are designated and listed according to the recommendations given in the 9th edition of BERGEY's "Manual of Systematic Bacteriology" (HOLT, 1984). The new denominations, deviating from those in the original publications or in former reviews, are *Bradyrhizobium* instead of *Rhizobium*, *Vibrio fisheri* instead of *Photobacterium fisheri*, *Vibrio harveyi* instead of *Beneckea harveyi*, *Rhodococcus rhodochrous* instead of *Corynebacterium* 7ElC and *Saccharopolyspora erythrea* instead of *Streptomyces erythreus*.

Since investigations into the occurrence of P-450 were restricted in the past to eubacteria only, its presence in archaebacteria or other bacteria can not be negated. A more widespread distribution of P-450 than indicated by the hitherto known examples may be supposed considering the fact that P-450 has been detected in quite diverse genera of gramnegative as well as grampositive bacteria and also from the more stochastic than systematic character of the distribution pattern presently known. More data are required to make it possible to establish whether the occurrence of P-450 correlates with taxonomical units. Hitherto, such relationships are indicated only in the case of some gram-positive species. Thus, a barbiturate-inducible P-450 has been detected in 11 of 12 tested strains of *Bacillus megaterium* (FULCO and RUETTINGER, 1987), and yet the 12th strain contained P-450 (however not induced by barbiturates). Likewise, three strains of *Streptomyces griseus* tested by SARIASLANI and KUNZ (1986) all produced a soybean flour-inducible P-450. Furthermore, our own investigations (unpublished) with various strains of *Rhodococcus rhodochrous*, not listed in Table 1, revealed the occurrence of n-hexane-inducible P-450 in five of seven strains. In opposite to this, the plasmid encodation of P-450 in *Pseudomonas* (RHEINWALD et al., 1973), and the fact that of numerous strains of *Acinetobacter* (ASPERGER et al., 1981) only a few were P-450-positive, argue against a speciescorrelated distribution of P-450 in these bacterial taxons.

Thus, the delineation of general rules for the occurrence or detection of bacterial P-450s appears not to be possible at all. Such a rule that predicts the obligate occurrence of P-450 in a large group of microorganisms seems to have been found for a P-450 catalyzing the lanosterol 14α-demethylation in yeasts, an essential step in the sterol biosynthesis (cf. also chapter 4 in this book). This P-450 is regulated by growth conditions such as oxygen tension but obviously not by exogenous inducers. The involvement of a bacterial P-450 in a biosynthetic pathway has hitherto been suggested only for the formation of erythromycin by *Saccharopolyspora erythrea* (cf. 3.9.) which has, however, to be regarded as a more specialized metabolic reaction than an essential one. Whether other biosynthetic reactions involve P-450 and can be found in a larger group of bacteria must be considered. The reports on obviously multiple and, some of them, constitutive forms of P-450 in the grampositive bacteria *Bacillus megaterium*, *Saccharopolyspora erythrea* and *Streptomyces griseolus* may point in this direction. Nevertheless, independently of the occurrence of constitutive, biosynthetic P-450s, there exists a multitude of oxygenatic biotransformations brought about by bacteria for which the enzymic mechanism still remains to be explored. Although, as mentioned above, all these reactions have not to be a priori P-450-dependent, it appears quite probable that still more bacterial P-450s will be detected in the future.

3. Description of individual cytochrome P-450 systems

3.1. Cytochrome P-450 in nitrogen-fixing bacteria of the genus *Bradyrhizobium*

As already mentioned, the detection of P-450 in bacteroids of *Bradyrhizobium japonicum* from nodules of soybean roots was the first description of a bacterial P-450. Occurrence of P-450 in further species of *Bradyrhizobium* was shown by the spectral demonstration of P-450 in bacteroids of *Bradyrhizobium* sp. (*Lupinus*) harvested from nodules of *Lupinus luteus* roots (Kretovich et al. 1972). The maximum of the specific content amounted to about 50 pmol/mg protein of cell-free extracts (Kretovich et al., 1974; Appleby, 1978). The P-450s of *Br. japonicum* bacteroids have been partially purified from the soluble protein fraction by means of ammonium sulfate precipitation (30–50% saturation), subsequent gel filtration on Sephadex G-75 and Sephacryl S-200, followed by chromatography on DEAE-cellulose (Appleby and Daniel, 1973). The anion exchange chromatography resolved P-450 reproducibly into three subspecies, P-450a, b and c, the molar ratio of which was about 1:1.5:3. Since, after addition of ligands and re-chromatography an interconversion of these forms with respect to their chromatographic behaviour was not achieved the occurrence of multiple forms of *Bradyrhizobium* P-450 was suggested (Appleby and Daniel, 1973). Similar P-450 heterogeneity was observed for *Bradyrhizobium* sp. (*Lupinus*) (Inozentsova et al., 1978).

Highly purified P-450c from *Br. japonicum* bacteroids with a specific content of 17.8 nmol P-450/mg protein has been obtained by means of chromatography either on hydroxylapatite (Appleby, 1978) or on phenobarbital-alkyl Sepharose 4B (Dus et al., 1976). Spectral properties, pI and M_r (cf. Tab. 12) resemble those of other P-450s. With respect to the amino acid composition, similarities with P-450_{LM2} from rabbit liver and P-450_{cam} have been claimed (Dus et al., 1976). Immunological cross-reactivities of all three P-450 subspecies have been shown by means of radioimmunoassay with anti-P-450_{cam} and anti-P-450_{LM2} antibodies (Dus et al., 1976). Differences between the subspecies in the optical absorption spectra of the oxidized forms (cf. Tab. 12), the ESR spectra (Peisach et al., 1972), and the ligand binding spectra revealed distinct spin-state equilibria prevailing in the individual forms (Appleby and Daniel, 1973). P-450a is a pure low-spin hemoprotein. In P-450b mostly the high-spin state predominates and P-450c is a mixed-spin form with a predominance of the high-spin state. A P-450 that is induced in free-living cells cultured under anaerobic conditions in the presence of nitrate as the terminal electron acceptor and which is eluted from DEAE-cellulose at the same ionic strength as the bacteroid P-450b was claimed to be a further subspecies of rhizobial P-450s. It was denominated P-450_{b1} since it was compared to P-450b almost pure high spin, and exhibited much lower binding affinities for alkyl amines (Appleby and Daniel, 1973), but a higher affinity to CO (Daniel and Appleby, 1972).

Generally, ligand binding to P-450 of both *Bradyrhizobium* species has been reported for a variety of compounds (Tab. 2). Type II spectra reflecting by the position of maxima and minima the original spin-state of the P-450 subspecies have been obtained with such compounds as alkyl amines, aniline and metyrapone. The observation that a variety of different substances induced spectral changes similar to type I spectral changes (Tab. 2) which are usually inter-

Table 2. Absorbance maxima and minima in difference spectra of cytochromes P-450 from *Bradyrhizobium* species induced by various ligands[1]

Ligand	P-450s from *Br. japonicum*[2]				P-450 from *Br. sp. (Lupinus)*[3]
	a	b	b_1	c	
Type I spectra					
Phenobarbital	[4]		[4]	[5] (0.145)	
Hexobarbital					384/430 (1.1)
Dimethylaniline					388/422 (5.9)
D(+)-Camphor					385/425 (1.6)
SKF 525-A					382/428 (0.31)
Type II spectra					
Octylamine	430/410 (4.0)	425/393 (0.43)	—	431/395—412 (7.0)	
Decylamine	432/412 (0.58)	426/393 (0.10)	422/390 (5.3)	432/412 (4.0)	
Aniline					427/402 (18.0)
Metyrapone					429/400 (3.7)

[1] Values given in brackets represent the spectroscopically determined dissociation constants (mM).
[2] According to Appleby and Daniel, 1973.
[3] According to Kretovich et al., 1974.
[4] No effects.
[5] Maxima and minima have not been cited in the literature.

preted as the binding of an enzyme substrate, indicates either a very low specificity of the substrate binding site of rhizobial P-450s or, what seems to be more probable, reflects unspecific distortions of the spectra of the P-450 preparations used in these studies. The high apparent K_s values (Kretovich et al., 1974) may support the latter interpretation.

Demonstration of P-450 ligand binding, together with the inhibition of nitrogenase activities as well as respiratory activities of bacteroids, have been used as arguments for the functional importance of rhizobial P-450 in the process of nitrogen fixation (Kretovich et al., 1974; Appleby et al. 1975). This hypothesis was supported by the fact that in free-living, aerobically grown cells, which cannot assimilate molecular nitrogen (Appleby, 1969b; Daniel and Appleby, 1972) or in bacteroids of a nitrogen fixation ineffective strain of *Bradyrhizobium* sp. (*Lupinus*) (Matus et al., 1973), only insignificant amounts of P-450 compared to active bacteroids (Appleby, 1969a) have been detected. Since an enzymic activity of rhizobial P-450 could not be discovered, models have been proposed that implicate the function of P-450 as an oxygen carrier or as terminal oxidase at levels of oxygen tension, low enough to avoid the inactivation of the oxygen-sensitive nitrogenase. Oxygen is, however, required for the generation of ATP consumed in the nitrogen-fixing process. At a low oxygen tension the ATP production in bacteroids from *Br. japonicum* was increased in the presence of leghemoglobin. This increase was more sensitive to inhibition by the P-450 ligand phenylimidazole than it was to the non-stimulated ATP formation in absence of leghemoglobin (Appleby et al., 1975). Hence, a more detailed model, with P-450 participating in a leghemoglobin-mediated system of oxidative phosphorylation active at very low oxygen tension, was proposed. In opposite to this, Inozentsova et al. (1979) refuse any effects of leghemoglobin on the inhibition of nitrogenas e activity by P-450 ligands.

Spectral evidence for the formation of oxygen-complexes of P-450a, b and c from *Br. japonicum* has been provided recently (Lambeir et al., 1985) by demonstrating oxygen-induced absorbance maxima at 417, 419, and 421 nm for the dithionite-reduced P-450s. Association constants in the range of $5 \cdot 10^5$ to $7 \cdot 10^5$ $M^{-1}s^{-1}$ and decay constants in the range of 1.6 to 4.8 s^{-1} have been determined. Although an enzymic reduction of P-450 by endogenous substrate could be observed with cellfree extracts (Appleby, 1969a), no information on electron transport components for rhizobial P-450s is available.

Taken together, the current data on the rhizobial P-450s are far from being able to give a conclusive answer as to what the function of the first isolated bacterial P-450 may be.

3.2. Induction of cytochrome P-450 in a guaiacol-degrading *Moraxella* species

The *Moraxella* species (strain GU2) in which P-450 has been detected is a bacterium isolated from garden soil on the basis of its ability to use the lignin degradation product guaiacol (Fig. 1) as sole source of carbon (Dardas et al., 1985). P-450 has been detected in high amounts (about 0.12 nmol/mg dry weight) in guaiacol-grown cells and recovered exclusively in the soluble protein fraction of cells disintegrated by means of a X-press. A further purification has not been reported.

Guaiacol-demethylating activity, the supposed function of the P-450, could not be demonstrated directly, neither with cell-free extracts nor with whole cells. This is why indirect evidence for the function of this P-450 in guaiacol oxidation has been elaborated to exclude gratuitous induction of P-450 by guaiacol. Thus, CO, in a mixture of 1:1 with oxygen, inhibited the whole cell oxidation of guaiacol indicating P-450 dependence. The detection of catechol in guaiacol-grown cultures and induction of catechol-1,2-dioxygenase by guaiacol made a demethylation of guaiacol as the first step of degradation via 1,2-catechol highly probable. Catechol oxidation was not inhibited by CO.

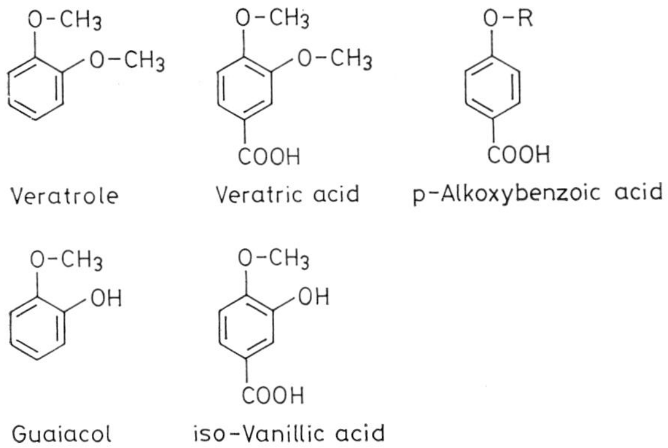

Veratrole   Veratric acid   p-Alkoxybenzoic acid

Guaiacol   iso-Vanillic acid

Fig. 1. Alkylaryl ether and derivatives of them functioning as substrates and inducers of bacterial cytochromes P-450. R = alkyl residues.

Hence, the demethylation should be the P-450-dependent step of guaiacol degradation. Furthermore, guaiacol induced type I difference spectra with cell-free extracts of *Moraxella* sp. GU2 indicating P-450 substrate binding. Similar spectra have been observed with the guaiacol homologs 2-ethoxyphenol, 2-propoxyphenol and 2-butoxyphenol and apparent spectral dissociation constants of 0.2, 4 and 7 μM compared to 0.15 μM for guaiacol have been determined. Metyrapone competed with guaiacol binding and produced reversed type I spectra with minima at 390 and maxima at 420 nm.

P-450 was not detected in cells grown on mineral media with succinate as the sole source of carbon but was besides by guaiacol also induced by 2-ethoxyphenol.

Altogether, these results are highly suggestive that the P-450 of this *Moraxella* strain is an alkoxyphenol dealkylating enzyme which resembles the P-450 found in *Nocardia* species NH1 (cf. 3.8.).

3.3. Hydrocarbon-inducible cytochrome P-450 of *Acinetobacter calcoaceticus*

Acinetobacter has been well studied in the past as a model organism for the assimilation of long-chain n-alkanes by bacteria (KLEBER et al., 1983; SINGER and FINNERTY, 1984). Whereas several strains utilize n-alkanes without the concomitant induction of P-450 (Tab. 3), occurrence of P-450 after growth on n-alkanes could be detected in some other strains (MÜLLER et al., 1978; ASPERGER et al., 1981; WYNDHAM, 1987). More detailed investigations have been carried out with the strain *A. calcoaceticus* EB 104.

Table 3. Occurrence of cytochrome P-450 in strains of the genus *Acinetobacter* grown on or induced by n-alkanes

P-450 non-detected	Ref.	P-450 detected	Ref.
A. calcoaceticus 69-V	a	*A. calcoaceticus* EB 102	c
A. species HO1	b	*A. calcoaceticus* EB 103	c
A. calcoaceticus CCM 2355	c	*A. calcoaceticus* EB 104	c
A. calcoaceticus CCM 5594	c	*A. calcoaceticus* EB 113	c
A. calcoaceticus CCM 5595	c	*A. calcoaceticus* EB 10C	c
A. calcoaceticus EB 106	c	*A. calcoaceticus* EB 10D	c
A. calcoaceticus EB 114	c	*A. calcoaceticus* EB 11	c
A. lwoffii CCM 5572	c	*A. calcoaceticus* A6E	d
A. lwoffii CCM 2376	c		
A. calcoaceticus ATCC 33305	d		
A. species AH60	d		
A. species DON2	d		

a) ASPERGER et al., 1978; b) ENSLEY and FINNERTY, 1980; c) ASPERGER et al., 1981; d) WYNDHAM, 1987.

Regulation

Inducibility of the P-450 from *A. calcoaceticus* Eb 104 was indicated by its absence in cells grown in minimal media with acetate, succinate, malate, arabinose, and ethanol respectively, as carbon source or grown on complex media such as yeast extract and nutrition broth (ASPERGER et al., 1984). None of these carbon sources, inclusively the non-growth substrates glucose and glycerole, inhibited the P-450 induction in the presence of n-hexadecane, indicating lack of catabolite repression. De novo protein synthesis was concluded from the suppression of P-450 induction in the presence of streptomycin, chloramphenicol or rifampicin as well as by the lack of immunoprecipitation of cell-free extracts of uninduced cells by anti-P-450 antibodies (Müller, 1984). The absence of cytochrome a in this bacterium, normally interfering with the spectral determination of P-450, together with the lack of catabolite repression facilitated extended experimental studies on the inducer specificity by cultivation of cells on complex media either in liquid culture or on agar slants.

P-450 of *Acinetobacter* is induced by a wide variety of aliphatic but also aromatic hydrocarbons (Asperger et al., 1985b; Kleber et al., 1985) as well as diverse derivatives of them (Tab. 4). The most effective inducing compound tested was biphenyl. But similar high contents of P-450 of about 0.4 nmol P-450/mg protein of whole cells have been obtained by dosage of n-hexane in an appropriate manner to cultures growing on nutrient broth. Polar substituents appeared to be ineffective with respect to inducing capability, since in no case induction has been observed with carboxylic acids, alcohols or amines, neither aliphatic nor aromatic ones, although these compounds were oxidized by *Acinetobacter*. Another discriminating factor consists in the spatial arrangement of the obligatory hydrophobic structural element necessary for induction. Thus, hydrophobic compounds that do not have a coplanar structural element of sufficient diameter, such as alicyclic compounds or strongly branched aliphatics, were unable to induce the P-450 in *Acinetobacter*. Otherwise, substituents that increase the polarity of a hydrophobic compound only insignificantly, did not or only partially spoil the inducer capacity, which was derived from positive results with olefines, ketones, ethers and haloalkanes (Tab. 4).

Altogether, the induction profile of P-450 in *Acinetobacter* resembles that of the liver-microsomal aryl hydrocarbon hydroxylase, the inducers of which are known as methylcholanthrene-type inducers (Conney, 1966; Ioannides and Parke, 1987). Opposite, barbiturates, such as phenobarbital did not induce the P-450 in *Acinetobacter*. Taking into account results obtained with different isomers of octane (Tab. 4) it may be assumed that a coplanar hydrophobic residue corresponding to the area of a n-propyl residue may be sufficient to bring about P-450 induction if the overall lipophilicity of the compound is retained.

Furthermore, the level of P-450 induction is controlled by the oxygen tension of the medium (Asperger et al., 1986). Decreased oxygen tension, below a level of about 3% of saturation, limiting the rate of growth of *A. calcoaceticus* EB 104 on n-hexadecane, resulted in an approximately 3-fold increase of the specific P-450 content.

Purification, molecular and spectral properties

Starting from hexadecane-grown cells, P-450 has been isolated in a highly purified form (19 nmol P-450/mg protein) with yields of about 18% (Mül-

Table 4. Induction profile for cytochrome P-450 of *Acinetobacter calcoaceticus* EB 104

Compound	P-450	Ref.
n-Alkanes (C_5-C_{20})	+	a
iso-Alkanes and cycloalkanes:		
2-Methylheptane	+	d
3-Methylheptane	+	d
4-Methylheptane	+	d
2,2,4-Trimethylpentane	—	c
Pristane	—	c
Cyclohexane	—	c
Decalin	—	d
Alkenes:		
1-Tetradecene	+	c
1-Decene	+	c
1,5-Hexadien	+	c
Phenylalkanes:		
1-Phenyltridecane	+	c
1-Phenyldodecane	+	c
1-Phenylheptane	+	c
Phenylethane	+	c
Toluene	+	c
1,2-Diphenylethane	+	c
Unsubstituted aromatic compounds:		
Benzene	—	b
Biphenyl	+	b
Indene	+	b
Tetralin	+	d
Naphthalene	+	b
Anthracene	—	b
Phenanthrene	+	b
Heterocyclic compounds:		
Pyridine	—	d
Furan	+	d
Tetrahydrofuran	—	d
Dioxane	—	d
Alkanols:		
1-ols (C_2; C_4; C_5; C_8; C_{12}; C_{16})	—	c
2-Tridecanol	—	d
Alkanoic acids:		
Monoacids (C_2-C_6; C_8; C_{10}; C_{12}; C_{14}; C_{16})	—	c
Diacids (C_{12}; C_{14}; C_{16})	—	c
Alkanones:		
Acetone	—	c
2-Hexanone	+	d
3-Hexanone	+	d
2-Dodecanone	+	c
7-Tridecanone	+	c
3-Hexadecanone	+	c
6-Hexadecanone	+	c
2,15-Hexadecanedione	+	c
Aliphatic ether and thioether:		
Diethylether	+	d
Dipropylether	+	d
Dihexylether	+	c
Diheptylether	+	c
Dioctylether	+	c
Di-iso-propylether	—	d
Diethoxyethane	—	c
Pentaoxapentadecane	—	c
Dioctylsulfide	+	d
Aliphatic amines:		
Hexylamine	—	
Decylamine	—	
Dioctylamine	—	
Halogenated alkanes:		
1-Bromodecane	+	d
1-Bromohexadecane	+	d
1,12-Dibromohexadecane	+	d
1,1,2-Trichloroethane	+	d
Various benzene derivatives:		
Phenol	—	d
Aniline	—	d
Nitrobenzene	—	d
Anisole	+	d
o-Vanilline	—	d
Benzylalcohol	—	d
Benzoic acid	—	d

Table 4 (continued).

Compound	P-450	Ref.	Compound	P-450	Ref.
Various compounds:			Camphor	—	c
			Phenobarbital	—	b
Triolein	—	d	Hexobarbital	—	b
Decylacetate	—	d	Pentobarbital	—	b
Cholesterol	—	d	Isosafrole	—	b

a) ASPERGER et al., 1984; b) ASPERGER et al., 1985b; c) KLEBER et al., 1985; d) ASPERGER, unpublished results; e) Inhibited growth totally, also if applied in the vapour phase.

LER et al., 1989). To increase the stability of P-450 and to recover it from sediments of ultracentrifugation, into which it distributed especially after growth on long-chain n-alkanes, purification was carried out in the presence of Triton X-100 (1 g/l). Extraction of 90,000 × g sediments was followed by chromatography on DEAE-cellulose, gel filtration with Sepharose CL-6B and chromatography on hydroxylapatite from which P-450 was eluted between 10 and 50 mM phosphate. A more convenient purification procedure (ASPERGER, unpublished) started from n-hexane induced cells grown on nutrient broth. From such cells P-450 was nearly exclusively recovered in the soluble protein fraction. By ammonium sulfate precipitation (30 to 50% saturation), gel filtration on Sephadex G-75, chromatography on DEAE-Sepharose and repeated gel filtration on Sephacryl S-300 it was purified to homogeneity. The buffers were supplemented with Tween 80 to increase P-450 stability. P-450 was concentrated finally and freed of Tween 80 by small-scale chromatography on DEAE-Sepharose. Predominant distribution of P-450 into the soluble protein fraction has also been observed with cells grown on minimal agar with n-nonane, applied in the vapor phase as carbon source (EREMINA et al., 1987).

The M_r, pI and the spectral maxima (cf. Tab. 12) were in the range of those of other P-450s. The amino acid composition (cf. Tab. 13) exhibited neither congruity with other bacterial P-450s nor striking differences (ASPERGER et al., 1983; MÜLLER et al., 1989). P-450 of *Acinetobacter* did not cross-react with antibodies against the P-450 of the n-alkane assimilating yeast *Candida maltosa* (KÄRGEL et al., 1985), and an antiserum against P-450 of *Acinetobacter* exhibited no cross-reactivities with $P\text{-}450_{cam}$ or P-450 from the alkane-grown *Candida maltosa* (MÜLLER, 1984).

Isolated P-450 was predominantly low-spin as judged by the Soret maximum of the oxidized state. Type II spectral changes with a shift from 417 to 420 nm were induced by the aliphatic amines hexyl, octyl and decyl amine or by benzylamine with apparent K_S-values of 5, 0.3, 0.03 and 2.5 mM. With imidazole the Soret maximum of ferric P-450 was shifted to about 424 nm. Type I spectral changes could not be achieved, neither with n-alkanes nor with other hydrophobic compounds.

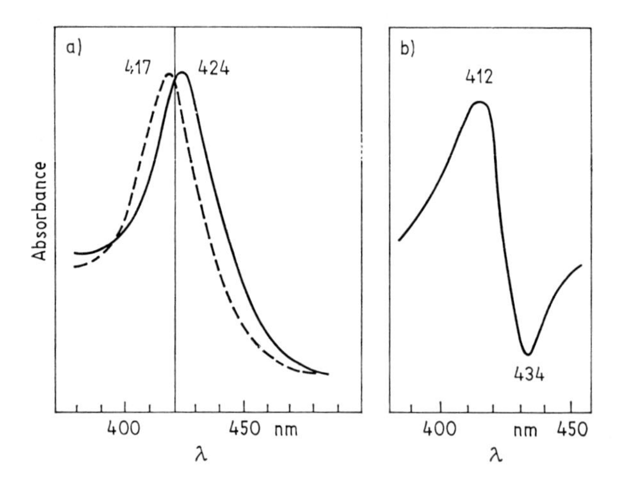

Fig. 2. Spectral features of ferric cytochrome P-450 from *A. calcoaceticus* EB 104. a) Soret bands of optical absolute spectra: (— — —) spectra at low concentration of P-450 (0.7 μM) in presence of non-ionic detergents or of serum albumin; (——) spectra of either low-concentrated P-450 in absence of detergent as well as albumin, or of high-concentrated P-450 (7 μM) in presence of Tween 80 (1 g/l), or after addition of imidazoles.
b) Difference spectra obtained by titration of ferric detergent-free P-450 either with non-ionic detergents (Emulgen 913, Tween 80, Triton X-100 and others) or with serum albumin.

Whereas absolute absorption spectra of purified, ferric P-450 monitored at low concentration (about 1 μM) in the presence of detergent exhibited the typical maximum at 417 nm, an anomalous absorption maximum at around 424 nm has been observed with detergent-free solutions or at increased concentration of P-450 (Fig. 2). The spectral red-shift could be reversed by addition of detergents or serum albumin and by dilution of P-450. Concomitantly with the spectral shift and influenced by the same factors an aggregation of P-450 occurs. With regard to the nearly equal spectral shift obtained by interaction of the 417 nm form of P-450 with imidazole, the observed spectral anomaly has been interpreted as an aggregation-induced conformational change of P-450 resulting in a replacement of the native 6th heme ligand by a histidyl residue (Asperger et al., 1988). The reversibility of this phenomenon indicates a quite flexible structure of P-450 from *Acinetobacter*.

Similar spectral anomalies have been described for a P-450 purified from human platelets and functioning as thromboxane synthase (Haurand and Ullrich, 1985) as well as for P-450 isolated from a mutant of *Saccharomyces cerevisiae* defective in lanosterol 14α-demethylation (Aoyama et al., 1987). In the latter case the spectral anomaly is claimed to be genetically determined, whereas for thromboxane synthase reversibility has been observed.

Only very low activities (about 0.7 nmol/min $\times$ mg protein) of alkane oxidation have been measured with cell-free extracts (Asperger, unpublished), although intact, resting cells oxidized hexadecane with a respiration rate of about 500 nmol/min $\times$ mg protein (Eremina et al., 1987). Nevertheless, evidence for the function of P-450 as an n-alkane hydroxylase has been provided by CO inhibition of activity, by similar subcellular distribution patterns for P-450 and activity, by simultaneous induction of hydroxylase activity as well as P-450 in response to biphenyl as inducer. Furthermore, the enzymic activity was restored with highly purified P-450 preparations, either obtained from n-hexadecane-grown or from n-hexane-induced cells. Reconstitution of activity required, beside the P-450, a ferredoxin and a NADH-dependent ferredoxin reductase, both isolated from *A. calcoaceticus* EB 104 (Asperger et al., 1985a).

Ferredoxin was detected when cell-free extracts from hexadecane-grown cells were adsorbed on DEAE-cellulose and a 0.5 M KCl eluate was further fractionated by means of ammonium sulfate precipitation and subsequent gel filtration on Sephadex G-75. The acidic, low-molecular protein mediating the reduction of mammalian cytochrome c in the presence of ferredoxin reductase from spinach was further purified by anion exchange chromatography and eluted at about 0.35 M KCl. The brownish protein with a M_r of about 9 kDa contained acid-labile sulfur and exhibited in the oxidized state an optical spectrum typical for [2 Fe–2S] ferredoxins (cf. Fig. 7) with maxima in the visible region at 415 and 460 nm, which disappeared after reduction by dithionite (Asperger et al., 1983). Furthermore, a NADH: acceptor oxidoreductase has been detected in the Sephadex G-75 eluate that catalyzed cytochrome c reduction in the presence of ferredoxin from *Acinetobacter*. Therefore, it has been defined as a ferredoxin reductase. Since other proteins that transfer electrons from NADH on to P-450 could not be detected, a P-450 system of *A. calcoaceticus* consisting of P-450, ferredoxin and a NADH-dependent ferredoxin reductase has been proposed (Asperger et al., 1985a). Taking into consideration, that the turnover number of the reconstituted system with hexadecane as substrate was quite low, it was supposed that additional factors are required to accomplish an optimal interaction between the extremely hydrophobic substrate and P-450. Substrate specificity still has to be explored in order to judge whether the broad inducer specificity finds any reflection in substrate conversion capacity. As detected by growth studies, P-450-containing strains of *Acinetobacter* were able to utilize middle-chain n-alkanes such as hexane up to decane as sole sources of carbon (Kleber et al., 1983; Eremina et al., 1987; Wyndham, 1987), whereas however other *Acinetobacter* strains could not grow on these hydrocarbons, although they were oxidized by resting cells of these strains, (Asperger and Aurich, 1977). It may be supposed that P-450-inducible *Acinetobacter* strains are more resistant to the toxic middle-chain n-alkanes than *Acinetobacter* strains that lack P-450.

3.4. Detection of cytochrome P-450 in luminescent bacteria of the genera *Vibrio* and *Photobacterium*

A functional role of P-450 as component of bacterial luciferases which oxidizes the fatty aldehydes necessary as substrates for in vitro bacterial bioluminescence had been proposed by Danilov (1979, 1987). The theoretical and experimental approach for this hypothesis is based on the inhibition of bioluminescence by a number of hydrophobic compounds that are known as substrates or inhibitors of P-450-dependent reactions.

Indeed, CO difference spectra of cell-free extracts from the luminescent bacteria *Vibrio fisheri* MJ-1 (Ismailov et al., 1979), *Vibrio harvey* and *Photobacterium phosphoreum* (Baranova et al., 1982) exhibited absorption bands at 450 nm. The P-450 content calculated on the basis of these spectra (Baranova et al., 1982) amounted to about 5 to 10 pmol/mg protein. P-450 was also detected in the luciferase preparations partially purified from the cytosolic proteins of *Vibrio fisheri* (Shumikhin et al., 1980; Danilov et al., 1985). The binding of the bioluminescence-quenching compounds to P-450, listed in Table 5, has been concluded from spectral effects that have been provoked by

Table 5. Compounds quenching bioluminescence of *Vibrio fisheri* and provoking P-450 type I difference spectra in cell-free extracts (a) or preparations of luciferase (b)

Compound	Preparation	Reference
Decanal	a	Danilov et al., 1981
Decane	a	Danilov et al., 1981
Camphor	a	Ismailov et al., 1979 Danilov et al., 1982
Phenobarbital	b	Danilov et al., 1985
Hexobarbital	b	Ismailov et al., 1981
Ethylmorphine	b	Ismailov et al., 1981
Dimethylaniline	b	Ismailov et al., 1981
Aminopyrine	b	Ismailov et al., 1981
Propylgallate	b	Ismailov et al., 1981

the addition of these compounds to cell-free extracts or to purified luciferase. Apparent dissociation constants that have been calculated on the basis of these spectral perturbations have been shown to correlate with the inhibition constants of bioluminescence quenching (ISMAILOV et al., 1981). Decanal reversed the inhibition competitively (DANILOV et al., 1981). These results and the inhibition of bioluminescence by metyrapone, SKF 525-A (MALKOV et al., 1982) and CO, respectively (DANILOV et al., 1985) were interpreted as an alkanal monooxygenase function of P-450 in luminescent bacteria. However, this model may be questioned, taking into consideration that luciferase preparations, described as highly purified, contained only 50 pmol P-450/mg protein (DANILOV et al., 1985). Assuming, the P-450 fragment can be assigned to a 50 kDa polypeptide, than the spectrally determined P-450 corresponds to only about 0.25% of the total protein. Furthermore, the strange type I difference spectra with extremely blue shifted maxima down to 380 nm, are mostly monitored in poorly fractionated cell-free preparations. Also the striking diversity of compounds which, beside the postulated P-450 substrate decanal, induce these spectra, argue for unspecific effects to be measured. But, apart from the validity of the proposed luciferase model, it is obvious that various species of the genera *Vibrio* and *Photobacterium* contain apparently constitutive P-450.

3.5. Cytochromes P-450 in *Bacillus megaterium*

In *Bacilli*, P-450 has been found hitherto in certain strains hydroxylating long-chain alkanoic acids or 3-oxo-Δ^4-steroids in the 15β-position. Both types of strains belong to the species *B. megaterium*.

3.5.1. The steroid-hydroxylating cytochrome P-450 system of *Bacillus megaterium* strain ATCC 13368

To investigate whether bacterial steroid hydroxylation in analogy to such reactions in higher organisms is also catalyzed by P-450-dependent enzymes, BERG et al. (1975) investigated the steroid 15β-hydroxylation system known to occur in *B. megaterium*. P-450 could be detected in amounts of about 8 pmol/mg protein in the cytosolic protein fraction in which progesterone-hydroxylating activity is found, too (BERG et al., 1975). The inhibition of steroid hydroxylase activity by typical inhibitors of P-450-dependent reactions, such as CO, SKF 525-A, imidazole and metyrapone (BERG et al., 1975) and the hydroxylation of progesterone by an enzyme system reconstituted from isolated P-450, an iron-sulfur protein and a NADPH-specific reductase, made the P-450 dependency of steroid 15β-hydroxylase from *B. megaterium* highly probable (BERG et al., 1979a, b).

The components

Highly purified P-450 (17 nmol/mg protein) has been obtained by means of chromatography on DEAE-cellulose, Ultrogel ACA-54, DEAE-Sepharose, Octyl-Sepharose, and hydroxylapatite (BERG et al., 1979a). The yield was about 3%. The final product was contaminated with about 20% P-420 but was electrophoretically homogeneous. Beside the M_r and pI (cf. Tab. 12) that were in the range of those of other P-450s, the amino acid composition has been determined (cf. Tab. 13). According to the ESR signals and the optical absorption bands (cf. Tab. 12) of the oxidized state, the isolated P-450 was in the low-spin form (BERG et al., 1979a, b). Ligand binding studies have not been reported.

The iron-sulfur protein, initially designated "megaredoxin" (BERG et al., 1976), was purified by means of similar chromatographic steps as those used for the purification of P-450. It was eluted from DEAE-Sepharose at 0.3 M KCl and from hydroxylapatite at 20 mM phosphate (BERG, 1982). With regard to an acid-labile sulfur to iron molar ratio of 0.9, ESR signals at 1.90, 1.93 and 2.06 g determined with partially purified preparations (BERG et al., 1976), a M_r of about 14 kDa and an acid-labile sulfur content of 1.8 mol/mol of electrophoretically homogeneous iron-sulfur protein, a [2Fe–2S] ferredoxin structure has been proposed (BERG, 1982). Whereas the ESR spectra and the amino acid composition resemble those of the [2Fe–2S] ferredoxins, putidaredoxin and adrenodoxin, that are components of the P-450 systems from *Pseudomonas putida* (GUNSALUS et al., 1974) and adrenal cortex mitochondria, respectively, the optical absorption spectrum (cf. Fig. 7) with a maximum of the oxidized form around 390 to 400 nm (BERG, 1982) corresponds more to iron-sulfur proteins containing [4Fe–4S] clusters (PALMER, 1975). The only partially purified reductase was NADPH-dependent and the activity was assayed by the reduction of dichlorophenolindophenol. The preparation contained proteins in a M_r range of 55 to 66 kDa and as prosthetic groups 4 mol FMN per mg protein. Nearly no FAD has been detected (BERG et al., 1976).

Substrate specificity

That the 15β-hydroxylation of progesterone may represent an additional activity of a rather unspecific P-450 should be excluded by the fact that from 12 non-steroidal potential P-450 substrates (Tab. 6) only aniline and imipramine were oxidized (BERG et al., 1979b; BERG and RAFTER, 1981). But, the rate of aniline hydroxylation did not exceed turnover numbers that have been observed for unspecific aniline hydroxylations catalyzed by non-P-450 hemoproteins (BERG and RAFTER, 1981).

Furthermore, as shown by studies with cell-free extracts, substrate specificity is restricted to such steroids (Tab. 6) that are characterized by a 3-oxo-Δ^4-

Table 6. Substrate specificity of the 3-oxo-Δ^4-steroid hydroxylase of *Bacillus megaterium* ATCC 13368

Compound	Activity[a] (%)	Enzyme system	Compound	Activity[a] (%)	Enzyme system
Steroid hydroxylation:			β-Sitosterol	—	c
Testosterone	20	b	Cholic acid	—	c
5α-Dihydrotestosterone	—	b	Deoxycholic acid	—	c
4-Androstene-3,17-dione	35	b	Lithocholic acid	—	c
Dehydroepiandrosterone	—	b			
5α-Androstan-3α,17β-diol	—	b	**Aliphatic hydroxylation:**		
5α-Androstan-3α,17β-diol-3.17-disulfate	—	b	Hexadecanoic acid, octadecanoic acid, prostaglandine $F_{2}\alpha$	—	c
Estrone	—	b			
Estradiol	—	b	**Aromatic hydroxylation:**		
Estriol	—	b	Biphenyl, benz(a)pyrene, lidocaine	—	c
Progesterone	100	b,c	Aniline	+	c
20α-Dihydroprogesterone	70	b			
17α-Dihydroprogesterone	75	b	**Dealkylation:**		
Pregnenolone	—	b	Aminopyrine, ethylmorphine, lidocaine, 7-ethoxyresorufine	—	c
5β-Pregnane-3α,20-diol	—	b			
Deoxycorticosterone	100	b			
Corticosterone	3	b			
Cholesterol	—	c	**N-oxidation:**		
Deoxycholesterol	—	c	Imipramine	+	c

a) Related to activity with progesterone; b) Cell-free extracts, BERG et al., 1976; c) Reconstituted system, BERG and RAFTER, 1981.

Table 7. Progesterone hydroxylation with cytochrome P-450 of *Bacillus megaterium* by means of peroxidatic reagents in the absence of electron transport proteins[a]

Reagent	Activity		Ratio of 15α/15β-hydroxylation
	(mM)	(pmol/min × nmol P-450)	
$NaIO_4$	2	346	1.3
$NaClO_2$	2	106	3.6
C_6H_5IO	5	630	3.0
$C_6H_5I(OAc)_2$	5	300	2.1
o-$NO_2C_6H_4IO$	5	910	2.2
m-$NO_2C_6H_4IO$	5	533	2.3
o-$NO_2C_6H_4I(OAc)_2$	5	813	2.4
m-$NO_2C_6H_4I(OAc)_2$	5	203	1.8
$C_6H_5I(OCOCF_3)_2$	5	593	2.4

a) Berg et al., 1979.

structure (Berg et al., 1976). That from 7 tested compounds possessing this structural features only progesterone was also hydroxylated at the 6β-position (20%) indicates a high regioselectivity for the 15β-position. Likewise, the hydroxylation reaction is governed by a strict stereospecificity for the β-position when taking place in presence of the natural electron transport system, whereas in case of peroxide catalyzed hydroxylations (Tab. 7) 15α-hydroxylations up to nearly 50% have been observed (Berg et al., 1979a).

Activity, reconstitution and regulation

Whereas the specific activities of only 2 nmol/30 min × mg protein for the hydroxylation of progesterone by cell-free extracts (Berg et al., 1975) appear to be very low, the turnover number of about 8 min^{-1} calculated from this value with regard to a specific P-450 content of 8 pmol/mg protein is comparable to those values reported for microsomal steroid hydroxylations (Coon and Inouye, 1985). Maximum turnover numbers of about 0.9 nmol/min × nmol P-450 have been measured during peroxidatic hydroxylation of progesterone by o-nitro-iodosobenzene (Tab. 7) in the presence of partially purified P-450 (Berg et al., 1979a). Hydroxylating activity (with $NaIO_4$ as oxygen donor) increased 2-fold if P-450 was supplemented with phosphatidylcholine in a molar ratio of 1:40 (Berg et al., 1979a) indicating that activity in reconstituted systems may be limited by sub-optimal incubation conditions. This is also obvious from the gap between the turnover number of 8 min^{-1} calculated from the activity of cell-free extracts and the turnover number of only 0.8 measured with reconstituted systems for the hydroxylation of progesterone.

Table 8. Substances without the capability to induce cytochrome P-450 of *Bacillus megaterium* strain ATCC 13368[a]

Steroids	Various compounds
Testosterone	Dodecanoic acid
Progesterone	Palmitic acid
17α-Hydroxyprogesterone	Phenobarbital
Corticosterone	β-Naphthoflavone
Deoxycorticosterone	16α-Cyanopregnenolone
11-Deoxycortisol	Corn oil
Cortisol	Thiamine + Nicotinic acid
Cholesterol	
Cholestenone	

a) BERG and RAFTER, 1981.

With regard to the obligatory requirement of ferredoxin for the reconstitution of enzymic activity (BERG et al., 1979a) the P-450 system of *B. megaterium* resembles the P-450_{cam} system (GUNSALUS et al., 1974). Opposite to the *Pseudomonas* system the ferredoxin of *B. megaterium* could be exchanged by adrenodoxin, and alternatively the *Bacillus* ferredoxin could substitute for the adrenodoxin in the 11β-hydroxylation system of adrenal cortex mitochondria. Similarly, the iron-sulfur protein reductases were interchangeable (BERG et al., 1979a).

Although the rather high specificity of the *B. megaterium* P-450 for the 15β-hydroxylation of 3-oxo-Δ^4-steroids may be considered as an argument for being a special steroid-metabolizing enzyme, the low level of this P-450 in the cell and the failure to detect a substrate-dependent induction (BERG and RAFTER, 1981) disagree with this kind of interpretation. The finding that P-450 is detected only in the stationary phase and that hydroxylating activity is multifold increased by incubating cells after exponential growth in carbon-free media may be interpreted as a catabolite repression that hinders the inducing effect of the substrates. But nevertheless, also in the absence of yeast and soybean extracts that have been used as carbon sources, the compounds listed in Table 8 did not increase the hydroxylase level obtained already without any added inducer. Therefore, the most likely induction mechanism for P-450 in *B. megaterium* ATCC 13368 appears to be a derepression caused by growth limitation whatever the triggering signal may be. The accumulation of an inducing-active metabolic product in the culture medium could be excluded (BERG and RAFTER, 1981). So far, it remains an open question as to what kind of physiological role the P-450 in this strain of *B. megaterium* may play. Among other strains of *B. megaterium* the strain ATCC 13368 appears to be rather unique, considering that from 12 tested ATCC-strains (FULCO and RUETTINGER, 1987) this was the only one in which the barbiturate-inducible P-450$_{BM-3}$ was not found. Otherwise, no progesterone hydroxylase activity has hitherto been detected in that strain of *B. megaterium* from which the P-450$_{BM-3}$ has been isolated (MATSON et al., 1977).

3.5.2. Fatty acid monooxygenase-linked cytochromes P-450 of *Bacillus megaterium*

Accidentally MIURA and FULCO (1974) detected a fatty acid (ω-n) hydroxylase in *B. megaterium* strain ATCC 14581. This enzyme was identified as a P-450-dependent monooxygenase (HARE and FULCO, 1975; MATSON et al., 1977). The mystery about the molecular organization of this monooxygenase could be resolved only recently by providing convincing evidence that this enzyme is made of a single polypeptide chain containing FAD, FMN as well as P-450-like-bound heme B and thus represents a catalytically selfconsistent P-450 protein (NARHI and FULCO, 1986; 1987). Another striking peculiarity of this enzyme is its inducibility by barbiturates but not by the enzyme substrates themselves (NARHI and FULCO, 1982).

P-450 multiplicity and molecular properties

Altogether, three different soluble P-450s having M_r's of 38, 46, and 119 kDa respectively, have been detected (SCHWALB et al., 1985). They have been designated P-450$_{BM-1}$, P-450$_{BM-2}$ and P-450$_{BM-3}$, the latter being the functionally active fatty acid monooxygenase. Whereas P-450$_{BM-2}$ has been only partially purified (4.1 nmol/mg protein), P-450$_{BM-1}$ has been obtained as an homogeneous protein (12 nmol P-450/mg protein). It has been characterized by its amino acid composition (cf. Tab. 13) and by its spectral features (cf. Tab. 12). Antibodies against P-450$_{BM-1}$ had no detectable affinity to P-450$_{BM-2}$ or to P-450$_{BM-3}$ (SCHWALB et al., 1985). P-450$_{BM-3}$ was separated from P-450$_{BM-1}$ by gel filtration, whereas P-450$_{BM-2}$ eluted at a somewhat higher ionic strength during anion exchange chromatography.

P-450$_{BM-3}$ has been purified to homogeneity from pentobarbital-induced *B. megaterium* (NARHI and FULCO, 1986) as well as from *Escherichia coli* transformed by a recombinant plasmid containing the *B. megaterium* cytochrome P-450$_{BM-3}$ gene (NARHI et al., 1988). Both isolated P-450s were identical in their molecular properties, spectral features (cf. Tab. 12) and functional activities. FAD, FMN and the heme occur in 1:1:1 molar ratio. They are distributed among two domains, which has been demonstrated by the characterization of corresponding fragments obtained by means of tryptic digestion in the presence of substrate followed by chromatography on a HPLC anion-exchange column (NARHI and FULCO, 1987). FAD and FMN were constituents of a 66 kDa fragment that contained the C-terminus of P-450$_{BM-3}$ and catalyzed the reduction of mammalian cytochrome c by NADPH. Three other discrete peptides, T-I, T-II and T-III, with M_r's of 55, 54 and 53.5 kDa respectively, exhibited P-450 spectra. The N-terminal amino acid sequence of T-I was identical to that of intact P-450$_{BM-3}$. T-I was able to bind the substrate as demonstrated by a type I spectral shift with myristate. This was not the case with T-II and T-III that had lost the 9 and 15, respectively, N-terminal amino acids, indicating the significance of the N-terminus for substrate binding. Reconstitution of activity with T-I and the 66 kDa fragment was, however, not accomplished.

The sequence Pro-Ala representing the N-terminus of P-450$_{BM-1}$ was not included within the N-terminal sequences determined for P-450$_{BM-3}$, T-I, T-II or T-III which makes it little probable that P-450$_{BM-1}$ may be a fragment of P-450$_{BM-3}$ identical with the P-450 domain. Therefore, the occurrence of multiple P-450s in *B. megaterium* should be taken into consideration.

Enzymic activities and substrate specificities

Although the P-450 of *B. megaterium* is known from the literature as a fatty acid ω-2 hydroxylase (MIURA and FULCO, 1974) hydroxy fatty acids, fatty alcohols, as well as fatty acid amides are also hydroxylated (Tab. 9). Beside hydroxylation unsaturated fatty acids undergo rather equally well epoxidation of the double bond (BUCHANAN and FULCO, 1978; RUETTINGER and FULCO, 1981). In case of palmitoleate, epoxidation amounted to about 50% in relation to hydroxylation. Other reactivities typical of P-450-dependent enzymes have not hitherto been reported.

Epoxidation has been detected by the identification of 9,10-epoxypalmitate and 9,10-dihydroxypalmitate when palmitoleate was incubated with cell-free extracts (BUCHANAN and FULCO, 1978) and was furthermore shown for several other unsaturated fatty acids. The ratio epoxidation to hydroxylation differed dependent on the substrate. It remained constant under various assay conditions or using different enzyme preparations, thus providing evidence for both activities being catalyzed by only one single enzyme (RUETTINGER and FULCO, 1981). This was finally confirmed by the highly purified P-450$_{BM-3}$ exhibiting both activities too (NARHI and FULCO, 1986). 9,10-Dihydroxypalmitate arises from 9,10-epoxypalmitate by the action of a hydratase present in preparations of partially purified P-450 (MICHAELS et al., 1980). ω-Hydroxylation of substrates was never found and the subterminal positions ω-1 to ω-3 were hydroxylated either in an equal ratio or with the predominance of the ω-2 position (MIURA and FULCO, 1974) indicating a unique regioselectivity of *B. megaterium* P-450. It was possible to substantiate

Table 9. Substrate specificity of the fatty acid monooxygenase of *Bacillus megaterium* ATCC 14581

Compound	Activity					Compound	Activity	
	a	b	c	d	e		a	f
Dodecanoic acid	7				+	Dodecanoyl amide	9	
Tridecanoic acid	23				+	Tetradecanoyl amide	10	
Tetradecanoic acid	57				+	Pentadecanoyl amide	7	
Pentadecanoic acid	100				+	Hexadecanoyl amide	2	
Hexadecanoic acid	63				+	Fatty acid methyl ester		
Heptadecanoic acid	38				+	(C_{12}—C_{18})	—	
Octadecanoic acid	7				+	Dodecanol	3	
cis-5-Tetradecenoic acid	39					Tridecanol	5	
cis-9-Hexadecenoic acid	110				+	Tetradecanol	7	
cis-10-Hexadecenoic acid		+				Pentadecanol	5	
trans-9-Hexadecenoic acid		+				Hexadecanol	1	
15-Hydroxy-9-cis-hexadecenoic acid		+				n-Alkanes (C_{12}—C_{18})	—	
cis-12-Octadecenoic acid	65					Secobarbital		—
cis-9-Octadecenoic acid	74					Phenobarbital		—
trans-9-Octadecenoic acid		+				Pentobarbital		—
12-Hydroxy-cis-9-octadecenoic acid		+				Methohexital		—
9-Hydroxystearic acid				+				
6-Hydroxystearic acid			+					
3-Hydroxystearic acid			+					

a) Ammonium sulfate precipitate of cell-free extracts, MIURA and FULCO, 1975; activity in % to pentadecanoic acid; b) Sephadex G-200 eluates, RUETTINGER and FULCO, 1981; c) Enzyme preparation as in b), MATSON and FULCO, 1981; d) Enzyme preparation as in b), MATSON et al., 1980; e) P-450$_{BM-3}$ NARHI and FULCO, 1986; f) Partially purified P-450, negative results were concluded from the failure to inhibit fatty acid monooxygenase activity, NARHI et al., 1983.

that all different hydroxylated fatty acids were produced by the identical enzyme (Ho and FULCO, 1976).

Considering that fatty acid esters or n-alkanes were not hydroxylated, the requirement of at least one polar group as a prerequisite for substrate binding has been suggested. The comparable good hydroxylation of in-chain hydroxy fatty acids and the competitive inhibition of palmitate hydroxylation by them has been interpreted as indicating the importance of a further non-hydrophobic binding region at the enzyme active site (MATSON and FULCO, 1981). This interpretation may be questioned considering that in the case of poorly water-soluble substrates the turnover is controlled not only by intrinsic enzymic parameters but by the partition of the substrate into the water phase as well. Thus, an optimum K_m (about 2 μM) for C_{15} and C_{16} fatty acids (NARHI and FULCO, 1986), the competitive inhibition of fatty acid hydroxylation by in-chain hydroxy fatty acids (MATSON and FULCO, 1981) but also the failure of n-alkanes and fatty acid methyl esters to be hydroxylated under the conditions employed (MIURA and FULCO, 1975) could resultant from both factors. It seems that the rate-limiting step of oxygenation at saturating levels of substrate is not affected by the structure of tested substrates or by the type of reaction, since all fatty acids, hydroxy fatty acids and unsaturated fatty acia had about the same V_{max} (RUETTINGER and FULCO, 1981; NARHI and FULCO, 1986). Progesterone was not hydroxylated (MATSON et al., 1977) showing the distinct nature of the P-450s from *B. megaterium* strain ATCC 14581 and from strain ATCC 13368 respectively (cf. 3.5.1.).

The catalytic activity of up to 4,600 is the highest value hitherto reported for P-450 enzyme systems (NARHI and FULCO, 1986). This high activity of the enzyme may be explained by its unique structure containing all functional elements of a P-450 enzyme system in a single polypeptide. Substrates induced type I spectral shifts (NARHI and FULCO, 1986) showing the conversion of the P-450 to the high-spin state. Since P-450 was not reduced by NADPH in absence of substrate, nor NADPH was oxidized with a significant rate, the reaction cycle starts as in other P-450 systems, obviously with a substrate-induced conversion to the high-spin state (NARHI and FULCO, 1986). Enzymic activity was not to be reconstituted by recombining the tryptic digestion fragments of P-450_{BM-3} although the P-450-containing fragment T-I was shown to bind the substrate and the reductase fragment was obviously also active since it reduced cytochrome c with nearly the same specific rates as in the case of authentic P-450_{BM-3}. Thus, function appears to require an exact positioning of functional domains (NARHI and FULCO, 1987) which may be disturbed by the loss of a small part of the polypeptide chain already during tryptic digestion. Otherwise, taking into consideration that in early publications (MIURA and FULCO, 1974; HARE and FULCO, 1975) the stimulation of fatty acid hydroxylation activity by bacterial ferredoxin has been reported, and considering that analytical data on the iron and the labile sulfur content of

Table 10. Induction profile for cytochrome P-450 of *Bacillus megaterium* ATCC 14581

Compound	Structural elements[1]				Induction	Ref.[2]
	R_1	R_2	R_3	X		
a) Pyrimidine derivatives						
Methohexital	allyl	1-methyl-2-pentenyl	CH_3	S	+	(a)
Thiamylal	allyl	1-methylbutyl	H	S	+	(a)
Secobarbital	allyl	1-methylbutyl	H	O	+	(a)
Talbutal	allyl	1-methylpropyl	H	O	+	(a)
Allobarbital	allyl	allyl	H	O	+	(a)
Aprobarbital	allyl	isopropyl	H	O	+	(a)
Thiopental	ethyl	1-methylbutyl	H	S	+	(a)
Pentobarbital	ethyl	1-methylbutyl	H	O	+	(a)
Amobarbital	ethyl	isopentyl	H	O	+	(a)
Butabarbital	ethyl	1-methylpropyl	H	O	+	(a)
Phenobarbital	ethyl	phenyl	H	O	+	(a)
5-Ethyl-5-p-tolylbarbiturate	ethyl	tolyl	H	O	+	(a)
Hexobarbital	methyl	1-cyclohexenyl	CH_3	O	+	(a)
p-Hydroxyphenobarbital	ethyl	p-hydroxyphenyl	H	O	—	(a)
Mephobarbital	ethyl	phenyl	CH_3	O	—	(a)
4-Methylprimidon	ethyl	tolyl	H	2H	—	(a)
Barbital	ethyl	ethyl	H	O	—	(a)
Barbituric acid	H	H	H	O	—	(a)
2-Thiobarbituric acid	H	H	H	S	—	(a)
b) Acylureas						
2-Phenyl-3-methylbutyrylurea	i-propyl	phenyl	H	O	+	(b)
2-Isopropyl-4-pentenoylurea	allyl	isopropyl	O	O	+	(b)
2-Phenyl-4-pentenoylurea	allyl	phenyl	H	O	+	(b)
2-Ethylhexanoylurea	ethyl	butyl	H	O	+	(b)
2-Phenylbutyrylurea	ethyl	phenyl	H	O	+	(b)
1-[2-Phenylbutyryl]-3-methylurea	ethyl	phenyl	CH_3	O	+	(c)

2-Phenylbutyrylthiourea	ethyl	phenyl	H	S	+	(b)
2-Phenyl-2-methylpropionylurea	methyl	phenyl + methyl	H	O	+	(b)
Phenylacetylurea	H	phenyl		O	—	(b)
Octanoylurea	H	hexyl		O	—	(b)
Dodecanoylurea	H	decyl		O	—	(b)
c) Acid amides						
2-Isopropyl-4-pentenamide	allyl	isopropyl			(+)	(b)
2-Phenylbutyramide	ethyl	phenyl			(+)	(b)
d) Acids						
2-Phenylbutyric acid	ethyl	phenyl			—	(b)
2-Ethylhexanoic acid	ethyl	butyl			—	(b)
e) Various compounds						
Caffeine; Theophylline; 5-Methyl-5-phenylhydantoin					(—)	(a)

[1]) cf. Fig. 3; [2]) References: a) KIM and FULCO, 1983; b) RUETTINGER et al., 1984; c) WEN and FULCO, 1985.

the P-450$_{BM-3}$ have not been provided, it should be investigated whether an iron-sulfur cluster being destroyed during tryptic digestion is present in the native P-450 of *B. megaterium*. On the other hand, being a NADPH-specific enzyme and containing FAD and FMN as prosthetic groups, P-450$_{BM-3}$ resembles quite well the microsomal P-450 systems that are not iron-sulfur protein-dependent (see Vol. 1 of this series).

Kinetic analysis of enzyme activity was hampered by a time-dependent inactivation caused by NADPH (Ruettinger and Fulco, 1981; Narhi and Fulco, 1986) resulting in a 10-fold lowered activity after prolonged incubation. Since the original activity could be restored after removing of NADPH by means of dialysis, a destruction of P-450 due to hydroperoxide formation by uncoupled NADPH oxidation has been ruled out (Narhi and Fulco, 1986). But, uncoupling of NADPH oxidation and product formation had been observed in other cases and found to be proportional to the hydrophobicity of the substrate (Ruettinger and Fulco, 1981). A K_m for NADPH has not yet been described. The monooxygenase activity was inhibited by CO and metyrapone (Matson et al., 1977) but surprisingly not by cytochrome c, although it is reduced in the presence of P-450$_{BM-3}$ with a turnover of about 1,300 (Narhi and Fulco, 1986).

Whether P-450$_{BM-1}$ also represents a fatty acid monooxygenase remains, up to now, an unresolved problem. Activity of purified P-450$_{BM-1}$ could not be tested, because an appropriate reducing system was not available. Although hydroxylation and epoxidation of palmitoleate by means of iodosobenzene diacetate has been accomplished, an enzymic reactivity similar to P-450$_{BM-3}$ remains questionable, since, for instance myristic acid was not hydroxylated (Schwalb et al., 1985).

Regulation

Progress in investigating P-450 of *B. megaterium* was markedly enhanced by the discovery of its phenobarbital-dependent induction (Narhi and Fulco, 1982). The constitutive fatty acid monooxygenase activity, of about 2 to

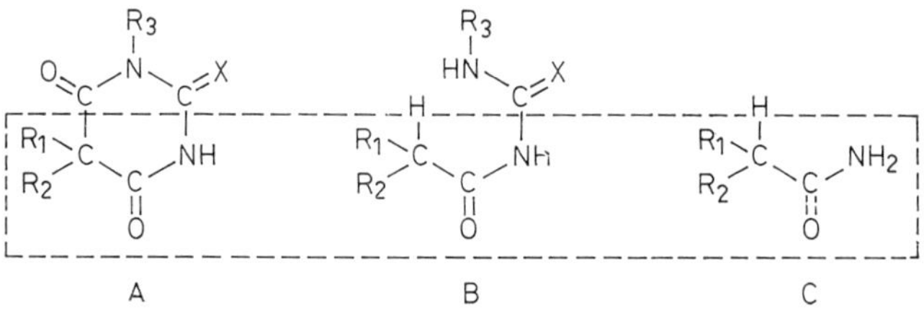

A B C

Fig. 3. Structures of compounds inducing cytochrome P-450 in *B. megaterium* ATCC 14581. A) Barbiturates (X = O) and thiobarbiturates (X = S); B) Acylureas (X = O) and thioacylureas (X = S); C) Acid amides. For R_1, R_2 and R_3 (cf. Tab. 10).

10 nmol/min/mg protein in cell-free extracts, as well as the specific P-450 content, of about 4 pmol/mg protein, were up to 30-fold increased. Later on, induction was accomplished by a variety of other barbiturates and by further compounds (Tab. 10) possessing as a common structural element an acylamide moiety (Fig. 3) such as acylureas, but also acid amides themselves. The substrates of fatty acid monooxygenase did not induce the P-450. On the other hand, none of the inducers was functionally active or was bound to P-450 (Narhi and Fulco, 1982; Narhi et al., 1983). Acylureas that are produced from corresponding barbiturates during autoclaving are in some cases more active inducers than the original barbiturates. Nevertheless, it was clearly established that also the barbiturates themselves are recognized as inducers (Fulco et al., 1983; Ruettinger et al., 1984). Enormous differences of inducing efficiency of more than 100-fold have been observed between the weak inducing acid amides and the most active barbiturate methohexital. One main factor which is responsible for inducing potency seems to be lipophilicity. This was deduced from the increasing inducing efficiency if the residues at the C_2 atom of the acyl moiety (cf. Fig. 3 and Tab. 10) were substituted by more hydrophobic ones or if urea oxygen was exchanged against sulfur. The higher inducing potency of acylureas, relative to the corresponding barbiturates and, furthermore, the capability of acid amides, opposite to the ineffectiveness of the corresponding carboxylic acids, to act as inducers (Kim and Fulco, 1983; Ruettinger et al., 1984) may be caused also by differences in hydrophobicity. The importance of regiospecific or steric interactions for inducer recognition is indicated by the failure of mephobarbital (Wen and Fulco, 1985), 4-methylprimidone or 5-methyl-5-phenylhydantoin (Tab. 10) to induce the fatty acid monooxygenase. The question whether the acylamide structure is required as an obligatory element of the inducer molecule is, however, not resolved. Other strongly hydrophobic compounds, sterically related to acylureas but without an acylamide group have not been tested. Therefore, such a postulation may not withstand a more detailed examination. Own experiments, carried out with a strain of *B. megaterium* not further characterized, revealed beside the expected induction of P-450 by pentobarbital also an induction by n-hexane, detectable by means of CO difference spectra monitored with whole cells (Asperger, unpublished).

With respect to the rather broad inducer specificity, the question has been posed whether the P-450 induction is dependent on inducer interaction with a specific receptor protein or whether it is a more colligative phenomenon (Wen and Fulco, 1985). Since induction experiments, such as for instance with phenobarbital, were normally carried out during exponential growth of cells on a medium containing casamino acids (Difco) as well as glucose, regulation by catabolite repression should be ruled out. Without an added inducer fatty acid monooxygenase was detected only in cells from the stationary phase of growth, increasing by elevating the content of casamino acids in the medium.

The reason for the latter effect remained unknown, since lipophilic extracts of the amino acid mixture did not affect the induction (NARHI and FULCO, 1982). A regulatory mechanism involving a putative repressor protein for the binding of the described inducers is indicated by the observations that in *Escherichia coli* carrying the recombinant plasmid with the P-450$_{BM-3}$ gene P-450 was expressed at a level similar to those in pentobarbital-induced *B. megaterium*, independent of whether an inducer was present or not (WEN and FULCO, 1987). The strongly inducer-dependent expression of P-450, if the cloned P-450$_{BM-3}$ gene was introduced back into *B. megaterium*, ruled out damage in the regulatory region of the cloned P-450$_{BM-3}$ gene as a reason for uncontrolled P-450 biosynthesis in *Escherichia coli*. In hybridization experiments with DNA of *B. megaterium*, nick-translated fragments of the recombinant plasmid hybridized none of the six plasmid bands, that were found with *B. megaterium* DNA, but the high-molecular DNA band, thus indicating that P-450$_{BM-3}$ is a chromosome-encoded protein (WEN and FULCO, 1987).

No unambiguous results exist on the regulation of P-450$_{BM-1}$ and P-450$_{BM-2}$. According to NARHI and FULCO (1982), 65% of phenobarbital-induced P-450 consist of the low-molecular form and 33% of pentobarbital-induced P-450 correspond to P-450$_{BM-1}$ and P-450$_{BM-2}$ (SCHWALB et al., 1985). Likewise, the discrepancies observed between the induction of P-450 and of fatty acid monooxygenase activity, as for instance in the case of mephobarbital which induced only P-450 but not activity (WEN and FULCO, 1985) also indicated the barbiturate-dependent induction of P-450s not identical with P-450$_{BM-3}$. Otherwise, results from Western blot analysis argue that only P-450$_{BM-3}$ is significantly induced by pentobarbital and that P-450$_{BM-1}$ and P-450$_{BM-2}$ obviously occur constitutively (FULCO and RUETTINGER, 1987).

Distribution among different strains

As it might be expected, with respect to the chromosomal encodation, pentobarbital-inducible P-450 is obviously distributed among various strains of *B. megaterium* (FULCO and RUETTINGER, 1987). In 11 of 12 ATCC strains of *B. megaterium* (cf. Tab. 1) significant levels of myristate hydroxylase as well as of P-450, have been determined after induction by pentobarbital. The only exception was the steroid 15β-hydroxylase-containing strain ATCC 13368 in which neither fatty acid monooxygenase nor P-450 was induced by pentobarbital. Furthermore, it was shown by means of Western blot analysis with antibodies against P-450$_{BM-1}$, P-450$_{BM-2}$ and P-450$_{BM-3}$ that, except strain 13368, P-450$_{BM-1}$ and P-450$_{BM-2}$ were constitutively present and not significantly induced by pentobarbital, whereas only cellfree extracts of pentobarbital-induced cells reacted immunopositively with anti- P-450$_{BM-3}$ antibodies. Opposite to the fact that P-450 has been detected in most of tested *B. megaterium* strains, in nine other *Bacilli* strains, comprising the seven species *B. licheniformis*, *B. pumilis*, *B. alvei*, *B. brevis*, *B. cereus*, *B. macerans* and *B. subtilis*, neither constitutive nor pentobarbital-inducible P-450s occurred.

3.6. Cytochrome P-450 of a cholesterol-utilizing *Corynebacterium*

The occurrence of P-450 in a *Corynebacterium* grown on cholesterol has been reported by YAMAMOTO et al. (1974). A M_r of 50 kDa has been determined from a 40% pure preparation. Absorbance maxima at 393, 510 and 645 nm have been observed in optical absolute spectra of the ferric form, indicating the dredominance of a P-450 in the high-spin state.

3.7. Alkane-inducible cytochrome P-450 of *Rhodococcus rhodochrous*

The reclassified alkane-oxidizing bacterium *Rh. rhodochrous* strain ATCC 19067, known in the past as *Corynebacterium sp. 7, ElC* was shown to contain P-450 when grown on n-octane as sole source of carbon (CARDINI and JURTSHUK, 1968; 1970). The classification of this, never purified P-450 as an n-alkane hydroxylase was based on its co-distribution with an essential part of cell-free n-octane-hydroxylating activity, on the inhibition of enzymic activity by CO and on the induction of P-450 by n-alkanes. Only minimum P-450 and hydroxylase activity was found in acetate-grown cells.

The octane hydroxylase has been suggested as consisting of at least two soluble components. After ammonium sulfate fractionation of the cytosolic protein fraction of sonically disrupted cells, activity was restored by reconstitution of the P-450-containing 25–40% ammonium sulfate precipitate with the fraction precipitating between 60–100% ammonium sulfate saturation. The fraction 40–60% was active by itself. The CO difference spectrum of the P-450-containing (approximately 0.3 nmol/mg protein) fraction exhibited an absorption maximum at 450 nm and an absorption shoulder at 425 nm. After exposure to deoxycholate, all P-450 was converted to P-420 (CARDINI and JURTSHUK, 1968). Data on the absolute absorption spectra and the spin-state of the P-450 are not available.

With respect to a high content of flavin and the diaphoretic activities with dichlorophenolindophenol, mammalian cytochrome c, and hexacyanoferrate (III) as electron acceptors, it has been suggested that the 60–100% ammonium sulfate fraction contained a NADH-dependent reductase as the second component of the octane hydroxylase. NADH-dependent reduction of P-450 was also stimulated by this protein fraction. But, considering that NADH: acceptor-oxidoreductase activities have been measured with all ammonium sulfate fractions, inclusively with the P-450 containing, it should not be ruled out that the 60–100% ammonium sulfate fraction contains a third component required for the reconstitution of an active hydroxylase. According to the high ammonium sulfate concentration at which this putative third component precipitates, an iron-sulfur protein may be suggested.

The oxidation products of n-octane have been identified as octan-1-ol and octanoic acid. A turnover number of 12 min^{-1} related to the spectrally determined P-450 can be calculated from the activity of the reconstituted system.

3.8. Cytochrome P-450 and p-alkoxybenzoate O-demethylase of *Nocardia sp. NH1*

Nocardia sp. NH1 is a bacterium that was isolated due to its capability to grow on p-alkoxybenzoates (cf. Fig. 1) as sole source of carbon (CARTWRIGHT et al., 1971). P-450 was detected monitoring optical spectra of the p-alkoxybenzoate O-dealkylating cytosolic protein fraction treated with CO.

The P-450 has been isolated to an apparent homogeneous state and purification has been demonstrated by several different methods (BROADBENT and CARTWRIGHT, 1974). But, the authors did not present any purification protocols, neither mentioned the specific P-450 content nor the heme content of the purified P-450. From 7.5 g cytosolic protein, about 45 mg P-450 have been isolated by following purification steps: ammonium sulfate precipitation (0 to 45% saturation), DEAE-Sephadex 50 chromatography, Sephadex G-100 chromatography and repeated chromatography on DEAE-Sephadex 50.

The M_r values of 42.5 kDa determined by SDS polyacrylamide gel electrophoresis and of 44 kDa calculated from the amino acid composition (cf. Tab. 13) correspond to those of other P-450s as do the pI and the spectral properties (cf. Tab. 12), too. According to the absorption maxima of the oxidized form at 417 nm and the ESR signals at 1.90, 1.97, 2.26 and 2.45 g, the isolated P-450 obviously occurs predominantly in the low-spin state. Although a type I spectral shift of the Soret band to 394 nm and the development of a band at 650 nm had been observed if iso-vanillate or 4-methoxybenzoate, but not if 2-, or 3-methoxybenzoate or camphor, were added, a high-spin shift caused by these substrates was not obvious from corresponding ESR spectra (BROADBENT and CARTWRIGHT, 1974).

Evidence for the function of P-450 as an alkoxybenzoate O-demethylase has been provided by the fact that purified P-450 oxidized 4-alkoxybenzoates in the presence of NADH when it was reconstituted with a protein fraction that had been separated from P-450 and did not oxidize these substrates by itself (BROADBENT and CARTWRIGHT, 1971; CARTWRIGHT and BROADBENT, 1974). Furthermore, P-450 was detected only in cell-free extracts of cells grown on the alkylaryl ethers 4-methoxy-, 4-ethoxy-, 4-n-propoxybenzoate or veratrate but not in cells grown on 4-hydroxybenzoate. These results make it improbable that the P-450 is involved in hydroxylations of the aromatic ring (CARTWRIGHT et al., 1971).

The nature of the electron transfer components has not been elucidated unambiguously, since they have been only partially purified (BROADBENT and CARTWRIGHT, 1971; CARTWRIGHT and BROADBENT, 1974). Starting from HUGHES press-disrupted cells, enzymic activity could be reconstituted with a protein fraction that has been eluted from gel filtration columns in a M_r range of about 60 to 80 kDa. This supported the hypothesis that the P-450 system of *Nocardia* NH1 consists of only two components. Otherwise, the activity of cell-free extracts prepared by sonication was dependent on a dialysable, low-molecular-weight protein component and dealkylating activity was also lost if the cell disruption was not carried out under exclusion of air (CARTWRIGHT et al., 1971). Both facts are indicative of the requirement of an iron-sulfur protein.

With regard to substrate specificity, the results obtained by respiratory measurements with either whole cells or with cell-free extracts show that the dealkylase of *Nocardia* is quite unspecific with respect to the length of the alkyl chain but is highly selective with respect to the position of the ether linkage relatively to the carboxylic group. Thus, m-alkoxybenzoates were not substrates whereas 4-n-alkoxybenzoates from C_1 to C_4, iso-vanillate and veratrate (cf. Fig. 1) were oxidized by induced cells or corresponding cell-free extracts.

3.9. Cytochrome P-450 of *Saccharopolyspora erythrea* linked to erythromycin biosynthesis

The only bacterial P-450 for which a biosynthetic function is proposed has been detected in *Saccharopolyspora erythrea*, a bacterium formerly known as *Streptomyces erythreus* (CORCORAN and VYGANTAS, 1982). This bacterium produces erythromycin A, a macrolide antibiotic. The biosynthesis of the erythromycin A precursor, erythronolide B, requires the introduction of a hydroxyl group in position 6 of the macrolide 6-deoxyerythronolide B (Fig. 4). P-450 from *Sa. erythrea* was purified by SHAFIEE and HUTCHINSON (1987) to homo-

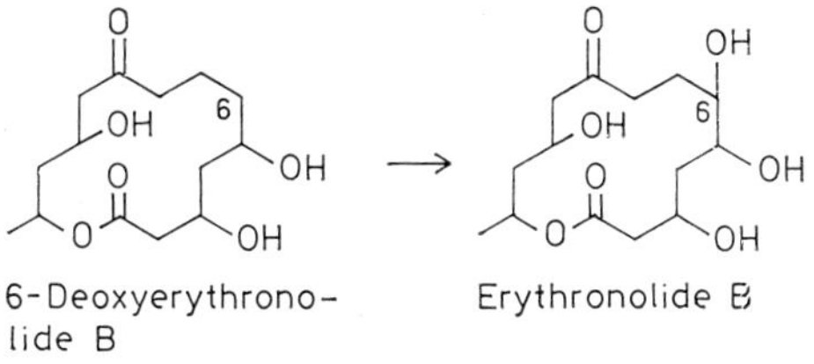

Fig. 4. Step of erythronolide B formation in *Sa. erythrea*.

geneity after CORCORAN and VYGANTAS (1982) had provided preliminary evidence for 6-deoxyerythronolide B hydroxylase being a P-450-dependent enzyme system.

Enzyme activity as well as P-450 (20 pmol/mg protein) reside in the soluble protein fraction of cell-free extracts. By means of ammonium sulfate precipitation (30–90% saturation), chromatography on DEAE-Sepharose and hydroxylapatite, preparative polyacrylamide gel electrophoresis and FPLC on Mono Q anion-exchange columns as the last step, two forms of P-450 have been isolated (SHAFIEE and HUTCHINSON, 1987). The resolution of P-450 into the two forms $P\text{-}450_I$ and $P\text{-}450_{II}$ has been achieved during chromatography on hydroxylapatite by elution with 0.01 and 0.5 M phosphate buffer, respectively. Each of the two P-450 forms, recovered in a ratio of 1:0.8, was purified according to the purification steps described above. Finally, purified $P\text{-}450_I$ contained 17.5 and $P\text{-}450_{II}$ 15.3 nmol P-450/mg protein. The total yield amounted to about 0.13 mg P-450 from 5.5 g cell protein and corresponded to a recovery of about 2.5% relative to P-450 in cell-free extracts. The homogeneity of the P-450 preparations was apparent from the $A_{280}:A_{416}$ ratio between 0.46 and 0.9, by single bands on SDS electropherograms and by single precipitation bands when tested against antisera to partially purified P-450. Absorption maxima (cf. Tab. 12), the M_r of 44.2 kDa and the pI of 4.6 were equal for both P-450 forms and even the amino acid composition (cf. Tab. 13) was quite similar. $P\text{-}450_I$ reacted both with anti-$P\text{-}450_I$ and with anti-$P\text{-}450_{II}$ antibodies prepared from highly purified proteins. Taking into account that the only definite difference between $P\text{-}450_I$ and $P\text{-}450_{II}$ consisted in their distinct elution from hydroxylapatite, major structural differences between the two proteins have been ruled out. An equilibrium between two structural states of one single protein has been considered as quite improbable since the hydroxylapatite elution parameters of the original forms have been retained after a variety of treatments (SHAFIEE and HUTCHINSON, 1987).

The hypothesis that P-450 of *Sa. erythrea* exerts deoxyerythronolide B 6-hydroxylase activity is supported by the O_2 dependence as well as CO inhibition of this reaction (CORCORAN and VYGANTAS, 1982), by the parallel enrichment of P-450 and enzymic activity, and also by the low apparent K_m values of about 10 μM for the substrate analogue (9R)-9-deoxo-9-hydroxy-6-deoxyerythronolide B (SHAFIEE and HUTCHINSON, 1987). Beside the natural substrate 6-deoxyerythronolide B, the chemically more stable reduction products (9R)- and (9S)-9-deoxo-9-hydroxy-6-deoxyerythronolide B are hydroxylated. Specific activities were quite similar to those with 6-deoxyerythronolide B when tested with partially purified P-450 (Forms I and II not resolved). However, V_{max} values with the (9S) epimer differed between $P\text{-}450_I$ and $P\text{-}450_{II}$ at a factor of about 2.5 (SHAFIEE and HUTCHINSON, 1987). Other reactivities or substrate specificities have not been reported. Only the low-spin form of P-450 could hitherto be observed. The turnover numbers calculated either for cell-free extracts (0.23×10^{-3} min^{-1}) or for purified preparations determined with ferredoxin and ferredoxin reductase from spinach as the reductase part (maximum of 7×10^{-3} min^{-1}) appear to be very low and may argue against the in vivo function of the *Saccharopolyspora* P-450 as an 6-deoxyerythronolide B hydroxylase. Since data on in vivo hydroxylation rates are not available, it is rather difficult to evaluate whether the low in vitro estimated activities are sufficient to support the erythromycin biosynthesis or whether the activity is drastically reduced by cell disruption due to the destruction of essential factors. One of the affected factors may be the reducing system. Hydroxylase activity has been shown to be NAD(P)H-dependent and to require at least two protein fractions beside the P-450s themselves. Furthermore, it was shown that ferredoxin and ferredoxin reductase from spinach were capable of substituting, respectively of increasing the native reductase activity (SHAFIEE and HUTCHINSON, 1987).

Recently, two NAD(P)H: acceptor oxidoreductases and an iron-sulfur protein, designated erythrodoxin, have been purified from *Sa. erythrea* and shown to reconstitute together with purified P-450$_I$ or P-450$_{II}$ the hydroxylase activity (SHAFIEE and HUTCHINSON, 1988). The electron-transferring proteins have been recovered from a 0.65 M KCl eluate of a DEAE-Sepharose column whereas P-450 was eluted already at 0.2 M KCl. By means of repeated DEAE-Sepharose chromatography, gel filtration and preparative polyacrylamide gel electrophoresis a 47 kDa reductase, active with NADH and NADPH, a 53 kDa reductase, specific for NADH, as well as an 27.5 kDa iron-sulfur protein have been obtained in an obvious homogeneous state, as judged by SDS gel electrophoresis. The iron-sulfur protein with the unusual high M_r, a pI < 4, and an acid-labile sulfur content of about 10 mol/mol protein exhibited an optical absorption spectrum (cf. Fig. 7) like [4Fe–4S] cluster-containing ferredoxins (PALMER, 1975) with a broad absorption maximum between 394 and 404 nm in the oxidized state. It stimulated the reduction of mammalian cytochrome c by the 47 kDa reductase. Both reductases contain FAD as prosthetic groups, exert diaphoretic activity with hexacyanoferrate-(III) or mammalian cytochrome c, and support the hydroxylation in the presence of "erythrodoxin" (SHAFIEE and HUTCHINSON, 1988).

Inhibition of the 6-deoxyerythronolide B hydroxylase, apart from that by CO, is obviously exerted by the sterol biosynthesis inhibitor 3-β-[2-(dimethylamino)ethoxy]androst-5-en-17-one since the addition of this compound to *Sa. erythrea* cultures resulted in the accumulation of 6-deoxyerythronolide B. This has been used for the production of the hydroxylase substrates (CORCORAN and VYGANTAS, 1982).

Information on the regulation of P-450 or the hydroxylase level is not available. Thus, it can not be judged whether the content of 20 pmol P-450 per mg cytosolic protein represents a constitutive or an induced level of P-450 at the time of cell harvest. Maximum 6-deoxyerythronolide B production has been observed in 6 to 12 h old cultures using a vegetative medium containing tryptone and yeast extract as the main carbon sources.

Still, more extended investigations on the substrate specificity and regulation of P-450 from *Sa. erythrea* are necessary to resolve the question of whether 6-deoxyerythronolide B hydroxylase activity has to be regarded as a main function of this P-450 or whether it represents a side reactivity only.

3.10. Cytochromes P-450 in *Streptomycetaceae*

3.10.1. Veratrole-demethylating cytochrome P-450 in *Streptomyces setonii*

Beside other methylaryl ether compounds the lignin-degrading actinomycete *S. setonii* 75Vi2 is also known to metabolize veratrole (cf. Fig. 1) (SUTHERLAND, 1986). Initial degradation steps are demethylation reactions, as shown by the accumulation of catechol in cultures of *S. setonii* 75Vi2 grown on a veratrole- and yeast extract-containing medium in the presence of the catechol 1,2-dioxygenase inhibitor 2,2'-dipyridyl. The presumptive participation of a P-450 in this reaction was supported by the observation of a typical P-450 absorption band in CO difference spectra of the cytosolic protein fraction from cells that had been grown in the presence of veratrole or guaiacol. Extracts from succinate-grown cells failed to give such spectra indicating the inducibility of the P-450 from *S. setonii* by alkylaryl ethers. Additional evidence for the function of the P-450 in veratrole and guaiacol demethylation has been provided by the demonstration of type I difference spectra of cell-free extracts with veratrole and guaiacol, respectively, and of type II difference spectra with metyrapone. Determination and inhibition of cell-free demethylating activity as well as attempts to purify the P-450 in order to gather more conclusive evidence on the enzymic function of the P-450 from *S. setonii* have not been reported up to now.

3.10.2. Isoflavonoid-induced cytochrome P-450 of *Streptomyces griseus*

Considering the high versatility of *S. griseus* with respect to the transformation of numerous multifunctional xenobiotics and natural compounds and stimulated by the detection of P-450 in other *Streptomycetes*, the strains *S. griseus* ATCC 13273, 10137 and NRRL B8090 have been tested for the occurrence of P-450 (SARIASLANI and KUNZ, 1986).

High levels of P-450, up to 0.12 nmol/mg protein, have been detected in the cytosolic protein fraction of all 3 strains grown on media that contained glycerol and yeast extract as carbon sources and were furthermore supplemented by soybean flour. P-450 was not detected in the membrane fraction. The main partion of P-450 precipitated from cell-free extracts between 35 and 45% saturation of ammonium sulfate. The P-450 content of cells grown on complex media such as sporulation broth or nutrient broth that had not been supplemented by soybean flour was very low or undetectable. Thus, P-450 induction in response to components of soybean flour has been assumed. Soybean oil and the known soybean constitutents daidzain and coumestrol failed to achieve an induction. But, when the isoflavonoid genisteine (Fig. 5) was added, high levels of P-450

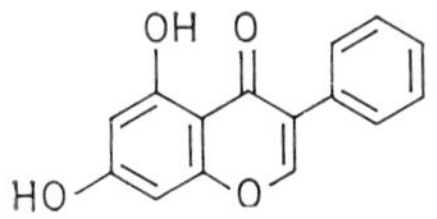

OH O

HO O

Fig. 5. Structure of genisteine, an inducer of cytochrome P-450 in *S. griseus.*

similar to those obtained in response to soybean flour were observed. Genisteine itself is not a constituent of soybean flour, but seems to be produced by the cells from its glucoside conjugate genistin that represents a major isoflavonoid compound in soybean flour. Up to now the inducing capability of genistin has not been tested.

3.10.3. Cytochromes P-450 of sulfonylurea herbicides-transforming *Streptomyces griseolus*

To investigate the breakdown of sulfonylurea herbicides (Fig. 6) the metabolism of several of these compounds by the soil bacterium *S. griseolus* (ATCC strain 11796) was studied (ROMESSER and O'KEEFE, 1986). Hydroxylation, dealkylation as well as deesterification reactions have been observed (O'KEEFE et al., 1987). CO difference spectra with the cytosolic protein fraction revealed up to 0.43 nmol P-450/mg protein in extracts from cells treated with herbicides, whereas in extracts from untreated cells a P-450 content of only 0.06 nmol/mg protein was obtained (ROMESSER and O'KEEFE, 1986).

Anion-exchange HPLC chromatography of protein fractions prepared by ammonium sulfate precipitation from cells grown without or in the presence of the herbicides sulfometuron methyl, chlorsulfuron, and chlorimuron ethyl, respectively, (Fig. 6) revealed the occurrence of one minor form of P-450 (P-450_{con}) in all cells, of at least three different P-450s in chlorimuron ethyl-treated cells and of two P-450s in cells treated with sulfometuron methyl or chlorsulfuron (O'KEEFE et al., 1988). The major form of P-450 (P-450_{SU1}) was present in all sulfonylurea-treated cells but notably absent in untreated cells.

Purification of P-450 from ammonium sulfate precipitates (40–60% of saturation) by means of gel filtration HPLC, anion exchange HPLC and a

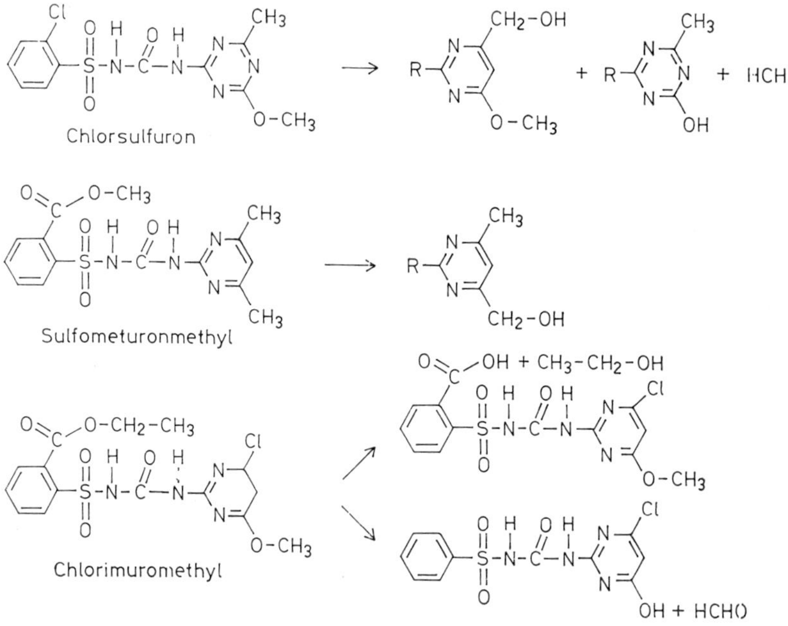

Fig. 6. Oxidation of sulfonylurea herbicides by cytochromes P-450 of *S. griseolus*.

second gel filtration yielded in a P-450$_{SU1}$ preparation containing 15 nmol P-450/mg protein. As judged by SDS gel electrophoresis, the protein was apparently homogeneous and had a M_r of about 46 kDa. Addition of 2 mM chlorsulfuron to the oxidized P-450 decreased the absorption at 418 nm concomitantly with a small increase of absorption at about 390 nm (O'KEEFE et al., 1988). Typical type I difference spectra with chlorsulfuron and chlorimuron ethyl, respectively, have been described for partially purified P-450 (40–60% $(NH_4)_2SO_4$ precipitates) from sulfometuron methyl-treated cells from which apparent K_s values of these substances of 0.74 and 0.22 mM, respectively, have been estimated (ROMESSER and O'KEEFE, 1986). P-450 binding of these compounds has not been observed with corresponding fractions of untreated cells. The CO complex of P-450$_{SU1}$ exhibited an absorption maximum of the Soret band at 449 nm which was similar to that of the partially purified P-450$_{SU2}$ from chlorimuron ethyl-induced cells but differed distinctly from the maximum at 451.5 nm of partially purified P-450$_{con}$.

That the herbicide-inducible P-450s are physiologically linked to the oxidation of the sulfonylureas has become apparent from the inhibition of activity by CO and from the obvious correlation between the induction of herbicide hydroxylation activity and the induction of P-450 (ROMESSER and O'KEEFE,

1986). Whereas cell-free extracts from herbicide-untreated cells did not hydroxylate the sulfonylureas, hydroxylation was achieved by the addition of purified P-450$_{SU1}$ to these extracts. Obviously, in contrast to the P-450$_{SU1}$ the electron transport system for P-450 reduction occurs constitutively (O'KEEFE et al., 1988). As reported by O'KEEFE et al. (1987), the P-450-reducing system consists of a NAD(P)H: iron-sulfur protein-oxidoreductase and an iron-sulfur protein containing four irons and four sulfurs. The still poorly characterized reductase has about equal affinity to NADPH and NADH. Furthermore, the reductase could be substituted by ferredoxin reductase from *spinach* or by putidaredoxin reductase from *Pseudomonas putida*.

The enzymic activities by which the P-450s of *S. griseolus* have been detected resemble those of xenobiotic metabolism in microsomes of higher organisms. As shown in Figure 6, sulfometuron methyl is transformed to a more polar product by hydroxylation of a methyl group bound to a pyrimidine ring of a complex compound. Chlorsulfuron was converted by both methyl group hydroxylation as well as O-demethylation and chlorimuron ethyl by O-demethylation or de-esterification. All three reactivities were demonstrated to be catalyzed by a single enzyme. More specifically, P-450$_{SU2}$ carried out only the O-demethylation of chlorimuron ethyl. None of the tested substrates were oxidized by P-450$_{CON}$ (O'KEEFE et al., 1987; ROMESSER and O'KEEFE, 1986). The distinct nature of these three P-450s has furthermore been established by the fact that antibodies against P-450$_{SU1}$ cross-reacted only very weakly with P-450$_{SU2}$ and P-450$_{CON}$. Anti-P-450$_{SU1}$ antibodies also failed to cross-react with P-450$_{CAM}$, several hemoproteins, and solubilized proteins of plant microsomes (O'KEEFE et al., 1987). The identity of the de-esterification with a P-450-catalyzed reaction is supported by concomitant, but mutually exclusive, O-demethylation and de-esterification of chlorimuron ethyl in a ratio of 1:1 by the purified P-450$_{SU1}$ (O'KEEFE et al., 1987). P-450-dependent de-esterification has also been reported to be carried out by microsomal P-450 of higher organisms (GUENGERICH, 1987).

To prove, whether the P-450 induction in *S. griseolus* in response to sulfonylureas represents a rather specific process or has to be considered as a more fortuitous detection, more extensive studies on the inducer specificity for these P-450s will be required. Nevertheless, the suppression of P-450 induction as well as that of sulfometuron methyl hydroxylase activity by chloramphenicol indicated a de-novo protein synthesis in response to sulfonylureas (ROMESSER and O'KEEFE, 1986). Obviously, the P-450 biosynthesis is not controlled by catabolite repression since all induction experiments have been carried out with cells growing on a complex medium. Influence of the growth phase on the induction of the individual P-450s still has to be explored.

That the sulfonylureas are appropriate substrates for the *S. griseolus* P-450s is apparent from the low K_m values of 0.77 mM and 0.06 mM estimated for sulfometuron methyl and chlorimuron ethyl, respectively (ROMESSER and

O'KEEFE, 1986), and from the quite high turnover number of about 10 to 20 min^{-1} (calculated from a specific activity of 7.8 nmol/min $\times$ mg protein of cell-free extracts) for the sulfometuron methyl hydroxylase. Nevertheless, the question about the actual physiological function of these P-450s able to convert such unusual substrates remains to be resolved.

4. Comparative evaluation of molecular and physiological properties of bacterial cytochromes P-450

The aim of this part of the paper is to give a condensed summary of the elementary data that characterize the bacterial P-450s described in more detail above. Comparative considerations focuse above all on the P-450_{cam} of *Pseudomonas putida* that represents the best characterized P-450 protein of all and which is very often cited as a synonym for bacterial P-450. Therefore, besides considering pecularities of individual P-450s, an attempt will be made to outline whether bacterial P-450s, with P-450_{cam} representing a general prototype of them, are clearly different from those of higher organisms or whether bacterial systems also reflect the functional and structural diversity of eucaryotic P-450s.

4.1. Cellular organization and molecular properties of diverse bacterial cytochrome P-450 systems

None of the eucaryotic P-450s, except the P-450 of the filamentous fungus *Fusarium oxysporum* (SHOUN et al., 1983), has been found to be a soluble protein. They are embedded either in the mitochondrial membrane or in membranes that are recovered from the microsomal fraction. Until now, corresponding membrane equivalents of the bacterial cell that carry P-450s have not been described and nearly all bacterial P-450s have been recovered from the cytosolic protein fraction. This applies also to the electron transfer components of bacterial P-450 systems so far described. The only observation where P-450 is mainly distributed into a particulate fraction concerns cell-free extracts of hexadecane-grown cells of *Acinetobacter* (MÜLLER et al., 1989). A membrane binding, however, could not be demonstrated unambiguously and several facts (cf. 3.3.) argue against a membrane-bound P-450 in vivo. So far, with respect to subcellular localization, all known bacterial P-450 systems resemble the system of *Pseudomonas putida.*

Many more difficulties exist concerning general conclusions about the molecular composition of bacterial P-450 systems. Apart from the fatty acid monooxygenase of *Bacillus megaterium,* only the reductase part for eight other bacterial P-450 systems has been investigated (Tab. 11). For six systems,

Table 11. Composition of bacterial cytochrome P-450 systems

Bacterium	Electron donor	Reducing proteins	Ref.
Pseudomonas putida ATCC 17453	NADH	[2Fe-2S] Ferredoxin and FAD-dependent reductase	a
Pseudomonas putida PpG 777	NADH	[2Fe-2S] Ferredoxin and FAD-dependent reductase	b
Acinetobacter calcoaceticus EB 104	NADH	[2Fe-2S] Ferredoxin and flavoprotein reductase	c
Bacillus megaterium (ATCC 13368)	NADPH	Iron-sulfur protein and FMN-dependent reductase	d
Bacillus megaterium (ATCC 14581)	NADPH	FAD- and FMN-containing single-molecule P-450 system	e
Nocardia sp. NH1	NADH	?	f
Rhodococcus rhodochrous	NADH	?	g
Saccharopolyspora erythrea	NAD(P)H	Iron-sulfur protein and one or two reductases	h
Streptomyces griseolus	NAD(P)H	[4Fe-4S] Iron-sulfur protein and a reductase	i

a) Gunsalus et al., 1974; b) Gunsalus, personal communication; c) Asperger et al., 1983; d) Berg, 1982; e) Narhi and Fulco, 1986; f) Cartwright and Broadbent, 1974; g) Cardini and Jurtshuk, 1970; h) Shafiee and Hutchinson, 1988; i) O'Keefe et al., 1987.

the participation of iron-sulfur proteins has been claimed. The available experimental data does not allow ruling out the involvement of such proteins in the other P-450 systems. The reductases are mostly NADH-dependent or accept both pyridine nucleotides. NADPH specificity is described for the *Bacilli* systems (Tab. 11). But, it has to be considered that, except for the *Pseudomonas* systems, the reductase are the most weakly characterized system components. Furthermore, it must be said, that striking differences are to be observed in respect of the nature of the ironsulfur proteins claimed to be constituents of the P-450 systems (Fig. 7). Typical [2Fe—2S] ferredoxin spectra comparable to the adrenodoxin of mitochondrial P-450 systems have been reported only for the *Pseudomonas* and *Acinetobacter* P-450s, whereas for the iron-sulfur proteins from grampositive bacteria either the Fe:S:protein ratio does not correspond to the type of the spectrum (Fig. 7d) or the stoichiometry as well as the optical spectra argue for the presence of

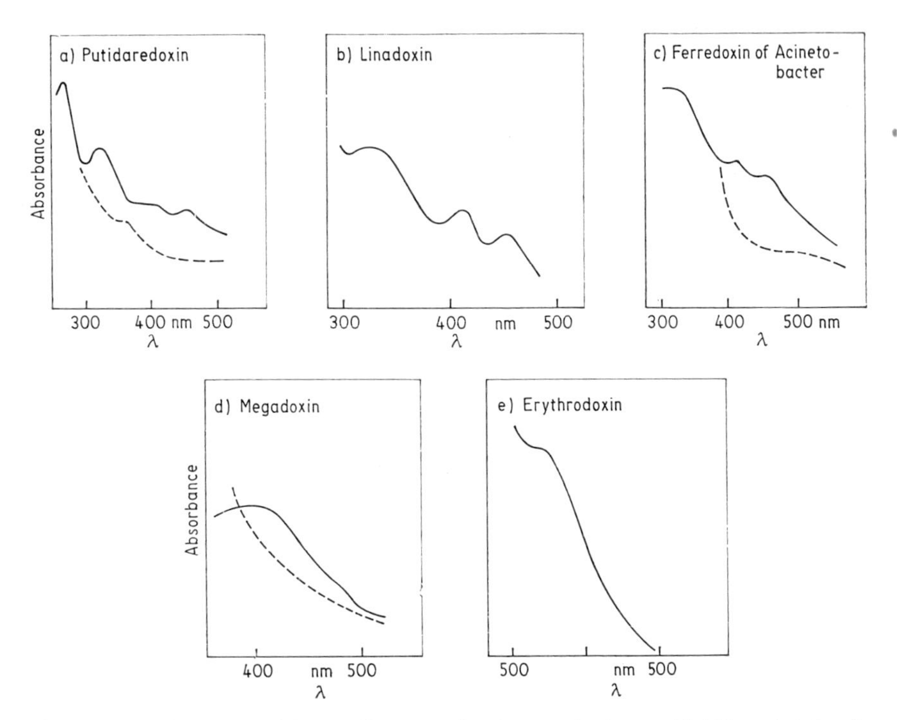

Fig. 7. Optical spectra of iron-sulfur proteins from cytochrome P-450-containing bacteria. (——) Oxidized state, (— — —) reduced state. References: a) GUNSALUS et al., 1974; b) GUNSALUS, personal communication; c) ASPERGER et al., 1983; d) BERG, 1982; e) SHAFIEE and HUTCHINSON, 1988.

[4Fe—4S] clusters (Fig. 7e) (PALMER, 1975). Altogether, little additional information such as ESR data or redox potentials is available to discuss this problem without speculating too much. So far, the only bacterial P-450 system resembling the microsomal systems in respect of its electron transfer chain is represented by the fatty acid monooxygenase from *Bacillus megaterium*. But, in this case also experimental proof for the absence of non-heme iron and acid-labile sulfur still has to be provided or the reconstitution of enzymic activity from proteolytic fragments has to be achieved.

Molecular data and spectral features of the P-450 proteins themselves are summarized in Table 12 and 13. Except P-450_{BM-3}, the M_r are in the range of those of P-450_{cam} and of P-450s from higher organisms (BLACK and COON, 1987). This proves to be true also for the absorbance maxima of optical spectra and, as so far determined, for ESR data (Tab. 12). Congruity between the amino acid compositions of different bacterial P-450s is not to be observed (Tab. 13). But, amino acid composition and pI of bacterial P-450s on the one hand and eucaryotic P-450s on the other hand clearly allows the recognition of

Table 12. Molecular and spectral properties of purified bacterial cytochromes P-450

Bacterium/P-450	M_r	pI	Maxima of optical absorption spectra[1]				ESR signals[2]	Ref.[3]
	(kDa)		ox	ox_S	red	CO	(g values)	
Pseudomonas putida (P-450$_{cam}$)	44	4.6	417; 535; 568	391; 540; 646	411; 540	447	1.91; 2.26; 2.46	i
Pseudomonas putida (P-450$_{lin}$)	45	4.83	417; 538; 568	392; 540; 646	411; 542	447		k, l
Bradyrhizobium japonicum								
P-450a			416; 532; 564		413; 543	449		b
P-450b			396; 525; 560; 645		412; 542	449	1.92; 2.24; 2.40	
P-450b$_1$			390; 646				1.72; 3.81; 7.90	
P-450c	46	4.9	417; 522; 566; 645		408; 543	447		
Acinetobacter calcoaceticus	52	4.7	417; 537; 563		408; 539	448		c
Bacillus megaterium (13368)	52	4.9	417; 534; 571		412; 544	450	1.92; 2.24; 2.41	d
Bacillus megaterium (14581)								
P-450$_{BM-1}$	38		414; 533; 567		410; 540	448		e
P-450$_{BM-3}$ and P-450$_{BM-3-1}$[4]	119		416; 525; 562	389	413; 540	447		f, m
Nocardia sp. NH1	42.5	4.94	417; 535; 567	394	412	448		g
Saccharopolyspora erythrea								
P-450$_I$ and P-450$_{II}$	44.2	4.6	416; 532; 567		409; 543	449		n
Streptomyces griseolus								
P-450$_{SU1}$	46		418			448.9		h
P-450$_{SU2}$						449.4		
P-450$_{CON}$						451.5		

[1] Absolute spectra, except CO compound maxima for *S. griseolus* and *Saccharopolyspora*, ox = ferric, ox_S = ferric substrate bound, red = ferrous, CO = ferrous CO bound, — = maximum absent. [2] For the oxidized substrate-free form. [3] References: cf. Table 13 for b, c, d, e, f, g, h and i) Gunsalus et al., 1974; k) Unger et al., 1986; l) Gunsalus, personal communication; m) Narhi et al., 1988; n) O'Keefe et al., 1988. [4] Expressed in *E. coli*.

Table 13. Amino acid composition of bacterial cytochromes P-450

Strain/P-450	Amino acid																			Ref.
	Asx	Thr	Ser	Glx	Pro	Gly	Ala	Val	Met	Ile	Leu	Tyr	Phe	His	Lys	Arg	Cys	Trp	Total	
Pseudomonas putida	36	19	21	55	27	26	34	24	9	24	40	9	17	12	13	24	6	1	397	a
Bradyrhizobium japonicum																				
P-450c	33	14	20	54	25	33	54	25	8	14	46	6	17	12	14	29	2	1	407	b
Acinetobacter calcoaceticus	45	22	21	54	21	27	33	32	10	20	31	6	19	11	24	20	2	2	400	c
Bacillus megaterium																				
ATCC 13368	49	24	34	55	25	27	25	27	7	27	44	11	17	9	28	15	10	1	435	d
Bacillus megaterium																				
ATCC 14581																				
P-450$_{BM-1}$	39	19	22	47	20	16	21	19	7	20	32	12	22	10	20	17	2	5	350	e
P-450$_{BM-3}$	99	53	60	143	53	120	93	71	23	48	99	41	45	24	63	44	12	9	1100	f
Nocardia sp. NH1	47	27	40	34	32	45	57	26	5	12	28	5	11	15	9	32	?	?		g
Saccharopolyspora erythrea																				
P-450$_{I}$	45	22	17	48	31	36	41	30	7	18	53	6	19	5	9	35	?	?		h
P-450$_{II}$	44	19	15	48	26	39	41	28	4	17	50	8	19	6	11	40	?	?		n

References: a) Haniu et al., 1982; b) Dus et al., 1976; c) Müller et al., 1989; d) Berg et al., 1979a; e) Schwalb et al., 1985; f) Narhi and Fulco, 1986; g) Broadbent and Cartwright, 1974; h) Shafiee and Hutchinson, 1987.

distinct structural features between these two groups. Hydrophobicity values, calculated as the percentage of the unweighted sum of Val, Met, Leu, Ile, Tyr, Phe, and Trp relative to the total number of residues, range between 19 up to 33.4%. This is significantly below the values of 35.5 up to 38.5% listed in the summary of BLACK and COON (1987) for eucaryotic P-450s. Furthermore, compared with the same eucaryotic P-450s, values of which are derived from sequence data, the total number of amino acid residues is decisively lower in the bacterial P-450s (Tab. 13). Whereas the pI values of several rat P-450s are all between seven and eight (SCHENKMAN et al., 1987), for all bacterial P-450s pI values between 4.6 and a maximum of five have been determined (Tab. 12). This difference is also reflected by the ratio of acidic to basic amino acid residues, amounting to about 0.7 to 0.8 for the membrane-bound P-450s but, to about one for the bacterial P-450_{cam} (BLACK and COON, 1987).

The lack of sequence data for bacterial P-450 under the current state of research, however, does not allow more informative comparisons with respect to phylogenetic relationships and makes it impossible at the present time to rank the bacterial P-450s into distinct P-450 gene families (NEBERT and GONZALEZ, 1987; NELSON and STROBEL, 1987).

4.2. Functional diversity and regulatory aspects of bacterial cytochromes P-450

Summarizing all the reactions claimed to be catalyzed by bacterial P-450s, it becomes evident that the procaryotic P-450s are also characterized by multireactivity and multisubstrate specificity. Furthermore, a broad spectrum of physiological importance is obvious. This comprises the utilization of foreign compounds as sole sources of carbon, the transformation of foreign natural and xenobiotic compounds, comparable to the detoxification reactions of microsomal P-450s (WISLOCKI et al., 1980) but also the biosynthesis of metabolic intermediates.

The functional role in degradative pathways is most obvious for the *Pseudomonas* P-450s (UNGER et al., 1986). Physiological relevance is also apparent for the P-450 in the alkane oxidizers *Rhodococcus* and *Acinetobacter*, for the alkylaryl ether-assimilating *Moraxella, Nocardia* and *Streptomyces*, for the cholesterol-degrading *Corynebacterium* or for the P-450 of *Saccharopolyspora*. However, the P-450s of unknown function such as those of *Rhizobiaceae, Vibrio* and *Streptomyces griseus*, as well as the true physiological function of P-450 in *Bacillus megaterium* or in *Streptomyces griseolus* remain to be explored. Although well defined enzymic activity has been reported for these P-450s, either the inducers are unknown, do not coincide with the enzyme substrate or, in the case of *Streptomyces griseolus*, the substrate is a man-made compound. Thus, it may be supposed that other, hitherto unknown, physiological functions underlie these enzyme activities. A deeper insight into the versatile metabolic processes of bacteria will help to elucidate these problems in the future and, vice versa, the study of bacterial P-450s may accelerate the unraveling of the mechanism and regulation of hitherto unexplored reactions in bacterial physiology.

Although the substrate specificities of bacterial P-450s remain rather poorly investigated, it has become obvious from the hitherto known examples that the reactions catalyzed by bacterial P-450s correspond to a good deal to the reaction diversity observed for P-450 of higher organisms. The reactivity spectrum, taking together all bacterial P-450s, comprises the hydroxylation of methyl and methylene groups in alkanes, terpenoids, steroids, macrolides or in aliphatic substituents of more complex compounds. Furthermore, epoxidations of isolated double bonds, O-dealkylations or de-esterifications occur and for the P-450_{cam} dehalogenations (CASTRO et al., 1985; MURRAY et al., 1985) have also been reported. Up to now there have been no examples of the hydroxylation of an aromatic ring or for N-demethylations.

Multifunctional and multisubstrate activity is not only distributed among different species or strains but also resides in individual P-450s. This is illustrated by the hydroxylation/epoxidation activity of P-450_{BM-3}, by the hydroxylations, O-dealkylations and de-esterification catalyzed by the P-450s from *Streptomyces griseolus*, by diverse reactions such as alcohol oxidation, epoxidation, oxidative and reductive dehalogenations, demonstrated also for P-450_{cam}, beside the hydroxylation of camphor (GELB et al., 1982; CASTRO et al., 1985; GUNSALUS et al., 1985; MURRAY et al., 1985) as well as by the hydroxylation of a variety of linalool structural analogs in different positions by the P-450_{lin} from *Pseudomonas putida* PpG 777 (BHATTACHARYYA, 1984). Therefore, the proposed strict substrate specificity of bacterial P-450s deduced from earlier studies with P-450_{cam} seems not to be a common principle for all bacterial P-450s.

Another point that must be paid attention to is the exciting high turnover number known for the P-450_{cam} system and also reported for a further bacterial P-450, the fatty acid monooxygenase from *Bacillus megaterium*. Compared to these enzymes, the rather low activities measured for other bacterial P-450 systems may be a consequence of suboptimal reaction conditions or of ignorance of the best fitting substrates for the corresponding P-450. More work on the identification and isolation of the appropriate electron transfer components of bacterial P-450 systems as well as on the elaboration of optimal reconstitution conditions will be required to resolve the question of the catalytic potency of bacterial P-450s.

High-rate substrate flow through a P-450-dependent reaction step is achieved sn bacteria not only by favourable intrinsic enzyme parameters but also by a itrong inducibility of most of the bacterial P-450s. P-450 induction up to levels of about 2% of whole cell protein has been observed. Since the levels of apparently constitutive occurring P-450s, such as from *Bradyrhizobia*, *Vibrio*, steroid hydroxylating *Bacillus megaterium* or *Saccharopolyspora*, are much lower, the existence of yet unknown inducers for these P-450s may be assumed. Such an assumption is supported by the fortuitous discovery of P-450_{BM-3} inducibility in response to the non-substrate phenobarbital. Catabolite repression of P-450 biosynthesis seems not to be widespreadly distributed among bacterial species. It has been demonstrated only for P-450_{cam} (Gunsalus et al., 1974), can be ruled out for *Acinetobacter*, *Bacillus megaterium* and *Streptomyces griseolus* (cf. 3.3., 3.5., 3.10.3.) or has not been investigated at all. However, much experimental data indicates an impact of the physiological state of the cell on the regulation of P-450. Thus, oxygen limitation increased the n-alkane-inducible P-450 in *Acinetobacter*. In *Bradyrhizobium*, significant amounts of P-450 were produced only under anaerobic conditions. An increase of P-450 after growth limitation (limiting factors not defined) has also been observed with *Bacillus megaterium* strains ATCC 13368 and 14581 (cf. 3.5.1., 3.5.2.). These regulatory effects may indicate an involvement of bacterial P-450 in hitherto unresolved physiological processes different to the degradation of foreign compounds or to the biosynthesis of secondary products.

For most of the bacterial P-450s, induction by exposure of the organism to foreign compounds has been described. A rather striking phenomenon is the broad inducer specificity observed for the P-450 from *Acinetobacter* and *Bacillus megaterium* (cf. 3.3., 3.5.2.). Indirect effects releasing the P-450 induction in response to such a great variety of exogenous compounds may be discussed considering the more or less strong inhibitory action of most of the tested hydrocarbon compounds on the microbial growth. Otherwise, inductions mediated by induced repressor binding proteins should also exhibit broad inducer specificity if the inducer recognition is based mainly on hydrophobic interactions. In these cases, binding will be controlled above all by partition parameters. The spatial extension of the molecules will be an additional discriminating factor, but altogether the selectivity is expected to be rather low. On this basis, a great variety of compounds having hydrophobic structural elements to fit into a more or less defined hydrophobic binding cave of the putative inducer receptor protein present in *Acinetobacter* or *Bacillus megaterium* would be able to act as inducers. Highly hydrophobic compounds of simple structure, as for instance the middle-chain n-alkanes, should be able to interact with many diverse hydrophobic inducer binding sites, thus representing universal inducers for enzymes participating in the transformation of hydrocarbons or compounds derived from them. Different fine structures of the inducer binding centers should be responsible for the facts that in the case of more voluminous compounds, residues without polar groups and having a coplanar structure resembling the methylcholanthrene-type inducers of microsomal P-450s are required for P-450 induction in *Acinetobacter*, whereas the fatty acid monooxygenase of *Bacillus megaterium* can also be induced by more polar compounds not having coplanar residues. At the present, it can not be said whether the principles discussed for the regulation of P-450 from *Bacillus megaterium* and *Acinetobacter* also apply to other bacterical P-450s, since only a few regulatory studies with a rather restricted number of test compounds, mostly narrowly related in structure, have been carried out with the other bacteria.

Since information on the genetics are available only for P-450_{cam} (UNGER et al., 1986) and P-450_{BM-3}, it also remains an open question whether the multiplicity of P-450 found in several bacteria (*Bradyrhizobium*, *Bacillus megaterium*, *Saccharopolyspora* and *Streptomyces griseolus*) is genetically encoded or originates from posttranslational modifications.

Thus, there remain a lot of problems to be unraveled in future research on bacterial P-450s, from the distribution of P-450 among diverse bacterial taxons, their functional diversity and their regulation up to practical exploitation of the biotransformation potency of bacterial P-450s.

5. References

ANTHONY, C., (1986), Adv. Microbiol. Physiol. **27**, 113–210.

AOYAMA, Y., Y. YOSHIDA, T. NISHINO, H. KATSUKI, U. S. MAITRA, V. P. MOHAN, and D. B. SPRINSON , (1987), J. Biol. Chem. **262**, 14260–14264.

APPLEBY, C. A., (1967), Biochim. Biophys. Acta **147**, 399–402.

APPLEBY, C. A., (1969a), Biochim. Biophys. Acta **172**, 71–87.

APPLEBY, C. A., (1969b), Biochim. Biophys. Acta **172**, 88–105.

APPLEBY, C. A. and R. M. DANIEL, (1973), in: Oxidases and Related Redox Systems, (T.E. KING, H.S. MASON and M. MORRISON, eds.), Univ. Park Press, Baltimore, 515–528.

APPLEBY, C. A., G. L. TURNER, and P. K. MACNICOL, (1975), Biochim. Biophys. Acta **387**, 461–474.

APPLEBY, C. A., (1978), Methods Enzymol. **52**, 157–166.

ASPERGER, O. and H. AURICH, (1977), Z. Allg. Mikrobiol. **17**, 419–427.

ASPERGER, O., H.-P. KLEBER, and H. AURICH, (1978), Acta Biol. Med. Ger. **37**, 191–198.

ASPERGER, O., A. NAUMANN, and H.-P. KLEBER, (1981), FEMS Microbiol. Lett. **11**, 309–312.

ASPERGER, O., R. MÜLLER, and H.-P. KLEBER, (1983), Acta Biotechnol. **3**, 319–326.

ASPERGER, O., A. NAUMANN, and H.-P. KLEBER, (1984), Appl. Microbiol. Biotechnol. **19**, 398–403.

ASPERGER, O., R. MÜLLER, and H.-P. KLEBER, (1985a), in: Cytochrome P-450, Biochemistry, Biophysics, and Induction, (L. VERECZKEY and K. MAGYAR, eds.), Elsevier Science Publishers, Amsterdam, 447–450.

ASPERGER, O., B. STÜWER, and H.-P. KLEBER, (1985b), Appl. Mikrobiol. Biotechnol. **21**, 309–312.

ASPERGER, O., A. A. SHARYCHEV, R. N. MATYASHOVA, B. LOSINOV, and H.-P. KLEBER, (1986), J. Basic Microbiol. **26**, 571–576.

ASPERGER, D., P. BERNDT, and H.-P. KLEBER, (1988), Abstr. 14th Intern. Congr. Biochem. Prague, Th: 125.

BARANOVA, N. A., G. S. YEREMINA, V. S. DANILOV, and N. S. YEGOROV, (1982), Dokl. Akad. Nauk SSSR **262**, 1001–1004.

BERG, A., K. CARLSTRÖM, J.-A. GUSTAFSSON, and M. INGELMAN-SUNDBERG, (1975), Biochem. Biophys. Res. Commun. **66**, 1414–1423.

Berg, A., J.-A. Gustafsson, M. Ingelman-Sundberg, and K. Carlström, (1976), J. Biol. Chem. **251**, 2831—2838.
Berg, A., M. Ingelman-Sundberg, and J.-A. Gustafsson, (1979a), J. Biol. Chem. **254**, 5264—5271.
Berg, A., M. Ingelman-Sundberg, and J. A. Gustafsson, (1979b), Acta Biol. Med. Ger. **38**, 333—344.
Berg, A. and J. J. Rafter, (1981), Biochem. J. **196**, 781—786.
Berg, A., (1982), Biochem. Biophys. Res. Commun. **105**, 303—311.
Bernhardt, F. H., E. Heymann, and P. S. Traylor, (1978), Eur. J. Biochem. **92**, 209—223.
Bhattacharyya, P. K., T. B. Samanta, A. J. Ullah, and I. C. Gunsalus, (1984), Proc. Indian Acad. Sci. Sect. A. **93A**, 1289—1304.
Black, S. D. and M. J. Coon , (1987), Adv. Enzymol. **60**, 35—87.
Broadbent, D. A. and N. J. Cartwright, (1971), Microbios **4**, 7—12.
Broadbent, D. A. and N. J. Cartwright, (1974), Microbios **9**, 119—131.
Buchanan, J. F. and A. J. Fulco, (1978), Biochem. Biophys. Res. Commun. **85**, 1254—1260.
Cardini, G. and P. Jurtshuk, (1968), J. Biol. Chem. **243**, 6070—6072.
Cardini, G. and P. Jurtshuk, (1970), J. Biol. Chem. **245**, 2789—2796.
Cartwright, N. J., K. S. Holdom, and D. A. Broadbent, (1971), Microbios **3**, 113—130.
Cartwright, N. J. and D. A. Broadbent, (1974), Microbios **10**, 87—96.
Castro, C. E., R. S. Wade, and N. O. Belser, (1985), Biochemistry **24**, 204—210.
Conney, A. H., (1966), Pharmacol. Rev. **19**, 317—366.
Coon, M. J. and K. Inouye, (1985), Ann. N. Y. Acad. Sci. **458**, 216—224.
Corcoran, J. W. and A. M. Vygantas, (1982), Biochemistry **21**, 263—269.
Daniel, R. M. and C. A. Appleby, (1972), Biochim. Biophys. Acta **275**, 347—354.
Danilov, V. S., (1979), Dokl. Akad. Nauk SSSR **249**, 477—479.
Danilov, V. S., A. D. Ismailov, Y. A. Malkov, and N. S. Egorov, (1981), Bioorg. Khim. **7**, 68—74.
Danilov, V. S., N. A. Baranova, A. D. Ismailov, and N. S. Yegorov, (1982), Eur. J. Appl. Microbiol. Biotechnol. **14**, 125—129.
Danilov, V. S., Y. A. Malkov, and N. S. Yegorov, (1985), Stud. Biophys. **105**, 157 to 165.
Danilov, V. S., (1987), Dokl. Akad. Nauk SSSR **294**, 1477—1480.
Dardas, A., D. Gal, M. Barrelle, G. Sauret-Ignazi, R. Sterjiades, and J. Pelmont, (1985), Arch. Biochem. Biophys. **263**, 585—592.
Dus, K., R. Goewert, C. C. Weaver, D. C. Arey, and C. A. Appleby, (1976), Biochem. Biophys. Res. Commun. **69**, 437—445.
Ensley, A. and W. R. Finnerty, (1980), J. Bacteriol. **142**, 859—868.
Eremina, S. S., O. Asperger, and H.-P. Kleber, (1987), Mikrobiologiya **56**, 764—769.
Fulco, A. J., B. H. Kim, R. S. Matson, L. O. Narhi, and R. T. Ruettinger, (1983), Mol. Cell. Biochem. **53/54**, 155—161.
Fulco, A. J. and R. T. Ruettinger, (1987), Life Sci. **40**, 1769—1775.
Garfinkel, D., (1958), Arch. Biochem. Biophys. **77**, 493—509.
Gelb, M. H., P. Malkonen, and S. G. Sligar, (1982), Biochem. Biophys. Res. Commun. **104**, 853—858.
Guengerich, F. P., (1987), J. Biol. Chem. **262**, 8459—8462.
Gunsalus, I. C., J. R. Meeks, J. D. Lipscomb, P. Debrunner, and E. Münck, (1974), in: Molecular Mechanisms of Oxygen Activation, (O. Hayaishi, eds.), Academic Press, New York, 559—613.
Gunsalus, I. C., P. K. Bhattacharyya, and K. Suhara, (1985), Curr. Top. Cell. Regul. **26**, 295—309.

HANIU, M., L. G. ARMES, K. T. YASUNOBU, B. A. SHASTRY, and I. C. GUNSALUS, (1982), J. Biol. Chem. **257**, 12664—12671.
HARE, R. S. and A. J. FULCO, (1975), Biochem. Biophys. Res. Commun. **65**, 665—672.
HAURAND, M. and V. ULLRICH, (1985), J. Biol. Chem. **260**, 15059—15067.
HO, P. P. and A. J. FULCO, (1976), Biochim. Biophys. Acta **431**, 249—256.
HOLT, J. G., (1984), Bergey's Manual of Systematic Bacteriology (9th edition) William and Wilkins, Baltimore.
INOZENTSOVA, I. A., S. S. MELIK-SARKISSIAN, and W. L. KRETOVICH, (1978), Dokl. Akad. Nauk SSSR **240**, 1468—1471.
INOZENTSOVA, I. A., S. S. MELIK-SARKISSIAN, and V. L. KRETOVICH, (1979), Dokl. Akad. Nauk SSSR **246**, 741—744.
IOANNIDES, C. and D. V. PARKE, (1987), Biochem. Pharmacol. **36**, 4197—4207.
ISMAILOV, A. D., N. S. YEGOROV, and B. S. DANILOV, (1979), Dokl. Akad. Nauk SSSR **249**, 481—485.
ISMAILOV, A. D., N. A. BARANOVA, V. S. DANILOV, and N. S. YEGOROV, (1981), Biokhimiya **46**, 234—239.
KÄPPELI, O., (1986), Microbiol. Rev. **50**, 244—258.
KÄRGEL, E., W.-H. SCHUNCK, P. RIEGE, H. HONECK, R. CLAUS, H.-P. KLEBER, and H.-G. MÜLLER, (1985), Biochem. Biophys. Res. Commun. **128**, 1261—1267.
KATAGIRI, M., B. N. GANGULI, and J. C. GUNSALUS, (1968), J. Biol. Chem. **243**, 3543—3546.
KIM, B. H. and A. J. FULCO, (1983), Biochem. Biophys. Res. Commun. **116**, 843—850.
KLEBER, H.-P., R. CLAUS, and O. ASPERGER, (1983), Acta Biotechnol. **3**, 251—260.
KLEBER, H.-P., R. MÜLLER, and O. ASPERGER, (1985), in: Environmental Regulation of Microbial Metabolism, (I. S. KULAEV, E. A. DAWES, and D. W. TEMPEST, eds.), Academic Press, London, 89—95.
KLINGENBERG, M., (1958), Arch. Biochem. Biophys. **75**, 376—386.
KRETOVICH, W. L., S. S. MELIK-SARKISSIAN, and W. K. MATUS, (1972), Biokhimiya **37**, 711—719.
KRETOVICH, W. L., S. S. MELIK-SARKISSIAN, M. V. RAIKCHINSTEIN, and A. I. ARCHAKOV, (1974), FEBS Lett. **44**, 305—308.
LAMBEIR, A. M., C. A. APPLEBY, and H. B. DUNFORD, (1985), Biochim. Biophys. Acta **828**, 144—150.
MALKOV, Y. A., V. S. DANILOV, and N. S. YEGOROV, (1982), Prikl. Biokhim. Mikrobiol. **18**, 76—80.
MATSON, R. S., R. S. HARE, and A. J. FULCO, (1977), Biochim. Biophys. Acta **487**, 487—494.
MATSON, R. S., R. A. STEIN, and A. J. FULCO, (1980), Biochem. Biophys. Res. Commun. **97**, 955—962.
MATSON, R. S. and A. J. FULCO, (1981), Biochem. Biophys. Res. Commun. **103**, 531—535.
MATUS, V. K., S. S. MELIK-SARKISSIAN, and W. L. KRETOVICH, (1973), Mikrobiologiya **42**, 112—118.
MAY, S. W. and A. G. KATOPODIS, (1986), Enzyme Microb. Technol. **8**, 17—21.
MICHAELS, B. C., R. T. RUETTINGER, and A. J. FULCO, (1980), Biochem. Biophys. Res. Commun. **92**, 1189—1195.
MIURA, Y. and A. J. FULCO, (1974), J. Biol. Chem. **249**, 1880—1888.
MIURA, Y. and A. J. FULCO, (1975), Biochim. Biophys. Acta **388**, 305—317.
MÜLLER, H.-G., H. HONECK, S. MAUERSBERGER, P. RIEGE, W.-H. SCHUNCK, H. TERYTZE, and E. KÄRGEL, (1978), XII. Int. Congr. Microbiol., München, Abstr. E. 51.
MÜLLER, H.-G., W.-H. SCHUNCK, P. RIEGE, and H. HONECK, (1984), in: Cytochrome P-450, (K. RUCKPAUL and H. REIN, eds.), Akademie-Verlag, Berlin, 337—369.
MÜLLER, R., (1984), Dissertation A, Karl-Marx-University, Leipzig.
MÜLLER, R., O. ASPERGER, and H.-P. KLEBER, (1989), Biomed. Biochim. Acta **48**, 243—254.

MURRAY, R. J., M. T. FISHER, P. G. DEBRUNNER, and S. G. SLIGAR, (1985), in: Metalloproteins Part I: Metal Proteins with Redox Roles, (P. M. HARRISON, ed.), Mcmillan, New York, 157–206.
NARHI, L. O. and A. J. FULCO, (1982), J. Biol. Chem. **257**, 2147–2150.
NARHI, L. O., B. H. KIM, P. M. STEVENSON, and A. J. FULCO, (1983), Biochem. Biophys. Res. Commun. **116**, 851–858.
NARHI, L. O. and J. A. FULCO, (1986), J. Biol. Chem. **261**, 7160–7169.
NARHI, L. O. and A. J. FULCO, (1987), J. Biol. Chem. **262**, 6683–6690.
NARHI, L. O., L.-P. WEN, and A. J. FULCO, (1988), Mol. Cell. Biochem. **79**, 63–71.
NEBERT, D. W. and F. J. GONZALEZ, (1987), Ann. Rev. Biochem. **56**, 945–993.
O'KEEFE, D. P., J. A. ROMESSER, and K. J. LETO, (1987), in: Phytochemical Effects of Environmental Compounds, (J. A. SAUNDERS, L. KOSAK-CHANNING, and E. E. COON, eds.), Plenum Publishing Corporation, New York, 151–173.
O'KEEFE, D. P., J. A. ROMESSER, and K. J. LETO, (1988), Arch. Microbiol. **149**, 406–412.
PALMER, G., (1975), Enzymes **12**, 1–56.
PEISACH, J., C. A. APPLEBY, and W. E. BLUMBERG, (1972), Arch. Biochem. Biophys. **150**, 725–732.
RHEINWALD, J. G., A. M. CHAKRABARTY, and I. G. GUNSALUS, (1973), Proc. Natl. Acad. Sci. USA **70**, 885–889.
ROMESSER, J. A. and D. P. O'KEEFE, (1986), Biochem. Biophys. Res. Commun. **140**, 650–659.
RUETTINGER, R. T. and A. J. FULCO, (1981), J. Biol. Chem. **256**, 5728–5734.
RUETTINGER, R. T., B.-H. KIM, and A. J. FULCO, (1984), Biochim. Biophys. Acta **801**, 372–380.
SARIASLANI, F. S. and D. A. KUNZ, (1986), Biochem. Biophys. Res. Commun. **141**, 405–410.
SCHENKMANN, J. B., L. V. FAVREAU, I. JANSSON, and J. E. MOLE, (1987), in: Drug Metabolism from Molecules to Man, (D. T. BENFORD, T. W. BRIDGES and G. G. GIBSON, eds.), Taylor and Francis, London, 1–13.
SCHWALB, H., L. O. NARHI, and A. J. FULCO, (1985), Biochim. Biophys. Acta **838**, 302–311.
SHAFIEE, A. and C. R. HUTCHINSON, (1987), Biochemistry **26**, 6204–6210.
SHAFIEE, A. and C. R. HUTCHINSON, (1988), J. Bacteriol. **4**, 1548–1553.
SHOUN, H., Y. SUDO, Y. SETO, and T. BEPPU, (1983), J. Biochem. **94**, 1219–1229.
SHUMIKHIN, V. N., V. S. DANILOV, Y. A. MALKOV, and N. S. YEGOROV, (1980), Biokhimiya **45**, 1576–1581.
SINGER, M. E. and W. R. FINNERTY, (1984), in: Petroleum Microbiology, (R. M. ATLAS, ed.), Macmillan Publishing Company, New York, 1–59.
SLIGAR, S. G. and R. J. MURRAY, (1986), in: Cytochrome P-450, Structure, Mechanism, and Biochemistry, (P. R. ORTIZ DE MONTELLANO, ed.), Plenum Press, New York, 290–296.
STIRLING, D. I., J. COLBY, and H. DALTON, (1979), Biochem. J. **177**, 361–364.
SUTHERLAND, J. B., (1986), Appl. Environ. Microbiol. **52**, 98–100.
UNGER, B. P., S. G. SLIGAR, and I. C. GUNSALUS, (1986), in: The Bacteria, Vol. X, (J. R. SOKATCH, ed.), Academic Press, Orlando, 557–589.
WEN, L.-P. and A. J. FULCO, (1985), Mol. Cell. Biochem. **67**, 77–81.
WEN, L.-P. and A. J. FULCO, (1987), J. Biol. Chem. **262**, 6676–6682.
WYNDHAM, R. C., (1987), Can. J. Microbiol. **33**, 1–5.
YAMAMOTO, A., S. TAKEMORI, and M. KATAGIRI, (1974), Seikagaku **46**, 534, cited in: MÜLLER et al., 1984.
YOSHIDA, Y., (1978), in: Cytochrome P-450, (R. SATO and T. OMURA, eds.), Kodanska, Academic Press, Tokyo, 194–202.

Chapter 2

Molecular Recognition by Cytochrome P-450$_{cam}$: Substrate Specifity, Catalysis and Electron Transfer

S. A. MARTINIS, J. D. ROPP, S. G. SLIGAR, and I. C. GUNSALUS

1. Introduction

The level of understanding in any biological system is directly proportional to one's ability to specifically perturb the integrity of the native system at the molecular level and monitor the effects of these alterations in a precise biological assay. In the case of enzyme structure-function investigations, one valuable approach is to selectively mutate the amino acid residue(s) of interest and evaluate the resultant characteristics and enzymatic activity of the mutant protein. The large number of genetically controllable proteins along with the relative ease of site-directed mutagenesis has made this technique of probing the structure-function relationships of proteins a routine procedure in many laboratories (STAYTON et al., 1989; OXENDER and FOX, 1987; GERLT, 1987; MOZHAEV et al., 1988). The current status in the application of genetic engineering to protein structure-function studies is exemplified by recent work involving cytochrome P-450$_{cam}$ (ATKINS and SLIGAR, 1988b; ATKINS and SLIGAR, 1989, 1990; ATKINS, 1988). In this system the use of site-directed mutagenesis has contributed to determining those residues at the enzyme's active site involved in substrate orientation, the molecular recognition of cytochrome P-450$_{cam}$ with its redox partners, as well as in the investigation of the unique thiolate heme ligation of cytochrome P-450$_{cam}$.

The wealth of information available on the reaction cycle, heme center spectroscopy, chemistry, and the tertiary structure of cytochrome P-450$_{cam}$ make it an ideal system for investigation. It is important to have such detailed information in order to facilitate rational design for experimental inquiry. Although it is not the intent of this review to outline all efforts on cytochrome P-450$_{cam}$ since its discovery in 1965 (HEDEGAARD et al., 1965), it is essential to the integrity of this chapter to summarize the background pertinent to the work reviewed herein; this will include a brief description of the notable features of the cytochrome P-450 family, followed by the cytochrome P-450$_{cam}$ reaction cycle, and finally will include a synopsis of the crystal structure results of cytochrome P-450$_{cam}$.

2. Background

The cytochromes P-450 are a diverse class of b-type heme-containing monooxygenases whose unique chemistry and spectral properties have been attributed to the mercaptide ligation of the iron protoporphyrin IX prosthetic group. Early investigators of this class of monooxygenase coined the term P-450 to denote a pigment with an absorption at 450 nm upon observation of a broad but intense absorption band at 450 nm after bubbling carbon monoxide into a dithionite reduced sample of mammalian microsomes (KLINGENBERG, 1958; GARFINKEL, 1958; OMURA and SATO, 1962, 1964a, b). Currently this enzyme family is found to be involved in a wide variety of biotransformations including such diverse examples as xenobiotic detoxification (EISEN, 1986) and steroid hormone biosynthesis (WATERMAN et al., 1986; JEFCOATE, 1986) in eucaryotes as well as the metabolism of environmental hydrocarbons in procaryotes (SLIGAR and MURRAY, 1986). The plethora of physiological biotransformations is the result of the wide range of chemical oxidations performed by the cytochromes P-450; these include aliphatic hydroxylation, N-, S-, and O-dealkylation, olefin and aryl epoxidation, C-C lyase activity, N- and S-oxygenation, and C-C desaturase activity (WHITE and COON, 1980; SLIGAR et al., 1984; WATANABE et al., 1981, 1982; TAKATA et al., 1980).

The cytochromes P-450 can be conceptually divided into two classes based on the components of their electron transport chains. The microbial and mitochondrial systems are distinguished by three protein components of the monooxygenase system. Specifically, NAD(P)H[1] serves as the source of the two electrons necessary to reduce an FAD-flavoprotein reductase which ultimately passes single electron equivalents to the cytochrome P-450 heme center via the iron-sulfur center of a redoxin protein. A generalized reaction cycle for cytochrome P-450_{cam} is shown in Figure 1. In contrast, for the microsomal monooxygenases, two electrons are passed, via NADPH, to the FAD prosthetic group of a more complicated FAD/FMN flavoprotein reductase; subsequent single electron transfer to the heme of cytochrome P-450 is mediated by the FMN prosthetic group. For a more detailed account of the reaction cycle of both eucaryotic and procaryotic cytochrome P-450 monooxygenases, the reader is directed to the following reviews: White and COON, 1980; SLIGAR and MURRAY, 1986; MURRAY et al., 1985; ORTIZ DE MONTELLANO, 1986.

Though the number of cytochrome P-450 isozymes presently under investigation is quite large, one in particular, cytochrome P-450_{cam}, has proven to be an excellent model in terms of the chemistry and physics of cytochrome P-450 catalysis, as well as in the biology and genetics of the regulation of the hydroxylase gene.

3. Camphor hydroxylation

Originally isolated from *Pseudomonas putida* (HEDEGAARD et al., 1965), cytochrome P-450_{cam} is a soluble enzyme responsible for the 5-exo hydroxylation of the monoterpene *d*-camphor. This hydroxylation reaction is the initial

[1] Standard abbreviations used throughout the text are as follows: NADH, nicotinamide adenine dinucleotide; NADPH, nicotinamide adenine dinucleotide phosphate; FAD, flavin adenine dinucleotide; FMN, flavin mononucleotide; UV, ultraviolet; ENDOR, electron nuclear double resonance; O_2, molecular dioxygen; H_2O_2, hydrogen peroxide; K_D, dissociation constant; k_{cat}, turnover number; K_M, Michaelis-Menten constant; ΔG_{spin}, the difference in the free energy of the ferric spin equilibrium; CO, carbon monoxide; V_{max}, maximal velocity; EPR, electron paramagnetic resonance; v_0, initial velocity; K^{app}, apparent dissociation constant; K_I, inhibitor dissociation constant.

= S
Fe^{3+} IIISH
(391 nm)
Fe^{2+} IIISH (408nm)
e_1
NADH
fp^o
Pd^r
O_2
NAD^+
fp^r
Pd^o
Fe^{3+} (417nm)
e_2
Fe^{3+} IIISH (418nm)
?
OH = S-OH
H
+
H_2O
Autodecay
Fe^{3+} IIISH + O_2 + e-

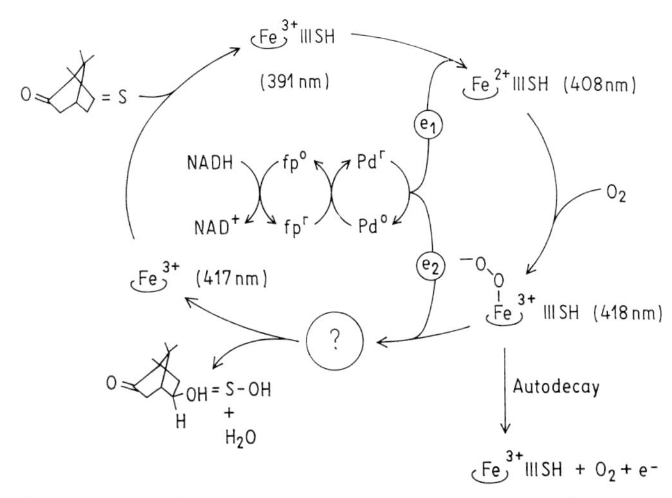

Fig. 1. A generalized reaction cycle of the cytochrome P-450_{cam} system. The abbreviations are as follows: S: substrate; NADH: reduced nicotinamide adenine dinucleotide; NAD^+: oxidized nicotinamide adenine dinucleotide; e: electron; fp: putidaredoxin reductase; and Pd: putidaredoxin. The intermediates of cytochrome P-450_{cam} may be characterized by a distinct optical spectrum; the maxima are recorded in parenthesis by the appropriate species.

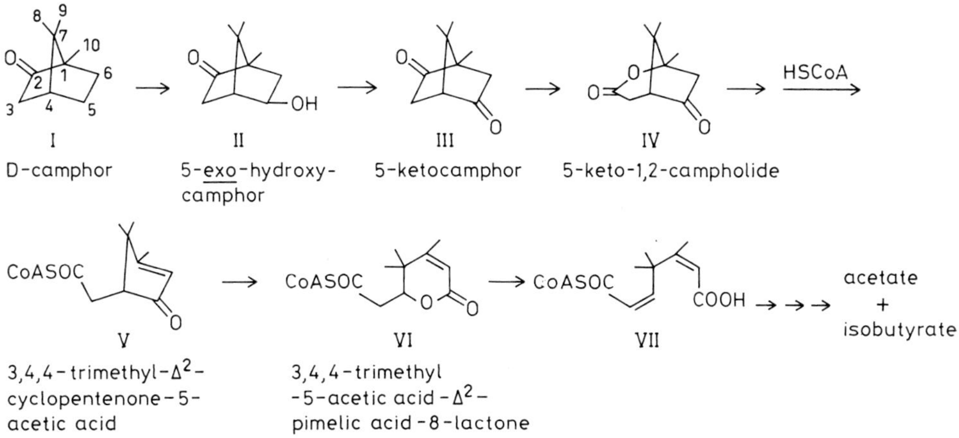

HSCoA
I
D-camphor
II
5-exo-hydroxy-camphor
III
5-ketocamphor
IV
5-keto-1,2-campholide
CoASOC
V
3,4,4-trimethyl-Δ^2-cyclopentenone-5-acetic acid
CoASOC
VI
3,4,4-trimethyl -5-acetic acid -Δ^2- pimelic acid -8-lactone
CoASOC
COOH
VII
acetate
+
isobutyrate

Fig. 2. Metabolic pathway for camphor degradation by *Pseudomonas putida*. Cytochrome P-450_{cam} catalyzes the regio- and stereospecific hydroxylation of d-camphor, the first step in the metabolic pathway.

step in the pathway which degrades camphor to isobutyrate and acetate (Fig. 2), thereby allowing the organism to utilize camphor as its sole source of carbon. Thorough investigation of cytochrome P-450_{cam} (Griffin et al., 1979) with techniques such as UV-visible spectroscopy (Gunsalus and Sligar,

1978; HANSON et al., 1976, 1977), magnetic circular dichroism (DAWSON and SONO, 1987; VICKERY et al., 1975; SHIMIZU et al., 1975), electron paramagnetic resonance (O'KEEFE et al., 1978; MASON et al., 1965; PETERSON, 1971; TSAI et al., 1970; PEISACH et al., 1971; LIPSCOMB, 1980), MÖSSBAUER spectroscopy (SHARROCK et al., 1973; MÜNCK and CHAMPION, 1975; CHAMPION et al., 1975), nuclear magnetic resonance (PHILSON et al., 1979; KELLER et al., 1972; GRIFFIN and PETERSON, 1975), ENDOR spectroscopy (LOBRUTTO et al.,

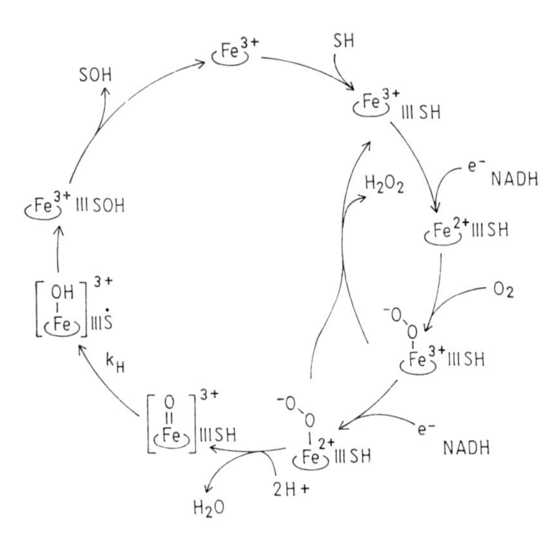

Fig. 3. Intermediates in the reaction cycle of cytochrome P-450_{cam}. Substrate binding of cytochrome P-450_{cam} elicits an increase in reduction potential allowing the input of NADH-derived electrons. Dioxygen binds at the ferrous center followed by the introduction of a second electron. Cleavage of the dioxygen bond results in an intermediate that has not been experimentally detected. Rather the indicated iron-oxo intermediates are postulated by analogy to the peroxidase family. These putative intermediate(s) hydroxylate substrate which is then released, regenerating the ferric resting state of the enzyme. In side reactions, autooxidation may occur producing hydrogen peroxide. SH and SOH refer to substrate and hydroxylated product, respectively.

1980), resonance Raman spectroscopy (BANGCHAROENPAURPONG et al., 1986; 1987; OZAKI et al., 1978; CHAMPION et al., 1978), differential scanning calorimetry (JUNG et al., 1985), and high pressure methodologies (MARDEN and HUI BON HOA, 1982; 1987; HUI BON HOA et al., 1988; FISHER et al., 1985) has provided a detailed model of the cytochrome P-450_{cam} reaction cycle.

A scheme for the following reaction cycle of the cytochrome P-450_{cam}-catalyzed camphor hydroxylation is presented in Figure 3. The enzyme's resting state is characterized by a ferric (Fe(III)) low-spin ($S = 1/2$) heme, which becomes predominately high-spin ($S = 5/2$) upon binding of the substrate camphor. As shown in Figure 1, this spin-state shift can be monitored spectrally by a Soret maxima shift from 417 to 391 nm. Coupled to this spin-state shift is an increase in the reduction potential from −300 mV to −170 mV (SLIGAR, 1976). This increase in potential makes the one electron reduction of cytochrome P-450$_{cam}$ by its electron transfer partner putidaredoxin, which has a reduction potential of −196 mV when bound to cytochrome P-450, thermodynamically favorable (SLIGAR and GUNSALUS, 1979). It has been proposed that some structural feature of the cytochrome P-450$_{cam}$-putidaredoxin complex, specifically the carboxy terminus of putidared oxin, facilitates on substrate hydroxylation in a manner unrelated to the required electron transfer process (SLIGAR et al., 1974; SLIGAR et al., 1980). Similar effects are observed in the adrenal mitochondrial systems, however, the generality of this "effector" function (LIPSCOMB et al., 1976) aside from electron transfer is not readily applicable to all cytochrome P-450 systems. The "effector" function of putidaredoxin will be addressed in a subsequent section of this chapter. Once reduced to the ferrous (Fe(II)) species, the camphor-bound, high-spin heme iron is able to bind molecular oxygen. Structural information from Mössbauer studies (CHAMPION et al., 1978) and resonance Raman spectroscopy (BANGCHAROENPAURPONG et al., 1986) demonstrate that this intermediate is best described as a superoxide anion bound to a ferric iron center, with the electron density assigned to the distal oxygen atom, not the heme iron. This intermediate has a room temperature half-life of about ninety seconds (LIPSCOMB et al., 1976) and its first order decay generates superoxide anion (SLIGAR et al., 1974a) and camphor-bound ferric enzyme in the autooxidation pathway. Steady-state analysis of the turnover of camphor by the native reconstituted system (PETERSON et al., 1972; HINTZ and PETERSON, 1981; HINTZ et al., 1982) indicates that this intermediate, denoted Fe(III)O-O$^-$, is the predominant species, and under single-turnover conditions is the last observable iron-oxygen intermediate species before the regeneration of the ferric resting state of the enzyme. These results and others (PEDERSON et al., 1977) propose the second electron transfer, also mediated by putidaredoxin via putidaredoxin reductase and NADH, to be the rate limiting step in the overall cycle. Alternatively, some reports have shown that under different conditions the first electron transfer can be the rate limiting step (BREWER and PETERSON, 1988). What remains consistent in all investigations, however, is that cytochrome P-450$_{cam}$ utilizes electron transfer rather than release of product to control its overall rate of reaction (HINTZ and PETERSON, 1981; HINTZ et al., 1982; BREWER and PETERSON, 1988; PEDERSON et al., 1977; HUI BON HOA, 1978). Figure 3 illustrates the existence of a nonproductive pathway which diverts electrons to the formation of hydrogen peroxide and camphor-bound ferric enzyme. As mentioned above, this may result from disproportionation of two equivalents of superoxide anion generated from the ferrous-dioxygen intermediate (SLIGAR et al., 1976; INGRAHAM and MEYER, 1985), or from the collapse of the two electron reduced oxygen bound species

$(Fe(II)O\text{-}O^-)$ prior to O-O bond scission. The precise mechanism of O-O bond cleavage has not been established. Investigations utilizing exogenous oxidants, such as peracids and aromatic peroxides to turnover cytochrome P-450_{cam} produced results suggestive of a homolytic cleavage mechanism of the peroxy bond (McCarthy and White, 1983a, b; Blake and Coon, 1981; White et al., 1980). However, the turnover of similar exogenous oxidants by metalloporphyrin model systems has indicated that either homolytic or heterolytic cleavage of the dioxygen bond may occur depending on the acidity of the oxidant (Lee and Bruice, 1985; Yuan and Bruice, 1985; Lee et al., 1988). Furthermore, the peroxidase family, which shares some of the catalytic capabilities of the cytochromes P-450 (Padbury and Sligar, 1985), operates via a heterolytic scission of the peroxy bond aided by a distal acid or base group (Holtzwarth et al., 1988; Poulos et al., 1980; Poulos and Kraut, 1980). Therefore, contradictory results supporting both homolytic and heterolytic mechanisms have hindered the understanding of oxygen activation and despite intense efforts, there has been no evidence to establish unequivocally the type of O-O bond cleavage mechanism when turnover is driven by the native reconstituted system and dioxygen.

No intermediates have been detected subsequent to O-O bond scission due to their transient nature. Rather, by analogy to the peroxidases (Marnett et al., 1986) and metalloporphyrin model systems (McMurray and Groves, 1986), the following scheme has been proposed to complete the catalytic cycle. Upon dioxygen bond cleavage, formation of a compound I-type species [Fe(V)O] has been postulated. In accordance with mechanisms suggested for the mammalian cytochromes P-450 (Groves and Subramanian, 1984), this intermediate carries out hydrogen abstraction from either the 5-*exo*- or 5-*endo*-face of the substrate by a radical mechanism (Gelb et al., 1982), resulting in a compound II-type intermediate, specifically a ferryl iron bound to a hydroxyl group (Fe(IV)OH). The hydroxyl group rebounds in a highly stereospecific manner, affording 5-*exo*-hydroxycamphor as the sole product and regenerating the ferric resting state of the enzyme.

4. Crystal structures

The crystallization of cytochrome P-450_{cam} (Poulos, et al., 1982) has provided an abundance of structural information, which has enhanced its role as a model system for the family of the cytochromes P-450. X-ray diffraction has produced high resolution crystal structures of substrate-free (Poulos et al., 1986), substrate-bound (Figure 4; Poulos et al., 1985, 1986b), and several inhibitor complexes (Poulos et al., 1987) of cytochrome P-450_{cam}. In short, the macromolecule is comprised of 40% alpha helices and 10% antiparallel beta sheet pairs which fold to form a triangular prism with dimensions of $60 \times 55 \times 30$ Å.

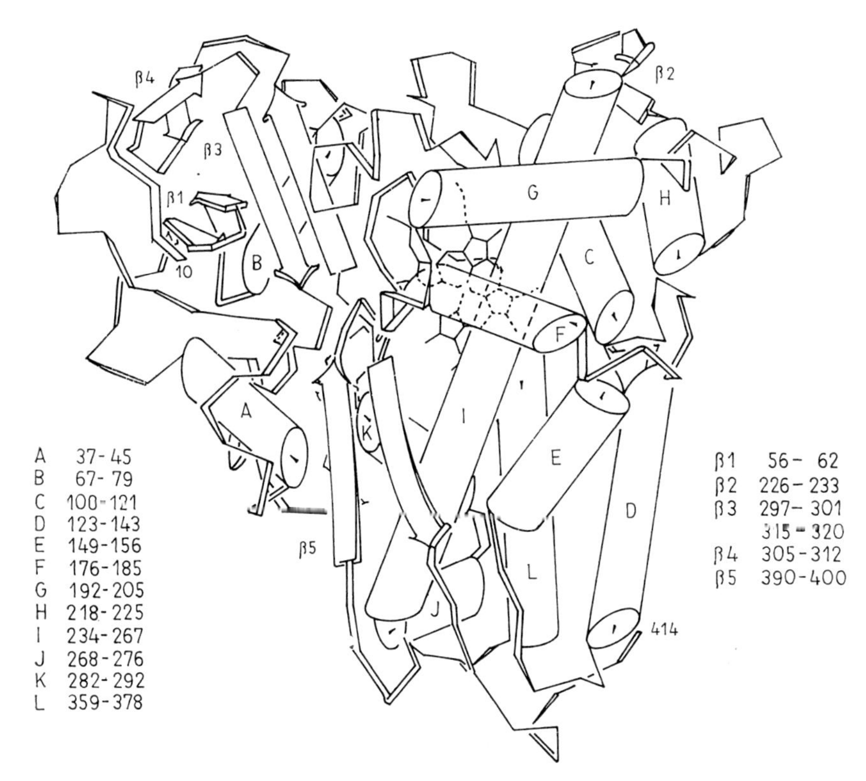

Fig. 4. Crystal structure of substrate-bound cytochrome P-450.

A comparison between the crystal structures of the camphor-free and -bound protein reveals highly conserved overall topologies. In the absence of an obvious camphor access channel, it has been postulated that the substrate's entry is facilitated by local dynamic fluctuations centralized around a small depression which lies directly over the camphor binding site. The most significant alteration induced by substrate binding occurs in the active site which, in the absence of camphor, is filled with an ordered network of six hydrogen bonded water molecules, one of which is liganded to the iron center. The end result is a low-spin ferric complex. Upon camphor binding, water is displaced producing a high-spin, penta-coordinated ferric center.

The resolved crystal structure reveals a hydrophobic active site which offers no acid or base groups to aid in catalysis. Specific interactions with amino acid side chains serve to anchor the camphor so that complete regio- and stereospecificity of hydroxylation is achieved (Fig. 5). One of the more striking features is active site Tyr 96, which forms a hydrogen bond to the carbonyl moiety of camphor. Two other residues, Phe 87 and Leu 244, lie in contact with the carbonyl, sterically stabilizing it. The *gem*-dimethyl group of Val 295 interdigitates between the 8,9 *gem*-dimethyl bridgehead substituent of camphor, influencing the substrate orientation through van der Waals forces. Finally, a number of other hydrophobic amino acids, including Val 247 and Leu 244, further define the camphor envelope, completely enclosing and orientating it in proximity to the heme, thus commanding highly specific hydroxylation.

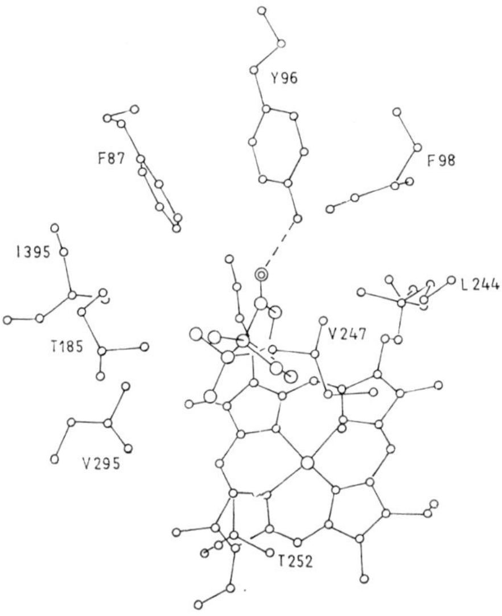

Fig. 5. Active site structure of the camphor-bound form of cytochrome P-450$_{cam}$.

From these well defined interactions, as illustrated by the x-ray crystal structure, it is clear that many forces are collectively utilized to bind, orient, and stabilize camphor. How much does each interaction contribute to molecular recognition? This is a question that may only be addressed by careful dissection of the system's components. The exploitation of substrate analogues (Fig. 6) and site-directed mutagenesis of cytochrome P-450$_{cam}$ has yielded a detailed analysis of substrate binding and catalysis that may serve as a model for future enzyme studies.

5. Metabolic switching: The oxidase function

As discussed earlier, the hydroxylation of camphor to exclusively produce 5-*exo*-hydroxycamphor functions via a tightly coupled system which requires two electrons ultimately derived from NADH. Norcamphor, which is essentially a camphor skeleton lacking the three methyl branches (Fig. 6), can also be metabolized by cytochrome P-450_{cam} affording three products: 3-*exo*, 6-*exo*, and 5-*exo*-hydroxynorcamphor (Tab. 1; ATKINS and SLIGAR, 1987). The distribution and overall yield of products is affected by isotopically labelling specific sites of hydroxylation with deuterium. Thus, deuteration of the five and six norcamphor positions enhances the relative production of 3-*exo*-hydroxynorcamphor. Likewise, utilization of the substrate norcamphor-3,3-d_2 results in a

SUBSTRATE ANALOGUES

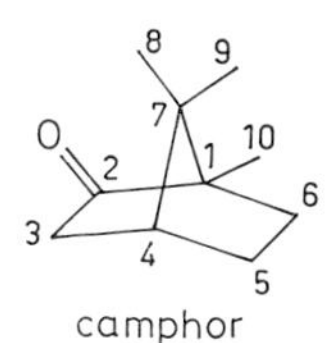

camphor

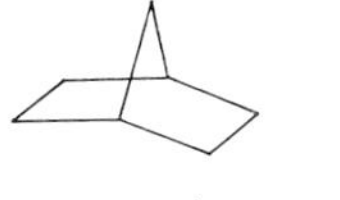
norcamphor

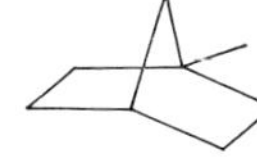
1-methyl norcamphor

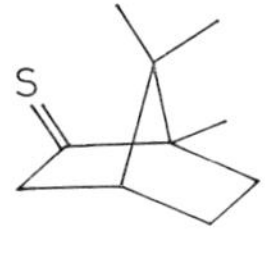

thiocamphor

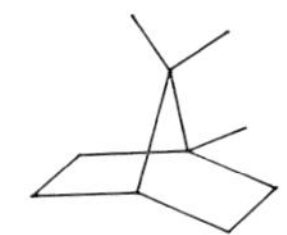
camphane

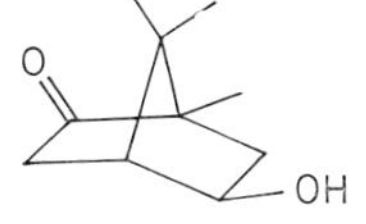

5-exo-hydroxycamphor

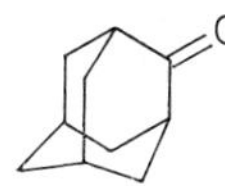

adamantanone

Fig. 6. Substrate analogues of cytochrome P-450_{cam}.

Table 1. Regiospecificity of norcamphor hydroxylation

Substrate [a]	**Relative percent of product**		
	3-*exo*-	5-*exo*-	6-*exo*-
Norcamphor	8%	45%	47%
5,6-d_2-Norcamphor	52%	27%	21%
3,3-d_2-Norcamphor	2%	49%	49%
3,3,5,6-d_4-Norcamphor	10%	44%	46%

[a] 5,6-d_2-norcamphor is 5,6-*exo*, *exo*-norcamphor-d_2 and the d_4-norcamphor is 5,6-*exo*, *exo*-norcamphor-3,3,5,6-d_4.

decreased yield of 3-*exo*-hydroxynorcamphor. Deuterium labelling at all of the norcamphor hydroxylation sites re-establishes the original relative product yield, but drastically lowers the total product yields.

In comparison to the decrease in the overall product yield, NADH and O_2 consumption as well as H_2O_2 production remain relatively consistent, presenting a striking unaccountability of electrons. This phenomenon has been pre-

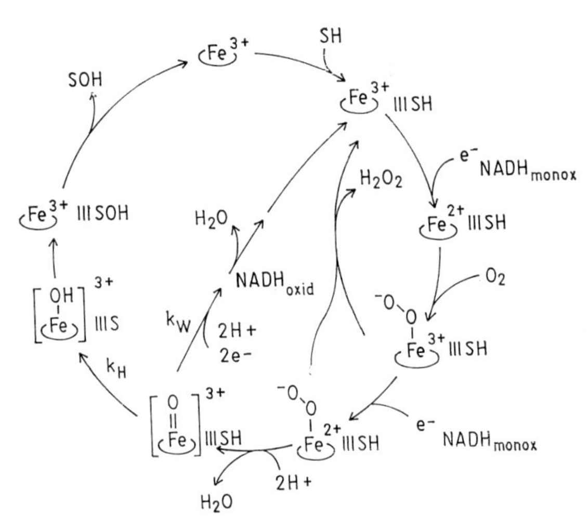

Fig. 7. Monooxygenase and oxidase cycles of cytochrome P-450_{cam}. The terms "$NADH_{mono}$" and "$NADH_{oxid}$" indicate points at which pyridine nucleotide reducing equivalents are utilized for the normal monooxygenase cycle and the further reduction of oxygen to result in oxidase stoichiometry, respectively. Hydrogen peroxide may be generated from dismutation of two equivalents of superoxide anion or directly from collapse of the two-electron reduced intermediate prior to O-O bond scission. This intermediate has not been experimentally detected. Reduction of the hydroxylating intermediate, $[FeO]^{3+}$, may occur as two one-electron reductions. SH and SOH refer to substrate and hydroxylated product, respectively.

viously documented based on the turnover of numerous substrates by the liver microsomal cytochromes P-450, including isozymes 2, 3a, 3b and 4 (Gorsky et al., 1987) and also in the 7-ethoxycoumarin O-dealkylation by rat hepatic phenobarbital- and 3-methylcholanthrene-induced cytochrome P-450 (Miwa et al., 1984; Harada et al., 1984). The tightly coupled monooxygenase reaction should theoretically have the stoichiometry of NAD(P)H to dioxygen to product of 1:1:1. Rather, in these results as well as the metabolism of deuterated norcamphor, a ratio of 2:1:1 is found (Atkins and Sligar, 1987). Thus, the electron equivalents consumed exceeds that used for product formation, posing the problem of determining the fate of the "extra" electrons.

Some of the cytochromes P-450 are known to form hydrogen peroxide in a side reaction and by autooxidation (Fig. 7). It can therefore be argued that if H_2O_2 is produced, it could serve as an intermediate subject to further reduction by analogy to catalase (JARIGAN et al., 1958) and chloroperoxidase (HAGER et al., 1972). However, the laboratories of GORSKY and SLIGAR carefully investigated the effects of this mechanism by adding exogenous H_2O_2, which was completely recovered. Analysis of autooxidation could not account for the excess of NADH-derived electrons consumed. Finally, the addition of catalase had no effect on any of the kinetic reaction parameters in the turnover of the enzymes.

GORSKY et al. (1987) proposed that the cytochromes P-450 are capable of exhibiting an oxidase activity. Thus, the stoichiometry of a four-electron reduction of dioxygen using two equivalents of NAD(P)H could account for the fate of the excess of electrons. Physiologically, the oxidase function is very significant. In the absence of an easily hydroxylatable substrate, the highly reactive intermediate, $[FeO]^{3+}$, could be destructive, either causing heme degradation or protein oxidation. With a continuous flux of electrons supplied by putidaredoxin, the enzyme can undergo measures to quench the $[FeO]^{3+}$ intermediate by the oxidase function to produce the ferric resting state of the enzyme and water, thus saving itself from self-destruction (ATKINS, 1988).

The isotope effects of the metabolism of norcamphor were also evaluated based on the kinetics of the steady state V_{max}, yielding interesting and highly informative results. Isotope effects of 1.22 and 1.16 were found for the kinetics of oxygen uptake and hydrogen peroxide production (ATKINS and SLIGAR, 1987), which is consistent with those reported by GELB et al. (1982) for the deuteration of camphor. The small magnitude of the isotope effects suggest significant masking, caused by the rate limiting step, which occurs prior to hydrogen abstraction; the rate limiting step in the turnover of camphor is the introduction of electrons into the cytochrome P-450 (HINTZ and PETERSON, 1981; HINTZ et al., 1982; BREWER and PETERSON, 1988; PEDERSON et al., 1977; HUI BON HOA, 1978). The kinetics of NADH consumption in the hydroxylation of norcamphor produced an inverse isotope effect of 0.77 (ATKINS and SLIGAR, 1987). The disparity in NADH consumption, oxygen uptake, and hydrogen peroxide formation indicates that NADH is utilized in a path that is independent from the latter two. ATKINS and SLIGAR proposed a modified cytochrome P-450_{cam} catalytic cycle that incorporates the oxidase function, which utilizes two additional electron equivalents derived from NADH, in a path that branches from the monooxygenase cycle at the $[FeO]^{3+}$ intermediate. Notably, this model can explain the observed inverse isotope effects on NADH consumption. Thus, in the case of deuterated norcamphor metabolism, there is an isotope-dependent redistribution of enzyme intermediates, resulting in a higher steady state concentration of a species that is substrate for a two electron reduction of NADH as well as a candidate to perform hydrogen abstraction.

5.1. Active site reconstruction

The uncoupled metabolism of norcamphor, which yields three hydroxylated products in comparison to the higly specific hydroxylation of camphor by cytochrome P-450$_{cam}$ prompted an investigation into the active site residues and their interaction with specific substrate moieties (ATKINS and SLIGAR, 1988b, 1989; ATKINS, 1988). Site-directed mutagenesis and substrate analogues were employed to focus on two spatially separated contact points between the substrate and key amino acids in the camphor binding site. Valine 295, which is interdigitated between the bridgehead methyl moieties of camphor, was substituted with an isoleucine to introduce steric bulk (ATKINS and SLIGAR, 1989). Valine 247 interacts by van der Waals forces with the 1-methyl group

Table 2. Active site reconstruction of cytochrome P-450$_{cam}$

System[a]	**% High Spin**	**Total yield hydroxy products/ equiv NADH**	**% 5-exo of total products**	**Yield 5-exo-/ NADH**
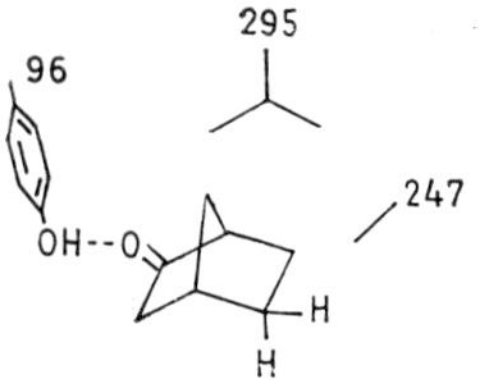 96, 295, 247, OH--O, H, H V247A/norc	44%	0.10	21%	0.02
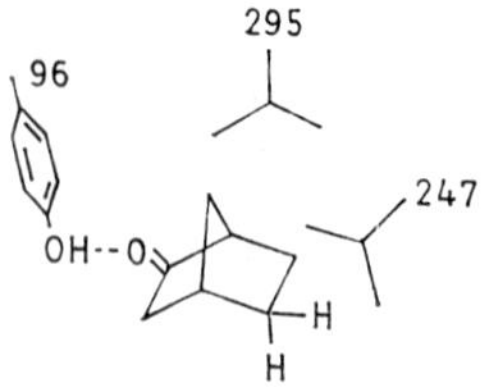 96, 295, 247, OH--O, H, H WT/norc	45%	0.12	45%	0.05
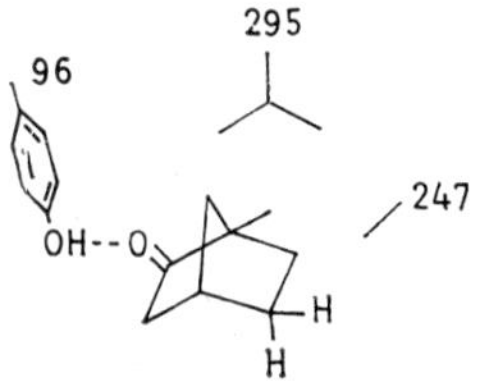 96, 295, 247, OH--O, H, H V247A/CH_3-norc	37%	0.40	73%	0.29

Table 2 (continued).

System[a]	% High Spin	Total yield hydroxy products/ equiv NADH	% 5-exo- of total products	Yield 5-exo-/ NADH
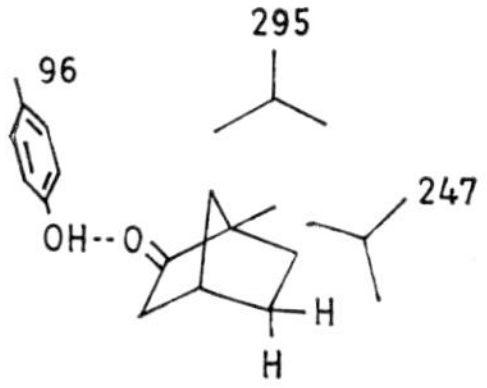 295, 96, 247, OH--O, H, H WT/CH_3-norc	48%	0.45	82%	0.37
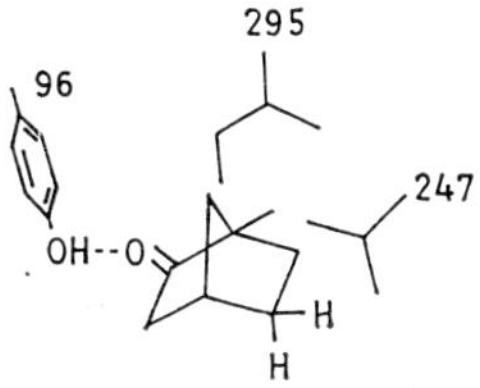 295, 96, 247, OH--O, H, H V295I/CH_3-norc	49%	0.45	90%	0.41
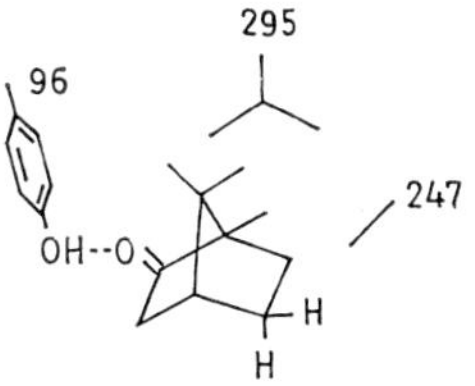 295, 96, 247, OH--O, H, H V247A/cam	95%	1.0	97%	0.97
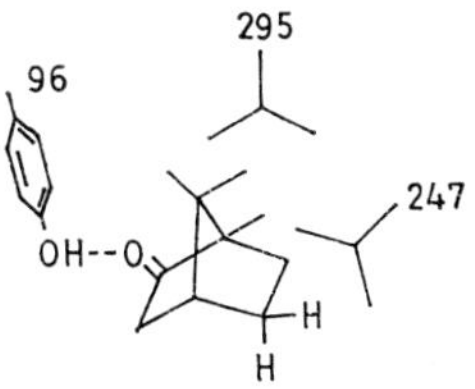 295, 96, 247, OH--O, H, H WT/cam	95%	1.0	100%	1.0

[a] the abbreviations used are: WT: wild-type; norc: norcamphor; CH_3-norc: 1-methyl-norcamphor; cam: camphor; V247A: the mutant with VAL-247 changed to Ala-247; V295I: the mutant with Val-295 changed to Ile-295.

of camphor. This contact was destroyed by mutating the protein residue to alanine. The substrate analogues, norcamphor and 1-methyl norcamphor were utilized to further probe these active site features of cytochrome P-450_{cam}. Combinations of these substrates and the wild-type and mutant proteins were characterized by determination of spin-state regulation, regiospecificity of hydroxylation, and binding affinity. Table 2 summarizes the active site reconstruction of cytochrome P-450_{cam}. The results are organized according to increasing efficiency of the enzyme-substrate complexes, which is based on yield of "correct" product as determined by hydroxylation at the five carbon position versus NADH utilization. As would be expected, maximal abolishment of protein-substrate contacts, for example in the case of the turnover of norcamphor by V247A[2], produced the lowest yield of total hydroxylated product per equivalent NADH, 0.10, indicating a significant amount of metabolic switching to the oxidase and/or peroxygenase function. Also, the combination affords the lowest percentage of the 5-hydroxy product, 21%. As Table 2 illustrates, increasing contact points which enhance binding, orientation, and stabilization of the substrate, influences higher degrees of specificity and more efficient utilization of NADH reducing equivalents to yield "correct product".

It is also interesting to note a series of results which illustrate the flexibility of the camphor binding pocket. Adamantanone is a bulky camphor derivative (Fig. 6), which binds in the active site of cytochrome P-450_{cam} with a similar K_D to that of camphor, 1.3 μM, and drives the high-spin ferric population to 97% (ATKINS, 1988). The introduction of further bulk at the valine 295 position by substitution with an isoleucine also affords a 97% high-spin population upon the binding of adamantanone. Furthermore, the wild-type and V295I specifically hydroxylate adamantanone at the 5-carbon position producing a single product (ATKINS, 1988; WHITE et al., 1984). The binding of camphor by both of these proteins also yields equivalent high-spin populations of 95% and exhibits complete regio- and stereospecificity of hydroxylation. V295I does possess a slightly higher K_D of 1.7 μM for camphor binding, which indicates a degree of independence of the spin-state regulation and binding equilibria as demonstrated previously in the literature (FISHER and SLIGAR, 1985b).

In summary, ATKINS has restructured the active site of cytochrome P-450_{cam} to successfully influence and control hydroxylation of alternative substrates. This investigation of molecular recognition has provided insight into the significance of the individual binding components in the camphor envelope. It has been documented that the 3-position of camphor, which is adjacent to the carbonyl (Fig. 6), supports a slightly more stable radical upon hydrogen

[2] Mutations are designated by the single letter abbreviation of the wild-type amino acid, its position number in the polypeptide chain, and the single letter abbreviation of the resultant mutant amino acid. For example, V247A represents a protein where valine at position 247 is changed to an alanine residue.

abstraction by the $[FeO]^{3+}$ intermediate (Collins et al., 1988). Wild-type, V295I and V247A all turn over norcamphor and 1-methyl norcamphor yielding a percentage of the 3-hydroxy derivative in the product distribution. The combination which has the greatest destruction of native protein-substrate contact points, V247A and norcamphor, affords 40% of the 3-*exo*-hydroxy-

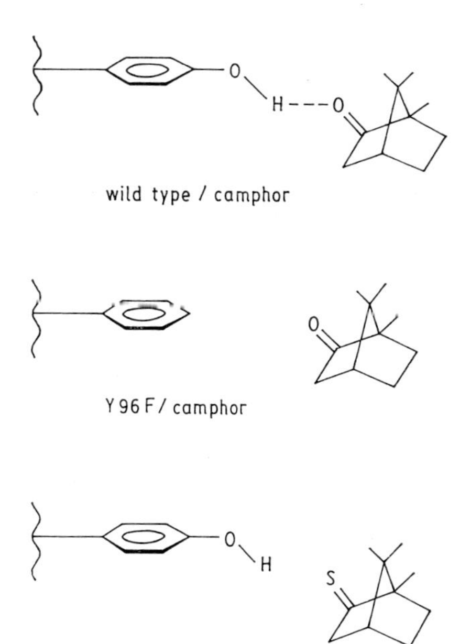

Fig. 8. Probes of the active site tyrosine 96 hydrogen bond in cytochrome P-450$_{cam}$. Complimentary changes are made to either the protein, by site-directed mutagenesis, or to the substrate. The substrate analogues used include thiocamphor and camphane.

norcamphor. Clearly, the active site of cytochrome P-450$_{cam}$ has evolved toi completely control and regulate the hydroxylation of its native substrate, camphor, for maximal efficiency in its metabolic function.

A final, distinct binding interaction between camphor and the cytochrome P-450$_{cam}$ active site merits attention. The side chain of tyrosine 96 extends into the substrate binding pocket offering a hydrogen bond to the carbonyl of camphor. This link anchors the camphor molecule, thus commanding complete regio- and stereospecificity. The role of this hydrogen bond was investigated by

Table 3. Binding and spin state parameters for hydrogen bonding probes of cytochrome P-450$_{cam}$

System	% HIGH SPIN	ΔG_{spin} (kcal/mol)	K_D (μM)	$\Delta G_{binding}$ (kcal/mol)
Wild-type/camphor	95%	−2.7	1.6 ± 0.3	−7.8
Y96F/camphor	59%	−0.21	3.3 ± 0.4	−7.4
Wild-type/ thiocamphor	65%	−0.36	3.0 ± 0.6	−7.5
Wild-type/camphane	46%	+0.9	46 ± 5	−5.9

$\Delta G = RT \ln$ [% high spin/% low spin], T = 25 °C
$\Delta G = RT \ln K_D$, T = 25 °C

both the application of site-directed mutagenesis and the utilization of substrate analogues (Fig 8; ATKINS and SLIGAR, 1988b). Specifically, in using site-directed mutagenesis, the hydrogen bond was deleted by substituting an isosterically equivalent amino acid, phenylalanine, for Tyr 96. Further modification was approached by using the substrate analogue, camphane (Fig. 6), which lacks the carbonyl, thus, also completely destroying the hydrogen bond. A second compound, thiocamphor was employed to explore weak hydrogen bonding capacities (VINOGRADOV et al., 1971).

As discussed earlier, cytochrome P-450$_{cam}$ is driven to a 95% high-spin complex in the presence of camphor. In comparison, Y96F only achieves a 59% high-spin conversion of the ferric center in the presence of saturating levels of camphor (ATKINS and SLIGAR, 1988a). Similarly, the spin-state populations elicited by an excess of thiocamphor and camphane with the native cytochrome P-450$_{cam}$ are 65% and 46% high- spin, respectively. Table 3 reflects that the wild-type/camphane interaction produces a drastically increased dissociation constant. When these combinations with wild-type are compared to Y96F, it should be noted that Y96F still contains two native residues, Phe 87 and Leu 244, which are in contact with the carbonyl group, whereas camphane would be incapable of anchoring by all three residues (POULOS et al., 1986). Thiocamphor, which is more isosteric with camphor as opposed to camphane, is characterized by a similar spin-state and binding affinity as the mutant protein. The change in the difference in the binding free energies of camphor to the native and mutant enzymes is approximately 0.5 kcal/mole, significantly less than a typical hydrogen bond energy of 2−5 kcal/mole (VINOGRADOV et al., 1971).

However, in a systematic exploration of the substrate binding site of tyrosyl-tRNA synthetase by FERSHT et al. (1985), complementary hydrogen bonding

Table 4. Isomeric distribution of hydroxylated products

System	Product distribution[a]		Total yield hydroxylated products[b]
Wild-type/camphor	5-*exo*-hydroxy	100%	100%
Y96F/camphor	5-*exo*-hydroxy	92%	
	5-*endo*-hydroxy	1%	
	6-*exo*-hydroxy	2%	100%
	9-hydroxy	< 1%	
	3-*exo*-hydroxy	4%	
Wild-type/ thiocamphor	5-*exo*-hydroxy	64%	
	6-*exo*-hydroxy	34%	98%
	3-*exo*-hydroxy	2%	
Wild-type/camphane	5-*exo*-hydroxy	90%	8%
	6-*exo*-hydroxy	10%	

[a] This is the distribution of products obtained, without regard for total yield. The numbering scheme for thiocamphor and camphane is identical to that for camphor so the 6- and 3-alcohols are analogous to those shown in Figure 6.
[b] Based on 100% yield expected from the NADH consumed.

and its effects on biological specificity were examined utilizing site-directed mutagenesis and substrate analogues, producing a similar difference in binding free energies. Notably, the investigation of a series of modified substrate-enzyme interactions allowed the categorization of various types of hydrogen bonds. Alterations to a charged donor or acceptor in a hydrogen bond results in weaker binding by approximately 3 kcal/mol, while substitution of an uncharged hydrogen bond donor or acceptor, which maintains that lack of charge decreases the binding free energy by only 0.5 to 1.5 kcal/mol. Therefore, a conservative replacement of phenylalanine for tyrosine would be expected to weaken binding by only a small free energy difference in binding of 0.5 kcal/mol. This energy difference can further be accounted for when other phenomenona associated with the substrate binding of cytochrome P-450_{cam} are observed. Briefly, energy contributions and differences may also originate from spin-state conversion which is based on water ligation to the heme iron. Enthalpy of other active site hydrophobic interactions, in particular Phe 87 and Leu 244 lie in contact with the camphor carbonyl reinforcing the duties of the tyrosine hydrogen bond. Finally, entropy associated with the ordered water molecules found in the substrate binding pocket introduces another thermodynamic influence. Collectively, these forces may serve to decrease the proposed theoretical energy difference expected for the removal of one hydrogen bond.

Alterations to the hydrogen bond interaction resulted in multiple product yields (Tab. 4; ATKINS and SLIGAR, 1988b). Also, investigation of NADH consumption found that the turnover of camphane by wild-type was highly inefficient, achieving only 8% product formation relative to NADH utilization, which reflects low biological specificity.

Biological specificity of the cytochrome P-450$_{cam}$ system can be quantitated by examining the k_{cat}/K_M. Assuming a minimal kinetic reaction scheme:

$$E + S \rightarrow ES \xrightarrow{k_2} E + P$$

where k_2 refers to the rate of formation of hydroxylated product from the ES complex. Normalization for yield of correct product relative to efficient NADH consumption affords an estimated k_{cat}. Since all kinetic determinations of NADH consumption have been performed with saturating concentrations of substrate and putidaredoxin, which are conditions where electron transfer is rate limiting, the rates of NADH consumption closely reflect the ferric, substrate-bound spin-state equilibrium. Thus, the measured K_D values may be used to approximate the relative K_M value under these V_{max} conditions. Table 5 presents the normalized values of k_2/K_D, displaying a five-fold decrease in specificity for the mutant enzyme and a seven-fold decrease for the wild-

Table 5. Kinetic parameters for hydrogen bonding probes of cytochrome P-450$_{cam}$

System	k_2, NADH consumption (nmoles NADH/ min × nmole P-450)	Relative[a] yield "correct isomer"	Corrected k_2 (nmoles/min × nmole P-450)	k_2/K_D[b], Relative Specificity $min^{-1}M^{-1} \times 10^{-6}$
Wild-type/ camphor	64 ± 3	1.0	64	40
Y96F/camphor	29 ± 2	0.92	27	8.2
Wild-type/ thiocamphor	36 ± 4	0.64	23	6.5

[a] Based on values reported in Table 4.
[b] As described in the text, k_2 is the relative rate of "correct" product formation, not the rate of hydrogen abstraction by the oxidizing intermediate. Also, the optically determined K_D is not a true "K_M".

type/camphane complex. This relatively large decrease in the efficiency of camphor hydroxylation is the additive result of small effects on k_2/K_D, and inefficient processing of the substrate to afford "incorrect" products, rather than an isolated effect on any one parameter.

In conclusion, analysis of the active site hydrogen bond interaction reveals that its disruption results in only minor changes in the macroscopic equilibrium binding constant, while ΔG_{spin} is significantly affected. This study emphasizes that the active site/camphor complex is optimally structured to afford maximal kinetic control of electron transfer by fine tuning the ferric spin equilibrium. In short, it provides some mechanistic insight into the kinetic control of electron transfer over camphor hydroxylation by cytochrome P-450$_{cam}$. Thus, the hydrogen bond of Tyr 96 effectively ensures proper orientation of camphor to shield the heme iron from solvent in order to maintain maximal gating of electron input.

Tyrosine 96 may also be used as a sensitive, natural probe to monitor the polarity of the active site by second derivative UV spectroscopy. The technique is based on a ratio (r_n) between two peak-to-trough second derivative absorbance differences which reflect the polarity of the microenvironment surrounding the protein's tyrosine residues (Ragone et al., 1984; Servillo et al., 1982; Ruckpaul et al., 1980). In the case of cytochrome P-450$_{cam}$, the percentage of tyrosine residues exposed to solvent is linearly dependent on the ferric high-spin concentration which may be varied by the binding of various substrate analogues (Fig. 9; Fisher and Sligar, 1985a).

As discussed previously in detail, Tyr 96 hydrogen bonds to the carbonyl moiety of camphor, expelling six water molecules from the active site. In the absence of camphor, active site Tyr 96 is left unshielded from the cluster of water molecules. This residue was conclusively implicated as the tyrosine responsible for the second derivative UV spectral changes based on site-directed mutagenesis (Atkins and Sligar, 1990). Substitution of Tyr 96 with phenylalanine produced a mutant cytochrome P-450$_{cam}$ which displayed a lack of spin-state dependent changes in the second derivative UV absorption spectrum. Therefore, these results demonstrate unequivocally that the microenvironment of Tyr 96 is primarily responsible for the second derivative UV spectrum.

This natural probe was employed to characterize the binding of product, 5-*exo*-hydroxycamphor, to cytochrome P-450$_{cam}$. 5-*exo*-hydroxycamphor binds with a K_D of 10 μM and drives the high-spin population to 26%. The predominantly low-spin population has been proposed to be caused by ligation of the product's hydroxyl group to the iron center. Results from the second derivative spectral results, indicate that the small fraction of high-spin enzyme present when the hydroxycamphor is bound does not result from a completely "dry" active site. Rather, based on the second derivative results, the large differences between substrate and product may be attributed to an alteration in solvent accessibility. Either the 5-hydroxy-camphor does not bind in the

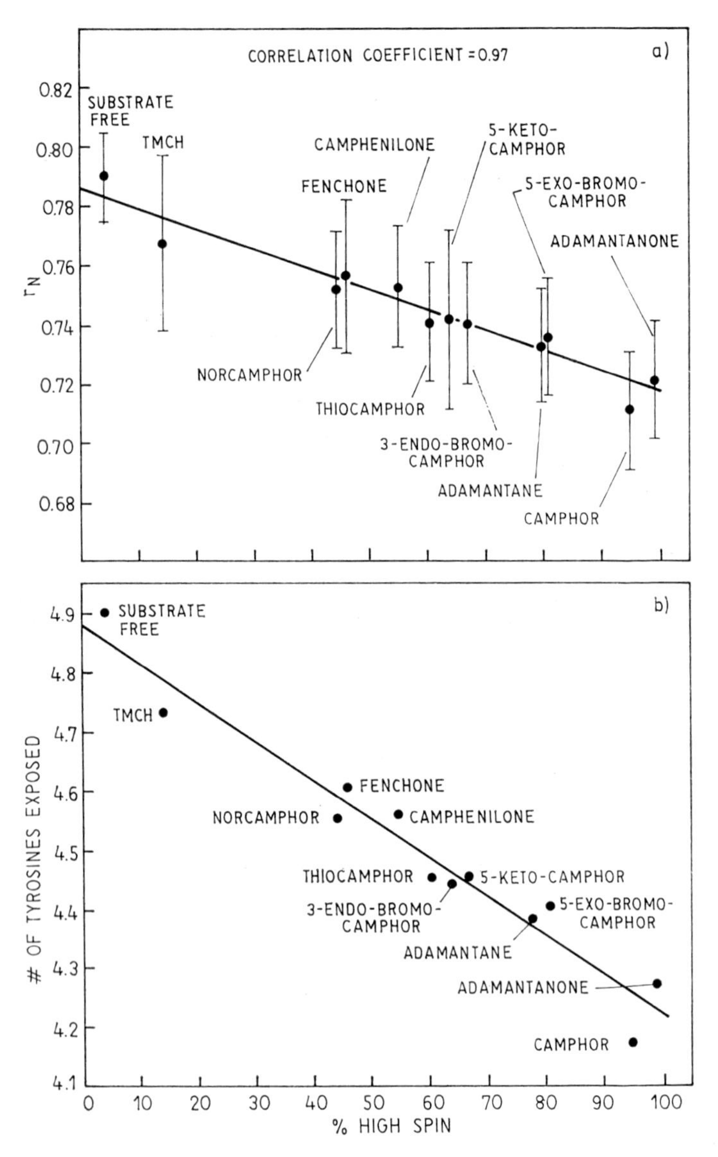

Fig. 9. Correlation between polarity of the tyrosine microenvironment (A) and tyrosine exposure to solvent (B) of cytochrome P-450_{cam} to ferric spin equilibrium upon binding of the indicated substrates based on second derivative UV spectroscopy.

same orientation as camphor or it is more polar, thereby attracting water to the active site. Alternatively, it is conceivable that the hydroxyl moiety of product may be sufficiently polar to affect the second derivative UV spectrum of Tyr 96. Most notably, the end result produces a predominantly low-spin species of cytochrome P-450$_{cam}$, which cannot accept electrons from the native reconstituted system due to its low redox potential. Thus, the protein-product complex induces a mechanism to turn off electron transfer in order to conserve reducing equivalents.

6. Axial ligand mutations: CYS 357

The cytochrome P-450s' unique and diverse range of activities has been partially attributed to the thiolate ligation of its heme prosthetic group. It is tempting to reason differences in the function of heme proteins based on their varied axial ligands, the histidine ligation of the oxygen carrying proteins, myoglobin and hemoglobin, the tyrosine ligation, such as in catalase, and, of course, the cysteinate ligation of the cytochrome P-450 monooxygenases. The advent of the cloning and high-level expression of cytochrome P-450$_{cam}$ (Koga et al., 1985; Unger et al., 1986) provides a genetic handle which may be effectively used to manipulate the protein to alter the heme ligation. In efforts to delineate enzymatic function relative to its heme ligation, the cytochrome P-450$_{cam}$ axial ligand, Cys 357, has been mutated to two other residues as evidenced by DNA sequencing. Specifically, histidine has been substituted in efforts to imitate the globin's ligation, and serine has been introduced to provide a residue that is isosteric to cysteine with only weak coordination possibilities (Unger, 1988).

It is interesting to note that both mutant proteins fail to incorporate heme during their synthesis or folding in *E. coli*, whereas wild-type is expressed as holoprotein. This phenomenon has been found in other endeavors pursuant to the modification of axial heme ligation. Specifically, the hexa-coordinated iron heme center of rat cytochrome b_5, which is axially liganded by histidines 39 and 63 has been modified at position 39 by replacement with a methionine residue (Sligar et al., 1987). The resultant protein, H39M, along with three other axial ligand mutants, H39V, H39L (Egeberg, unpublished results), and H63M (von Bodman et al., 1987) have been expressed as the apo-protein. The proteins have been purified as the apo-protein and reconstituted with heme, affording the holoprotein for further characterization. Yet another example of heme ligand mutations may be demonstrated by sperm whale myoglobin. The histidine axial ligand of sperm whale myoglobin has been substituted by cysteine and tyrosine in attempts to mimic the activities of cytochrome P-450 and catalase respectively (Egeberg et al., 1988). Both mutant proteins were expressed essentially as apo-protein, purified, and reconstituted. The resultant holoproteins were found to be unstable. Also, site-directed mutagenesis of the proposed axial ligand of rat liver cytochrome P-450, the invariant Cys 456, was pursued (SHIMIZU, et al., 1988). Replacement with tyrosine and histidine failed to yield holoprotein. Collectively, these examples illustrate the importance of the native axial ligand, in the linked processes of polypeptide chain folding and heme incorporation. Reconstitutions allow for further characterization, but physical spectroscopy studies or x-ray crystallography are necessary to verify proper ligation by the desired amino acid residue.

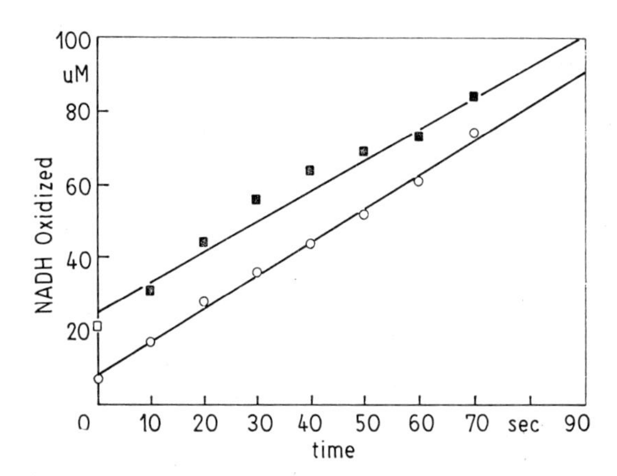

Fig. 10. NADH-oxidation rates of cytochrome P-450$_{cam}$ (○) and the *E. coli* expressed mutant C357H (■).

In the case of cytochrome P-450$_{cam}$, reconstitution of purified C357H with heme yielded holoprotein with optical spectra that bore resemblance to both the native protein and myoglobin. The Soret for the oxidized species was found to be at 417 nm, which is identical to that of substrate-free cytochrome P-450$_{cam}$, but contrasts myoglobin's Soret at 407 nm. The reduced spectra displayed a band at 406 nm for the mutant as compared to 408 and 433 nm for wild-type tytochrome P-450$_{cam}$ and myoglobin respectively. However, CO difference speccroscopy produced a normal spectrum with a maximum at 414 nm which is more closely aligned with myoglobin's band at 423 nm than the characteristic hyper-spectrum of cytochrome P-450$_{cam}$ at 446 nm (HANSON et al., 1976, 1977).

The reduction potential for C357H was measured to be −280 mV in the presence of camphor, which differs substantially from substrate-bound cytochrome P-450$_{cam}$ at −170 mV. This potential would mandate a loss of activity as it is lower than that of the protein's native redox partner, putidaredoxin, which has a reduction potential of −196 mV (SLIGAR, 1976). In fact, efforts to elicit camphor hydroxylation with the mutant protein in the native reconstituted system and also with exogenous oxidants, including *m*-chloroperbenzoic acid, phenyl peracetic acid, and iodosobenzene were unsuccessful and indicated a complete loss of native activity. It is interesting to note, however, that the native electron-transfer chain was still able to consume electrons derived from NADH at a rate similar to the native system (Fig. 10). Also, the mutant was found to be capable of peroxidizing the substrate pyrogallol with a K_M of 4.8 mM and a V_{max} of 166 μmol pyrogallol/μmol heme/min, 10 fold higher than the wild-type enzyme. The mutant protein binds oxygen and was characterized by a fast autooxidation with a half-life of 150 sec, which is similar to that of 200 sec of the native protein.

In summary, the native axial ligands of not only cytochrome P-450$_{cam}$, but other heme proteins, have proven to be essential in the structural integrity and function of these metalloproteins. They appear to be important in the mechanism of heme incorporation which would indicate that they also play a role in protein folding. In the case of cytochrome P-450$_{cam}$, without structural evaluation of the reconstituted protein to ascertain heme ligation, it is difficult to determine the exact contribution of the substituted thiolate ligand residue to the loss of activity. However, this initial investigation does reflect that the mutant is able to bind O_2 and accept NADH-derived electrons, although it is not capable of camphor hydroxylation which may be in part due to the change in reduction potential.

7. Putidaredoxin

Putidaredoxin, an iron sulfur protein, is responsible for the input of the first and second reducing equivalents to cytochrome P-450$_{cam}$. In its absence, the dioxy-ferrous substrate-bound form of cytochrome P-450$_{cam}$ autooxidizes rapidly, releasing superoxide without product formation (Lipscomb et al., 1976). If Fe(II)O_2 is combined with either oxidized or reduced putidaredoxin, the complex rapidly decomposes to ferric cytochrome P-450$_{cam}$ with the concomitant formation of 5-*exo*-hydroxycamphor. The dissociation constants of the putidaredoxin complexes with fluorescein-isothiocyanate labelled cytochrome P-450$_{cam}$ were measured by fluorescence methodologies (Sligar and Gunsalus, 1976). Putidaredoxin was found to bind cytochrome P-450$_{cam}$, quenching the fluorescent label in a one-to-one stoichiometry. The K_D for the oxidized putidaredoxin complex is 3 μM at 283 K, while the reduced putidaredoxin binds more tightly with a K_D of 0.5 μM.

When the oxidized or reduced species of putidaredoxin is added to photochemically reduced, oxygenated cytochrome P-450$_{cam}$, the complex rapidly breaks down to Fe(III) concurrent with 5-*exo*-hydroxycamphor formation (Lipscomb et al., 1976). Notably, regardless of redox state, putidaredoxin induces the camphor hydroxylase activity of cytochrome P-450$_{cam}$ in substoichiometric concentrations, implicating its role as an effector in addition to a provider of electrons.

Investigations to delineate this mechanism have been directed toward the carboxy terminus of putidaredoxin, in particular the terminal tryptophan and penultimate glutamine (SLIGAR et al., 1974). Proteolytic cleavage of these two residues from putidaredoxin results in a protein (*des*-trp-gln-Pd)[3] which maintains an intact iron sulfur center as evidenced by EPR spectroscopy. Its optical spectrum is identical to the native protein with the exception of the UV region which reflects the loss of one tryptophan. In addition, the reduction potential was equivalent to that measured for the native putidaredoxin, reflecting conservation of the catalytic core. In contrast to native putidaredoxin, *des*-trp-gln-Pd exhibited an increase in its dissociation constant to 50 μM describing its interaction with cytochrome P-450$_{cam}$. Furthermore, its effector activity diminished significantly. Addition of sub-stoichiometric concentrations of putidaredoxin to oxyferrous cytochrome P-450$_{cam}$ results in a second-order process which induces camphor hydroxylation with a rate constant of 51,000 $M^{-1}sec^{-1}$. While substitution of the native iron-sulfur protein with *des*-trp-gln-Pd lowers this rate constant to 970 $M^{-1}sec^{-1}$.

The role of the tryptophan residue in the camphor hydroxylation system is currently under investigation utilizing site-directed mutagenesis technologies. The cloning and expression of putidaredoxin (KOGA et al., 1989) has enabled conservative substitution of the aromatic amino acids, Phe and Tyr for Trp, as well as nonaromatic mutations including Val, Leu, Lys, and Asp (DAVIES et al., 1990). In addition to these substitutions, the terminal residue, Trp 106 has been deleted; thus the protein is expressed as a polypeptide chain with only 105 residues (*des*-trp-Pd)[3]. The resultant purified mutant proteins were reconstituted into the native electron transfer system, which includes NADH, putidaredoxin reductase, and cytochrome P-450$_{cam}$. Rates of NADH consumption were measured under conditions where they were limited by the electron transfer process from putidaredoxin to cytochrome P-450$_{cam}$. The results are recorded in Table 6. As would be expected, the wild-type protein displayed the highest velocity of 47 μM NADH/min. The substitution of Trp 106 with aromatic amino acid residues, Phe and Tyr, produced diminished rates of 26 and 16 μM NADH/min respectively. The steady-state kinetic parameters, K_{app} and turnover number were determined by varying wild-type and mutant putidaredoxin concentrations and analyzing the data assuming a single site for multiprotein association. Values of K_{app} were determined to be 5 μM for the wild-type and 14 μM for both W106F and W106Y, indicating tight complexation between putidaredoxin and cytochrome P-450$_{cam}$. Turnover numbers were determined to be 41 sec^{-1} for wild-type, 38 sec^{-1} for W106F, and 20 sec^{-1}

[3] Abbreviations for the proteins in the camphor hydroxylase system are as follows: Pd: putidaredoxin; *des*-trp-Pd: putidaredoxin with the tryptophan C-terminal residue deleted; *des*-trp-gln-Pd: putidaredoxin with the tryptophan C-terminus and penultimate glutamine proteolytically removed.

Table 6. Steady-state kinetics of NADH consumption of the reconstituted cytochrome P-450$_{cam}$ electron transfer chain utilizing carboxy-terminal putidaredoxin mutants

Putidaredoxin	v_0(+cam) (μM/min)	v_0(−cam) (μM/min)	Activity[a] (%)
Wild-type	47	0.11	100
W106F	26	5.2	80
W106Y	16	6.7	58
W106V	6.2	5.0	~0
W106L	3.6	3.1	~0
W106D	5.9	5.7	~0
W106K	5.2	5.1	~0
des-trp	5.2	5.1	~0

[a] Percentage electron transfer rate above autooxidative background:

$$\frac{v_0(+\text{cam}) - v_0(-\text{cam})}{v_0(-\text{cam})}$$

for W106Y. Perhaps most significantly, the nonaromatic replacements and the *des*-trp-Pd mutant resulted in drastically reduced velocities which were essentially equivalent to the respective background NADH consumption rates in the absence of camphor. Thus, the aromatic side chain of tryptophan at the carboxy-terminus of putidaredoxin is strongly implicated in the mechanism of electron transfer and camphor hydroxylation from a diprotein complex.

8. Putidaredoxin-cytochrome P-450$_{cam}$ association

The lack of a putidaredoxin crystal structure has hampered attempts to provide a structural description of the complex between putidaredoxin and cytochrome P-450$_{cam}$ and the molecular details of the electron transfer event. Recent experiments by Stayton et al. (1989), however, have implicated a putative binding site for putidaredoxin based on competition studies between putidaredoxin and cytochrome b_5 in their association with cytochrome P-450$_{cam}$. Although cytochrome P-450$_{cam}$ and cytochrome b_5 are not physiological electron transfer partners, certain mammalian cytochromes P-450 are known to interact with membrane bound cytochrome b_5 (Hildebrandt and Estabrook, 1971; Bonfils et al., 1981; Pompon and Coon, 1984). The lack of a spectral handle, however, had previously stymied attempts to measure a binding constant and it is only recently that the association of cytochrome b_5

and cytochrome P-450_{cam} has been monitored via a genetically engineered cytochrome b_5 labelled with the environmentally sensitive fluorophore acrylodan (STAYTON et al., 1988). In these experiments STAYTON et al. (1988) generated mutations of the soluble portion of rat liver cytochrome b_5 (VON BODMAN et al., 1986) that contain cysteine residues in place of surface threonine residues. Rat liver cytochrome b_5 contains no native cysteine residues so the mutation facilitated the use of the sulfhydral selective fluorophore acrylodan.

The involvement of a conserved set of surface carboxylates on cytochrome b_5 with a variety of electron transfer proteins including cytochrome P-450_{cam} is well documented (STAYTON et al., 1988; RODGERS et al., 1988; DAILEY and STRITTMATTER, 1979; TAMBURINI and SCHENKMAN, 1985). The interaction of cytochrome b_5 and cytochrome P-450_{cam} was found to be ionic-strength dependent, suggesting an electrostatic contribution to the binding free energy. It has been proposed that the association of cytochrome b_5 with cytochrome P-450_{cam} occurs via the anionic surface of cytochrome b_5 and an analogous cationic surface on cytochrome P-450_{cam} (STAYTON et al., 1989). Computer graphic modelling, based on the high-resolution crystal structures of these proteins, and optimized for an orientation which utilizes the conserved carboxylates of cytochrome b_5, have produced a good electrostatic fit based on charged pair interactions for the docking of cytochrome b_5 and cytochrome P-450_{cam} (STAYTON et al., 1989). A binding competition study was performed to asses the involvement of this cationic site on P-450_{cam} with the association of putidaredoxin. Amino acid sequence data reports that putidaredoxin contains a series of localized carboxylate groups in its primary structure (TANAKA et al., 1974) which have been proposed to form a negatively charged patch on the surface of the folded protein. Furthermore, based on modification experiments, this cluster of acidic amino acids was found to be important in establishing the putidaredoxin-putidaredoxin reductase complex (GEREN et al., 1986). Adrenodoxin, an iron sulfur protein involved in the reduction of the adrenal methylene hydroxylase, also contains a primary cluster of carboxylates that has been implicated in its association with both its reductase and cytochrome P-450_{scc} (LAMBETH and KRIENGSIRI, 1985).

The results of the binding competition investigation shown in Figure 11 clearly indicate competitive binding between cytochrome b_5 and putidaredoxin with cytochrome P-450_{cam}. An inhibitor constant of 0.8 μM corresponding to a dissociation constant for putidaredoxin and cytochrome P-450_{cam} is indicated. Under similar conditions, the binding constant for cytochrome b_5 and cytochrome P-450_{cam} was determined to be 1.1 μM (STAYTON et al., 1989). These results, combined with those of GEREN et al. (1986) suggest that the cluster of carboxylic acids in putidaredoxin may form an anionic binding site on the surface of the protein which contributes to the free energy of binding to cytochrome P-450_{cam} in a manner similar to that observed in cytochrome b_5.

Current investigation into the validity of the electrostatic association of cytochrome P-450$_{cam}$ and putidaredoxin is currently in progress. Mutations have been made in both cytochrome P-450$_{cam}$ and putidaredoxin to reverse the charge of the proposed ionic interaction (STAYTON, unpublished results).

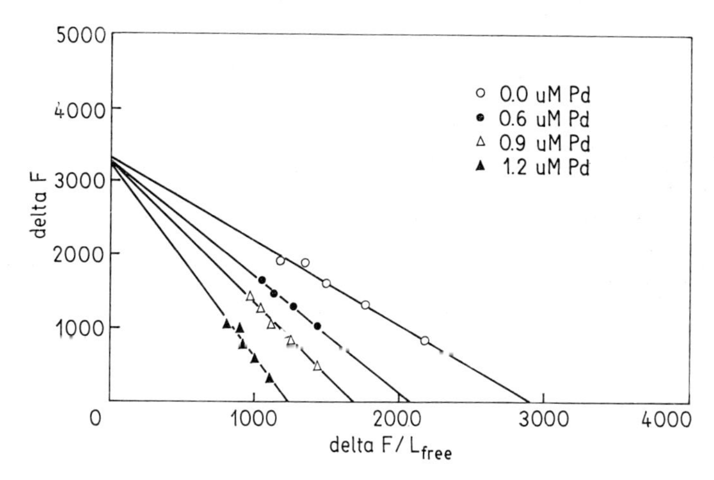

Fig. 11. Binding competition EADIE-HOFFSTEE plot of cytochrome b_5 and putidaredoxin in the association with cytochrome P-450$_{cam}$. The ordinate represents the change in fluorescence and the abscissa is the change in fluorescence over the free ligand concentration. Putidaredoxin additions were varied while the concentrations of cytochrome b_5 and cytochrome P-450$_{cam}$ remained constant. The convergence of the linear plots at the ordinate axis demonstrates competitive binding.

An interesting note should be included in this section on the findings in the computer graphics docking model mentioned earlier. It was found that a conserved phenylalanine in the cytochromes P-450, specifically Phe 350 in P-450$_{cam}$, lies in the path between the cytochrome b_5 heme and the cytochrome P-450$_{cam}$ heme. It is tempting to suggest from this computer modelling investigation, along with the competitive binding data for putidaredoxin and cytochrome b_5, and also from experimental evidence presented by DAVIES et al. (1990) that the electron transfer path between the iron-sulfur center of putidaredoxin and the heme of cytochrome P-450$_{cam}$ is mediated by the carboxy-terminal tryptophan of putidaredoxin and the phenylalanine 350 of cytochrome P-450$_{cam}$.

9. Summary

Site-directed mutagenesis of cytochrome P-450$_{cam}$ and its redox partners has served as a valuable tool in the exploration of the various mechanisms involved in camphor hydroxylation by cytochrome P-450$_{cam}$. Substitution of specific active site residues have altered molecular recognition of the substrate, potentially enabling reconstruction of the active site to facilitate hydroxylation of alternative substrates. Axial ligand mutations have revealed that native coordination is significant not only in catalysis and activity, but also in protein folding and heme incorporation. Putidaredoxin's long history as an effector as well as a source of electrons to cytochrome P-450_{cam} has been well supported by mutations in its carboxy terminus. Specifically, the aromaticity of the terminal residue has been precisely implicated as essential in camphor hydroxylation. Also, the complexation of putidaredoxin and cytochrome P-450_{cam} has been examined utilizing another effector, cytochrome b_5. Based on the crystal structures of cytochromes b_5 and P-450_{cam}, computer graphics have been utilized to illustrate a localization of positive charges on the surface of cytochrome P-450_{cam}, which define a potential binding site for redox partners. Finally, mechanistic studies of norcamphor metabolism have provided results which have further defined a more detailed catalytic cycle of the cytochrome P-450s.

Acknowledgement

The authors would like to thank Dr. W. M. Atkins, Dr. J. L. Beck, Dr. P. S. Stayton and M. Davies for communication of unpublished results and/or valuable editorial assistance. This work is supported by grants from the National Institutes of Health GM31756 and GM33775 and by the Biotechnology Research and Development Corporation.

10. References

Atkins, W. M. and S. G. Sligar, (1987), J. Am. Chem. Soc. **109**, 3754–3760.
Atkins, W. M., (1988), Ph. D. thesis, University of Illinois, Urbana, IL.
Atkins, W. M. and S. G. Sligar, (1988a), Biochemistry **27**, 1610–1616.
Atkins, W. M. and S. G. Sligar, (1988b), J. Biol. Chem. **283**, 18842–18849.
Atkins, W. M. and S. G. Sligar, (1989), J. Am. Chem. Soc. **111**, 2715–2717.
Atkins, W. M. and S. G. Sligar, (1990), Biochemistry **29**, 1271–1275.
Bangcharoenpaurpong, O., A. K. Rizos, P. M. Champion, D. Jollie, and S. G. Sligar, (1986), J. Biol. Chem. **261**, 8089–8092.
Bangcharoenpaurpong, O., P. M. Champion, S. A. Martinis, and S. G. Sligar, (1987), J. Chem. Phys. **87**, 4273–4284.
Blake, R. C. and M. J. Coon, (1981), J. Biol. Chem. **256**, 12127–12133.
Bonfils, C., C. Balny, and P. Maurel, (1981), J. Biol. Chem. **256**, 9457–9465.
Brewer, C. B. and J. A. Peterson, (1988), J. Biol. Chem. **263**, 791–798.
Champion, P. M., J. D. Lipscomb, E. Münck, P. Debrunner, and I. C. Gunsalus, (1975), Biochemistry **14**, 4151–4158.
Champion, P. M., I. C. Gunsalus, and G. C. Wagner, (1978), J. Am. Chem. Soc. **100**, 3743–3751.
Collins, J. R. and G. H. Loew, (1988), J. Biol. Chem. **263**, 3164–3170.
Dailey, H. A. and P. Strittmatter, (1979), J. Biol. Chem. **254**, 5388–5396.
Davies, M. D., L. Qin, J. L. Beck, K. S. Suslick, H. Koga, T. Horiuchi, and S. G. Sligar, (1990), J. Am. Chem. Soc. **112**, 7396–7398.

DAWSON, J. H. and M. SONO, (1987), Chem. Rev. **37**, 1257–1273.
EBLE, J. S. and J. H. DAWSON, (1984), J. Biol. Chem. **259**, 14389–14395.
EGEBERG, K. D., B. A. SPRINGER, S. G. SLIGAR, D. MORIKIS, and P. M. CHAMPION, (1988), Symposium on Oxygen Binding Heme Proteins Structures, Dynamics, Functions, and Genetics, Pacific Grove, CA, 1988, PII-7.
EISEN, H. J., (1986), in: Cytochrome P-450: Structure, Mechanism, and Biochemistry, (P. R. ORTIZ DE MONTELLANO, ed.), Plenum Press, New York, pp. 315–344.
FERSHT, A. R., J.-P. SHI, J. KNILL-JONES, D. M. LOWE, A. J. WILKINSON, D. M. BLOW, P. BRICK, P. CARTER, M. M. Y. WAYE, and G. WINTER, (1985), Nature **314**, 235–238.
FISHER, M. T. and S. G. SLIGAR, (1985a), Biochemistry **26**, 4707–4803.
FISHER, M. T. and S. G. SLIGAR, (1985b), J. Am. Chem. Soc. **107**, 5018–5019.
FISHER, M. T., S. F. SCARLATA, and S. G. SLIGAR, (1985), Arch. Bioch. Biophys. **240**, 456–463.
GARFINKEL, D., (1958), Arch. Biochem. Biophys. **77**, 493–509.
GELB, M. H., D. C. HEIMBROOK, P. MALKÖNEN, and S. G. SLIGAR, (1982), Biochemistry **21**, 370–377.
GEREN, L., J. TULS, P. O'BRIEN, F. MILLETT, and J. A. PETERSON, (1986), J. Biol. Chem. **261**, 15491–15495.
GERLT, J. A., (1987), Chem. Rev. **87**, 1079–1105.
GORSKY, L. D., D. R. KOOP, and M. J. COON, (1984), J. Biol. Chem. **259**, 6812–6817.
GUNSALUS, I. C. and S. G. SLIGAR, (1978), in: Advances in Enzymology, (A. MEISTER, ed.), John Wiley & Sons, Inc., New York, pp. 1–44.
GRIFFIN, B. W. and J. A. PETERSON, (1975), J. Biol. Chem. **250**, 6445–6451.
GRIFFIN, B. W., J. A. PETERSON, and R. W. ESTABROOK, (1979), in: The Porphyrins, Vol. VII, (D. DOLPHIN, ed.), Academic Press, New York, pp. 333–375.
GROVES, J. T. and D. V. SUBRAMANIAN, (1984), J. Am. Chem. Soc. **106**, 2177–2181.
HAGER, L. P., D. L. DOUBEK, R. M. SILVERSTEIN, J. H. HARGIS, and J. C. MARTIN, (1972), J. Am. Chem. Soc. **94**, 4364–4366.
HANSON, L. K. S. G. SLIGAR, I. C. GUNSALUS, M. GOUTERMAN, and C. R. CONNELL, (1976), J. Am. Chem. Soc. **98**, 1976–1977.
HANSON, L. K., S. G. SLIGAR, and I. C. GUNSALUS, (1977), Croatica Chemica Acta **49**, 237–250.
HARADA, M., G. T. MIWA, J. S. WALSH, A. Y. H. LU, (1984), J. Biol. Chem. **259**, 3005–3010.
HEDEGAARD, J., and I. C. GUNSALUS, (1965), J. Biol. Chem. **240**, 4038–4043.
HILDEBRANDT, A. and R. W. ESTABROOK, (1971), Arch. Biochem. Biophys. **143**, 66–79.
HINTZ, M. J. and J. A. PETERSON, (1981), J. Biol. Chem. **256**, 6721–6728.
HINTZ, M. J., D. M. MOCK, L. L. PETERSON, K. TUTTLE, and J. A. PETERSON, (1982), J. Biol. Chem. **257**, 14324–14332.
HOLZWARTH, J. F., F. MEYER, M. PICKARD, and H. B. DUNFORD, (1988), Biochemistry **27**, 6628–6633.
HUI BON HOA, G. E. BEGARD, P. DEBEY, and I. C. GUNSALUS, (1978), Biochemistry **17**, 2835–2839.
HUI BON HOA, G., C. DI PRIMO, I. DONDAINE, S. G. SLIGAR, I. C. GUNSALUS, and P. DOUZOU, (1988), Biochemistry **28**, 651–656.
INGRAHAM, L. L. and D. L. MEYER, (1985), in: Biochemistry of Dioxygen, (E. FRIEDEN, ed.), Plenum Press, New York, pp. 7–10.
JARNIGAN, R. C. and J. H. WANG, (1958), J. Am. Chem. Soc. **80**, 786–789.
JEFCOATE, C. R., (1986), in: Cytochrome P-450: Structure, Mechanism, and Biochemistry, (P. R. Ortiz de Montellano, ed.), Plenum Press, New York, pp. 387–428.
JUNG, C., P. BENDZKO, O. RISTAU, and I. C. GUNSALUS, (1985), in: Cytochrome P-450:

Biochemistry, Biophysics, and Induction, (L. Vreeczky and K. Magyar. eds.), Budapest, 1985, Akadeniai Kiado, Budapest, pp. 19—27.
Keller, R. M., K. Wüthrich, and P. G. Debrunner, (1973), Proc. Natl. Acad. Sci., USA **69**, 2073—2075.
Klingenberg, M., (1958), Arch. Biochem. Biophys. **75**, 376—386.
Koga, H., B. Rauchfuss, and I. C. Gunsalus, (1985), Bioch. Biophys. Res. Comm. **130**, 412—417.
Koga, H., E. Yamaguchi, K. Matsunaga, H. Aramaki, and T. Horiuchi, (1989), J. Biochem. **106**, 831—836.
Lambeth, J. D. and S. Kriengsire, (1985), J. Biol. Chem. **260**, 8810—8816.
Lee, W. A. and T. C. Bruice, (1985), J. Am. Chem. Soc. **107**, 513—514.
Lee, W. A., L.-C. Yuan, and T. C. Bruice, (1988), J. Am. Chem. Soc. **110**, 4277—4283.
Lipscomb, J. D., S. G. Sligar, M. J. Namtvedt, and I. C. Gunsalus, (1976), J. Biol. Chem. **251**, 1116—1124.
Lipscomb, J. D., (1980), Biochemistry **19**, 3590—3599.
LoBrutto, R., C. P. Scholes, G. C. Wagner, and I. C. Gunsalus, (1980), J. Am. Chem. Soc. **102**, 1167—1170.
Marden, M. C. and G. Hui Bon Hoa, (1982), Eur. J. Bioch. **129**, 111—117.
Marden, M. C. and G. Hui Bon Hoa, (1987), Arch. Bioch. Biophys. **253**, 100—107.
Marnett, L. J., P. Weller, and J. R. Battista, (1986), in: Cytochrome P-450: Structure, Mechanism, and Biochemistry, (P. R. Ortiz de Montellano, ed.), Plenum Press, New York, pp. 29—78.
Mason, H. S., T. Yamano, J.-C. North, Y. Hashimoto, and P. Sakagishi, (1965), in: Oxidases Related Redox Systems, (1964), Proc. Symp., Vol. 2, pp. 879—899.
McCarthy, M. B. and R. E. White, (1983a), J. Biol. Chem. **258**, 9153—9158.
McCarthy, M. B. and R. E. White, (1983b), J. Biol. Chem. **258**, 11610—11616.
McMurry, T. J. and J. T. Groves, (1986), in: Cytochrome P-450: Structure, Mechanism, and Biochemistry, (P. R. Ortiz de Montellano, ed.), Plenum Press, New York, pp. 1—28.
Miwa, G. T., J. S. Walsh, and A. Y. H. Lu, (1984), J. Biol. Chem. **259**, 3000—3004.
Mozhaev, V. V., I. V. Berezin, and K. Martinek, (1988), in: Critical Reviews in Biochemistry, (G. P. Fasman, ed.), CRC Press, Inc., Boca Raton, pp. 235—281.
Münck, E. and P. M. Champion, (1975), Ann. N. Y. Acad. Sci. **244**, 142—162.
Murray, R. I., M. T. Fisher, P. G. Debrunner, and S. G. Sligar, (1985), in: Metalloproteins, Part I: Metal Proteins with Redox Roles, (P. M. Harrison, ed.), Macmillan Press, New York, pp. 157—206.
O'Keefe, D. H., R. E. Ebel, and J. A. Peterson, (1978), J. Biol. Chem. **250**, 7405—7415.
Omura, T., and R. Sato, (1962), J. Biol. Chem. **237**, 1375—1376.
Omura, T., and R. Sato, (1964a), J. Biol. Chem. **239**, 2370—2378.
Omura, T., and R. Sato, (1964b), J. Biol. Chem. **239**, 2379—2385.
Ortiz de Montellano, P. R., ed., (1986), Cytochrome P-450: Structure, Mechanism, and Biochemistry, Plenum Press, New York.
Oxender, D. L., and C. F. Fox, eds., (1987), Protein Engineering, Alan R. Liss, Inc., New York.
Ozaki, Y., T. Kitagawa, Y. Kyogoku, Y. Imai, C. Hashimoto-Yutsudo, and R. Sato, (1978), Biochemistry **17**, 5826—5831.
Padbury, G. and S. G. Sligar, (1985), J. Biol. Chem. **260**, 7820—7823.
Pederson, T. C., R. H. Austin, and I. C. Gunsalus (1977), in: Microsomes and Drug Oxidations, (V. Ullrich, ed.), Pergamon Press, New York, 275—283.
Peisach, J., W. E. Blumberg, S. Ogawa, E. A. Rachmilewitz, and R. Oltzik, (1971), J. Biol. Chem. **246**, 3342—3355.
Peterson, J. A., (1971), Arch. Bioch. Biophys. **144**, 678—693.

PETERSON, J. S., YUZURU, I., and B. W. GRIFFIN, (1972), Arch. Bioch. Biophys. **149**, 197–208.

PHILSON, S. B., P. G. DEBRUNNER, P. G. SCHMIDT, and I. C. GUNSALUS, (1979), J. Biol. Chem. **254**, 10173–10179.

POMPON, D. and M. J. COON, (1984), J. Biol. Chem. **259**, 15377–15385.

POULOS, T. L. and J. KRAUT, (1980), J. Biol. Chem. **255**, 8199–8205.

POULOS, T. L., S. T. FREER, R. A. ALDEN, S. L. EDWARDS, U. SKOGLAND, K. TAKIO, B. ERIKSSON, N. XUONG, T. YONETANI, and J. KRAUT, (1980), J. Biol. Chem. **255**, 575–580.

POULOS, T. L., M. PEREZ, and G. WAGNER, (1982), J. Biol. Chem. **257**, 10427–10429.

POULOS, T. L., B. C. FINZEL, I. C. GUNSALUS, G. C. WAGNER, and J. KRAUT, (1985), J. Biol. Chem. **260**, 16122–16130.

POULOS, T. L., B. C. FINZEL, and A. J. HOWARD, (1986a), Biochemistry **25**, 5314 to 5319.

POULOS, T. L., B. C. FINZEL, and A. J. HOWARD, (1986b), J. Mol. Biol. **195**, 687–700.

POULOS, T. L., and A. J. HOWARD, (1987), Biochemistry **26**, 8165–8174.

RAGONE, R., G. COLONANA, C. BALESTRIERI, L. SERVILLO, and G. IRACE, (1984), Biochemistry **23**, 1871–1875.

RODGERS, K. K., T. C. POCHAPSKY, and S. G. SLIGAR, (1988), Science **240**, 1657–1659.

RUCKPAUL, K., H. REIN, D. P. BALLOU, and M. J. COON, (1980), Biochim. Biophys. Acta **626**, 41–56.

SERVILLO, L., G. COLONNA, C. BALESTRIERI, R. RAGONE, and G. IRACE, (1980), Anal. Biochem. **126**, 251–257.

SHARROCK, M., E. MÜNCK, P. G. DEBRUNNER, V. MARSHALL, J. D. LIPSCOMB, and I. C. GUNSALUS, (1973), Biochemistry **12**, 258–265.

SHIMIZU, T., M. NOZAWA, Y. HATANO, Y. IMAI, and R. SATO, (1975), Biochemistry **14**, 4172–4178.

SHIMIZU, T., K. HIRANO, M. TAKAHASHI, M. HATANO, and Y. FUJII-KURIYAMA, (1988), Biochemistry **27**, 4138–4141.

SLIGAR, S., J. LIPSCOMB, P. DEBRUNNER, and I. C. GUNSALUS, (1974a), Bioch. Biophys. Res. Comm. **61**, 290–296.

SLIGAR, S. G., P. G. DEBRUNNER, J. D. LIPSCOMB, M. J. NAMTVEDT, and I. C. GUNSALUS, (1974b), Proc. Nat. Acad. Sci., USA **71**, 3906–3910.

SLIGAR, S. G., (1976), Biochemistry **15**, 5399–5406.

SLIGAR, S. G., and I. C. GUNSALUS, (1976), Proc. Nat. Acad. Sci., USA **73**, 1078–1082.

SLIGAR, S. G., and I. C. GUNSALUS, (1979), Biochemistry **18**, 2290–2295.

SLIGAR, S. G., K. A. KENNEDY, and D. C. PEARSON, (1980), Proc. Natl. Acad. Sci., USA **77**, 1240–1244.

SLIGAR, S. G., M. H. GELB, and D. C. HEIMBROOK, (1984), Xenobiotica **14**, 63–86.

SLIGAR, S. G., and R. I. MURRAY, (1986), in: Cytochrome P-450: Structure, Mechanism, and Biochemistry, (P. R. ORTIZ DE MONTELLANO, ed.), Plenum Press, New York, pp. 429–503.

SLIGAR, S. G., K. D. EGEBERG, J. T. SAGE, D. MORIKIS, and P. M. CHAMPION, (1987), J. Am. Chem. Soc. **109**, 7896–7897.

STAYTON, P. S., M. T. FISHER, and S. G. SLIGAR, (1988), J. Biol. Chem. **263**, 13544 to 13548.

STAYTON, P. S., T. L. POULOS, and S. G. SLIGAR, (1989), Biochemistry **28**, 8201–8205.

STAYTON, P. S., B. A. SPRINGER, W. M. ATKINS, and S. G. SLIGAR, (1989), in: Metal Ions in Biological Systems, Vol. 8, (H. SIGEL ed.), Marcel Dekker, Inc., New York, in press.

TAMBURINI, P. P. and J. B. SCHENKMAN, (1985), Proc. Nat. Acad. Sci., USA **84**, 11–15.

Takata, T., M. Yamazaki, K. Fujimori, Y. H. Kim, S. Oae, and T. Iyanagi, (1980), Chem. Lett., The Chem. Soc. of Jpn., 1441—1444.

Tanaka, M., M. Haniu, K. T. Yasunobu, K. Dus, and I. C. Gunsalus, (1974), J. Biol. Chem. **249**, 3689—3710.

Tsai, R., C. A. Yu, I. C. Gunsalus, J. Peisach, W. Blumberg, W. H. Orme-Johnson, and H. Beinert, (1970), Proc. Natl. Acad. Sci., USA **66**, 1157—1163.

Unger, B. P., S. G. Sligar, and I. C. Gunsalus, (1986), J. Biol. Chem. **261**, 1158—1163.

Unger, B. P., (1988), Ph. D. thesis, University of Illinois, Urbana, IL.

Vickery, L., A. Salmon, and K. Sauer, (1975), Biochim. Biophys. Acta **386**, 87—98.

Vinogradov, S. N. and R. H. Linell, (1971), in: Hydrogen Bonding, Van Nostran Reinhold, New York, pp. 123—124.

von Bodman, S. B., M. A. Schuler, D. R. Jollie, and S. G. Sligar, (1986), Proc. Nat. Acad. Sci., USA **83**, 9443—9447.

Watanabe, Y., T. Numata, T. Iyanagi, and S. Oae, (1981), Bull. Chem. Soc. Jpn. **54**, 1163—1170.

Watanabe, Y., S. Oae, and T. Iyanagi, (1982), Bull. Chem. Soc. Jpn. **55**, 188—195.

Waterman, M. R., M. John, and E. R. Simpson, (1986), in: Cytochrome P-450: Structure, Mechanism, and Biochemistry, (P. R. Ortiz de Montellano, ed.), Plenum Press, New York, pp. 345—386.

White, R. E. and M. J. Coon, (1980), Ann. Rev. Biochem. **49**, 315—356.

White, R. E., S. G. Sligar, and M. J. Coon, (1980), J. Biol. Chem. **255**, 11108—11111.

White, R. E., M. B. McCarthy, K. D. Egeberg, and S. G. Sligar, (1984), Arch. Bioch. Biophys. **228**, 493—502.

Yuan, L.-C. and T. C. Bruice, (1985), J. Am. Chem. Soc. **107**, 512—513.

Chapter 3
Cytochromes P-450 in Alkane-Assimilating Yeasts

H.-G. Müller, W.-H. Schunck, and E. Kärgel

1. Introduction

Monooxygenases of cytochrome P-450 type have been detected in bacteria, fungi, plants, invertebrates, and vertebrates. All cytochromes P-450 so far known are distinguished by a common structural characteristic – the thiolate (S^-) ligand to the heme iron. Due to this unusual environment of the heme group cytochromes P-450 are enabled to activate molecular oxygen for the hydroxylation of many lipophilic compounds.

This unique ability led in process of phylogenetic development to the evolution of multiple cytochromes P-450 distinguished by different catalytic functions and biological importance. We can be sure that many cytochromes P-450 are still waiting yet to be detected, mainly in the kingdom of bacteria and fungi. Based upon our present knowledge it can be supposed that all cytochromes P-450 both, known and yet unknown have a common ancestor.

The yeast *Saccharomyces cerevisiae* belongs to the first organisms in which cytochrome P-450 was identified. As early as 1964 it was found by LINDENMAYER and SMITH in this yeast grown on glucose. However, these investigators were not yet able to understand the function. Some years later LEBEAULT et al. (1971) and GALLO et al. (1971) demonstrated the cytochrome P-450 formation in strains of *Candida tropicalis* grown on n-alkanes and ascribed the alkane hydroxylation to this heme enzyme. At almost the same time, HEINZ et al. (1970) obtained evidence for the involvement of cytochrome P-450 in the hydroxylation of fatty acids by a *Torulopsis* strain producing biotensides. With this, the three physiological states which favour cytochrome P-450 formation in various yeasts were thus already identified: the growth on fermentable sugars, the growth on n-alkanes, and the production of glycolipids. However, only a limited number of yeast species has the both latter capacities.

For about 20 years it seemed to be quite sure that cytochrome P-450 systems can consist of either three or two proteins. In the former case the supply of electrons from reduced nicotinamide adenine dinucleotide to cytochrome P-450 as terminal oxidase proceeds via a specific reductase and an iron sulfurprotein. Such cytochrome P-450 systems have been found to be characteristic of the mitochondrial cytochromes P-450 which hold important positions in steroid hormone synthesis in mammalia and of several bacterial systems. In cytochrome P-450 systems consisting of two proteins a NADPH cytochrome P-450 reductase containing one FAD and one FMN per molecule transfers the electrons from NADPH directly to the cytochrome P-450 as terminal oxidase. They are typical of the endoplasmic reticulum of the eucaryotic cell.

Recently an unusal bacterial cytochrome P-450 was found in strains of *Bacillus megaterium* (NARHI and FULCO, 1986; FULCO and RUETTINGER, 1987), which unifies both functions – e-transfer and oxygen activation – in one protein. It contains heme, FAD and FMN as prosthetic groups.

In yeasts only the microsomal 2-protein type of cytochrome P-450 monooxygenase systems has been found up to now.

This review summarizes the advances in knowledge on cytochromes P-450 occuring in alkane-utilizing yeasts achieved since about 1980. Earlier literature was cited only exceptionally. It was considered in great detail some years ago by ourselves (MÜLLER et al., 1984), furthermore by KÄPPELI (1986) and by KING and WISEMAN (1987). Knowledge about bacterial cytochromes P-450 was likewise summarized by MÜLLER et al. (1984) and more recently in an extended form by SLIGAR and MURRAY (1986). Furthermore, current-state overviews about other microbial cytochromes P-450 are given in this volume and in other parts of this series.

2. Occurrence and functions of cytochromes P-450 in yeasts

At present nearly 30 yeast species of different genera which contain hemoproteins of this type are known (Tabl. 1). Hitherto six different enzymatic reactions could definitely be connected with cytochrome P-450 in yeasts:

a) The demethylation of lanosterol in position 14

This reaction is an essential step in the biosynthesis of ergosterol, which is an indispensable constituent of yeast membranes. Doubtless, it was the cytochrome P-450 found by LINDENMAYER and SMITH (1964) which catalyzes this reaction. Later, YOSHIDA and co-workers investigated this enzyme system from *S. cerevisiae* very thoroughly (YOSHIDA and AOYAMA, this volume). Semianaerobic growth on glucose favours the biosynthesis of this cytochrome P-450 in baker's yeast.

The importance of this demethylation reaction in the biosynthesis of ergosterol and the significance of this sterol for the membrane structure of fungi allows the expectation of the wide-spread occurrence of this cytochrome P-450 in fungi (cp. Tab. 1).

b) The Δ^{22}-desaturation of ergosta-5,7-dien-3-β-ol

A report of HATA et al. (1983) supplied evidence that this reaction of the ergosterol biosynthesis is likewise catalyzed by a cytochrome P-450 monooxygenase system. In their studies they used mutants of *S. cerevisiae* defective in 14α-demethylase reaction, but able to catalyze the mentioned desaturation reaction, which is sensitive to CO, SKF 525-A, and metyrapon. Further results are necessary to ensure the identity of this cytochrome P-450 and to elucidate its properties. Obviously, a very low level in *S. cerevisiae* complicates the investigations. (For details see the contribution of YOSHIDA and AOYAMA in this volume).

c) The hydroxylation of benzo(a)pyrene

The enzymatic attack of polycyclic aromatic hydrocarbons by distinct mamma-

Table 1. Cytochromes P-450 in yeasts

Species	Detected cytochrome P-450 dependent functions	References
Brettanomyces anomalus	not identified[1]	Kärenlämpi et al., 1980
Candida maltosa	n-alkane-hydroxylation	Honeck et al., 1982
	fatty acid hydroxlation	Blasig et al., 1988
Candida tropicalis	n-alkane hydroxylation	Duppel et al., 1973
	fatty acid hydroxylation	Duppel et al., 1973
Candida pulcherima	induced by n-alkanes	Takagi et al., 1980
Candida boidinii	not identified	Sharyshev 1988, personal communication
Candida silvanorum	n-alkane hydroxylation	Sharyshev and Anderegg 1988, personal communica-cation
Debaryomyces hansenii	not identified[1]	Kärenlampi et al., 1980
Debaryomyces formicarius	n-alkane hydroxylation	Sharyshev et al., 1986, unpublished
Hansenula anomala	not identified[1]	Kärenlampi et al., 1980
Kluyveromyces fragilis	not identified[1]	Kärenlampi et al., 1980
Pichia guilliermondii	induced by n-alkanes	Tittelbach et al., 1976
Pichia fermentans	not identified	Kärenlampi et al., 1980
Saccharomyces cerevisiae	lanosterol demethylation	Aoyama and Yoshida 1978
	hydroxylation of benzo(a)-pyren	Woods and Wiseman 1980
	Δ 22-desaturation of ergosta-5.7-dien-3-βol	Hata et al., 1983
Saccharomyces uvarum	lanosterol demethylation	Sanglard et al., 1986
Saccharomyces bayanus	not identified[1]	Kärenlampi et al., 1980
Saccharomyces chevalieri	not identified[1]	Kärenlampi et al., 1980
Saccharomyces italicus	not identified[1]	Kärenlampi et al., 1980
Schizosaccharomyces pombe	not identified[1]	Poole et al., 1974
Schizosaccharomyces japonicum	not identified[1]	Kärenlampi et al., 1980
Torulopsis spec.	fatty acid hydroxylation	Heinz et al., 1970
Torulopsis candida	induced by n-alkanes	Ilchenko et al., 1980
Torulosis glabrata	not identified[2]	Kleber et al., 1988
Torulopsis dattila	not identified[2]	Kleber et al., 1988
Torulopsis apicola	not identified[2]	Kleber et al., 1988
Trichosporon cutaneum	not identified[1]	Laurila et al., 1984
Trichosporon adeninovorans	induced by n-alkanes	Mauersberger 1988, unpublished
Yarrowia lipolytica[3]	induced by n-alkanes, hydroxylation of fatty acids	Delaisse et al., 1981

[1] formed during growth on fermentable sugars, probably resembling the cyt P-450_{14DM} of *S. cerevisiae*.
[2] induced during glycolipid biosynthesis.
[3] formerly genus *Candida*.

lian liver microsomal cytochromes P-450 has been very thoroughly substantiated now (YANG and BAO, 1987). AZARI and WISEMANN (1982) described the purification of a cytochrome P-450 form from *S. serevisiae* grown on 20% glucose, termed as cytochrome P-448, which is able to catalyze the hydroxylation of benzo(a)pyrene to various metabolites such as 3-hydroxybenzo(a)-pyrene, 9-hydroxybenzo(a)pyrene, and 7,8-dihydroxybenzo(a)pyrene. The involvement of cytochrome P-450 was established by reconstitution experiments using purified yeast cytochrome P-448, NADPH-cytochrome P-450 reductase, and phospholipid. But, the reaction rate is very low compared to 3-methylcholanthrene induced cytochrome P-450 forms of rat liver, only 0,14 nmole 3-hydroxybenzo(a)pyrene/nmole cytochrome P-448/hour was formed (AZARI and WISEMAN, 1982; KING et al., 1984). Further results are necessary to prove that this reaction in aromatic hydrocarbon metabolization — or another one — is the essential function of this cytochrome P-450 isolated by AZARI and WISEMAN. Regrettably, no laboratory made any direct comparison between the cytochromes P-450 purified by the groups of YOSHIDA and of WISEMAN, respectively. Therefore, some questions remain open. Cloning and sequencing of the gene of the cytochrome P-450 (P-448) isolated by the latter group would confirm its identity.

d) The terminal hydroxylation of long chain n-alkanes

This reaction is the first step in alkane catabolism in many microorganisms which can utilize these hydrophobic compounds as the only source of carbon and energy (BÜHLER and SCHINDLER, 1984; MÜLLER et al., 1984). Approximately 20% of the nearly 500 yeast species possess this ability (SCHAUER and SCHAUER, 1986). To our knowledge, in all yeasts checked hitherto the property of alkane utilization always has been found to be connected with the alkane-inducible cytochrome(s) P-450. Therefore, we suppose the occurrence of alkane-hydroxylating cytochrome(s) P-450 in all yeasts capable of growing on n-alkanes. Subterminal hydroxylation of n-alkanes has not yet been detected in yeasts, although enzymes responsible for the subsequent reactions seem to occur in *C. maltosa* as shown by investigations with hexadecane-(2)-ol (SCHUNCK et al., 1987, a, b).

e) The terminal hydroxylation of fatty acids

Various investigators (DUPPEL et al., 1973; GALLO et al., 1973, 1976; BERTRAND et al., 1980; MARCHAL et al., 1982; BLASIG et al., 1988) studying the alkane hydroxylation by *C. tropicalis, Y. lipolytica*, or *C. maltosa*, have demonstrated that microsomal fractions isolated after growth on n-alkanes can hydroxylate both n-alkanes and fatty acids in ω-position. In many cases lauric acid was used as the test substrate to determine the activity of the "alkane-hydroxylating cytochrome P-450". However, recent results suggest that at least two different cytochrome P-450 forms are induced by growth on n-alkanes which may differ distinctly in their substrate specificities against n-alkanes and fatty acids.

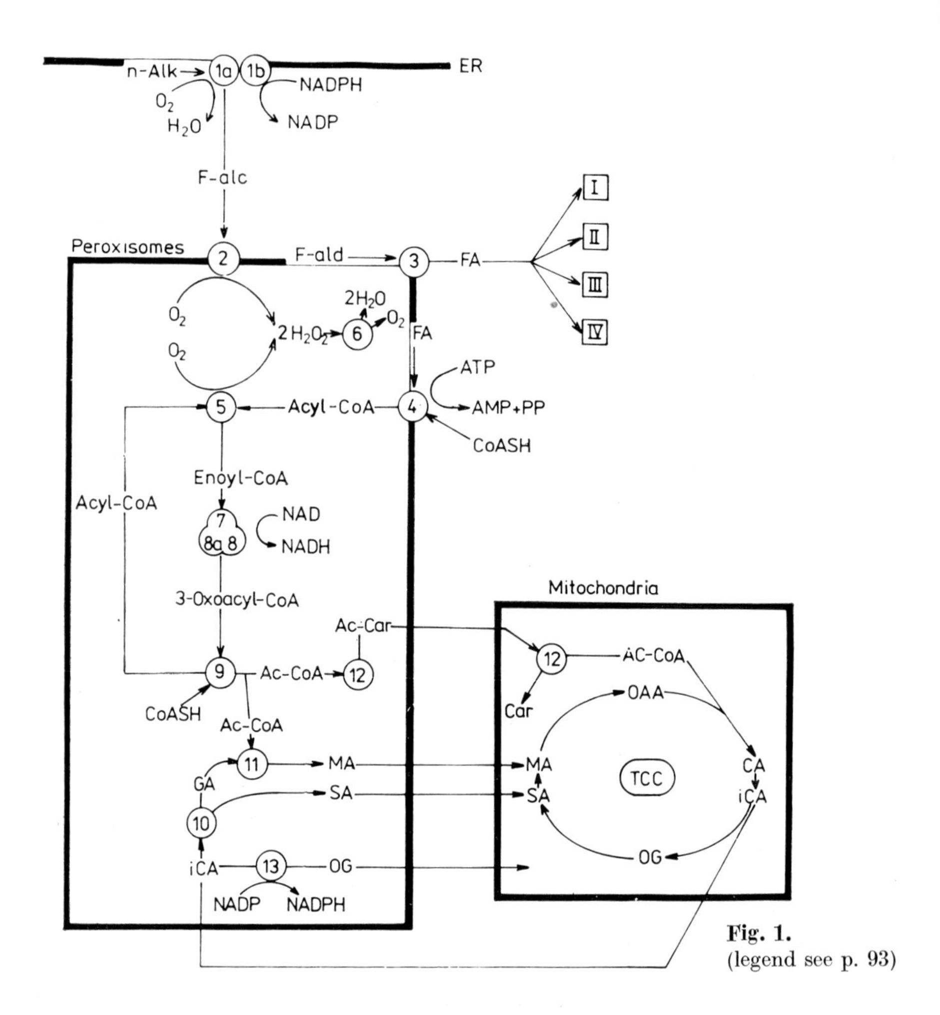

Fig. 1.
(legend see p. 93)

f) The subterminal hydroxylation of fatty acids

A fatty acid hydroxylation in $\omega-1$ position was demonstrated in a *Torulopsis* strain producing glycolipids by HEINZ et al. (1970) . The reaction was NADPH- and O_2-dependent and CO-sensitive, indicating the involvement of a cytochrome P-450. Hydroxylated fatty acids are needed for the synthesis of the biosurfactant. Whether any other biological importance can be ascribed to this cytochrome P-450 remains to be clarified.

In addition, several other reactions being typical of mammalian liver cytochromes P-450 have been ascribed to yeast cytochromes P-450, e. g. the demethylation of benzphetamine (DUPPEL et al., 1973) by microsomes of *C. tropicalis* or the dealkylation of aminopyrine, p-nitroanisole and caffeine (SAUER

Fig. 1. Monoterminal n-alkane oxidation and its subcellular localization in yeasts. (The scheme is confined to the main pathway, adjacent reactions were mostly omitted. The squares symbolize other transformations of fatty acids as key intermediates without specification of their intracellular localization).

1a = cytochrome P-450 monooxygenase
1b = NADPH — cytochrome P-450 reductase
2 = fatty alcohol oxidase
3 = fatty aldehyde dehydrogenase
4 = fatty acyl-CoA synthetase
5 = fatty acyl-CoA oxidase
6 = catalase
7 = enoyl-CoA hydratase
8 = L-3 hydroxyacyl-CoA dehydrogenase
8a = hydroxyacyl epimerase
9 = 3-oxoacyl-CoA thiolase
10 = isocitrate lyase
11 = malate synthase
12 = carnitine acetyltransferase
13 = NADP dependent isocitrate dehydrogenase

I = elongation of fatty acids
II = desaturation of fatty acids
III = lipid biosynthesis
IV = ω-hydroxylation

n-AlK = n-alkanes
F-alc = fatty alcohol
F-ald = fatty aldehyde
FA = fatty acid
Ac-CoA = acetyl-CoA
Ac-Car = acetyl carnitine
Car = carnitine
GA = glyoxylate
iCA = isocitrate
CA = citrate
MA = malate
OAA = oxalacetate
OG = 2-oxo-glutatarate
SA = succinate
TCC = tricarbonic acid cycle

---

et al., 1982) by microsomes of *S. cerevisiae var uvarum*, the O-deethylation of ethoxycoumarin, or the hydroxylation of biphenyl in *C. tropicalis* and *Trichosporon cutaneum* (Käppeli et al., 1985). However, an attribution of these reactions to a cytochrome P-450 enzyme doesn't seems to be possible only with reservations (cf. chapter 4, this volume).

3. Cytochrome P-450 and alkane catabolism — its pathway and subcellular organization

Nearly 100 yeast species, mainly belonging to the genera *Candida*, *Torulopsis*, *Pichia*, and *Yarrowia*, are able to grow alternatively either on carbohydrates or on long-chain hydrocarbones as the only source of carbon and energy. Frequently, comparable growth rates can be achieved with both substrates. The different properties of both groups of compounds — the former rich in oxygen and, therefore, hydrophilic, the latter, without oxygen and strongly apolar — require different pathways for uptake and metabolism.

As so far investigated, a cytochrome P-450 enzyme system holds a key position in n-alkane catabolism in all alkane-utilizing yeasts, for it is catalyzing the first enzymatic step of alkane conversion to a more hydrophilic intermediate. For a better understanding of this central role of the alkane-hydroxylating cytochrome P-450 a short overview of the whole catabolic pathway will be presented (cp. Fig. 1).

3.1. Alterations in yeast cells during growth on n-alkanes

The induction of cytochrome P-450 by long-chain n-alkanes is included in a complex physiological adaptation of the yeast cells to alkane utilization.

Genetic analysis indicates that the function of more than 80 genes may be required to bring about the phenotype of alkane assimilation (studies with *Y. lipolytica*, Bassel and Mortimer, 1982, 1985). Among them, at least 26 genes are linked to alkane uptake and oxidation to fatty acids. Transition from glucose to n-alkane utilization produces characteristic modifications in the yeast cell both at the biochemical and morphological level.

- Chemical and structural alterations at the cell surface which are related to hydrocarbon transport (Osumi et al., 1975a; Meissel et al., 1976; Käppeli et al., 1978, 1984; Fischer et al., 1982; Röber and Reuter, 1984).
- Induction of alkane-hydroxylating cytochrome P-450(s) and NADPH-cytochrome c (P-450) reductase (Gilewicz et al., 1979; Mauersberger et al., 1984, 1987; Loper et al., 1985).
- Induction of fatty alcohol and fatty aldehyde oxidizing enzymes (Mauersberger et al., 1984, 1987; Krauzova et al., 1985, 1986; Kemp et al., 1988).
- Increased formation of peroxisomes (Osumi et al., 1975b) and induction of peroxisomal β-oxidation and glyoxylate cycle enzymes (Tanaka et al., 1982, 1988).

Microbial n-alkane utilization starts with the uptake of the apolar substrate by the cell. Despite many efforts the basic principle of this process has not yet been identified. Most findings agree with the suggestion that the uptake of n-alkanes by the yeast cell is a passive process, which is facilitated by special hydrophobic structures. (For details of the state of knowledge see BÜHLER and SCHINDLER, 1984).

3.2. Cytochrome P-450-catalyzed terminal hydroxylation of n-alkanes as initiating reaction

As already mentioned the first enzymatic step of hydrocarbon assimilation by yeasts is the hydroxylation by a cytochrome P-450 enzyme system in terminal position (Fig. 2) leading to the fatty alcohol as the first intermediate of alkane catabolism. This enzyme system is membrane-bound and consists of a NADPH-cytochrome P-450 reductase transferring electrons, and a cytochrome P-450 acting as hydroxylase in a typical monooxygenase reaction. Both enzymes were obtained in a highly purified state, both of *C. maltosa* (RIEGE et al., 1981; HONECK et al., 1982; SCHUNCK et al., 1983) and of *C. tropicalis* (BERTRAND et al., 1980; LOPER et al., 1985) (For details see section 4). Successful reconstitution experiments with the highly purified enzymes (HONECK et al., 1982) and the selective inhibition by CO of the first step of alkane degradation under in-vivo conditions (SCHUNCK et al., 1987a, b) provided evidence leaving no doubt that the initiating reaction is catalyzed by cytochrome P-450.

In contrast to yeasts, in various filamentous fungi a subterminal hydroxylation, frequently starting quite distant from a terminus, is evident (REHM and REIFF, 1981). However, the enzymes responsible for the first reaction are as yet unknown in these cases. In bacteria, both a cytochrome P-450 dependent (CARDINI and JURTSCHUK, 1970; ASPERGER et al., 1984) and a non-heme iron

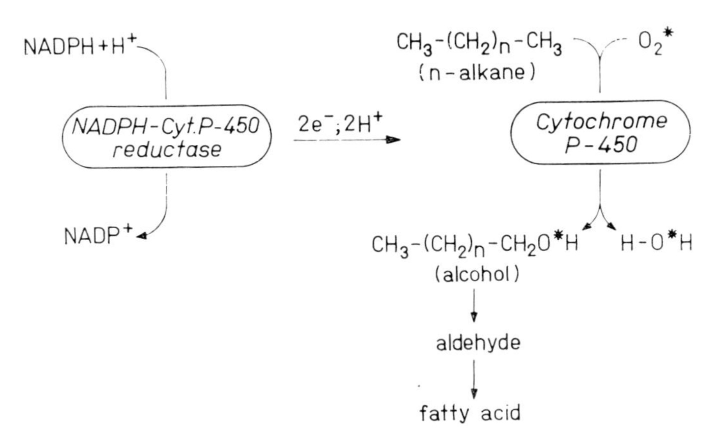

Fig. 2. Reaction scheme of the alkane-hydroxylating cytochrome P-450 system of yeasts. The following reactions are catalyzed by a fatty alcohol oxidase and a NAD-dependent fatty aldehyde dehydrogenase.

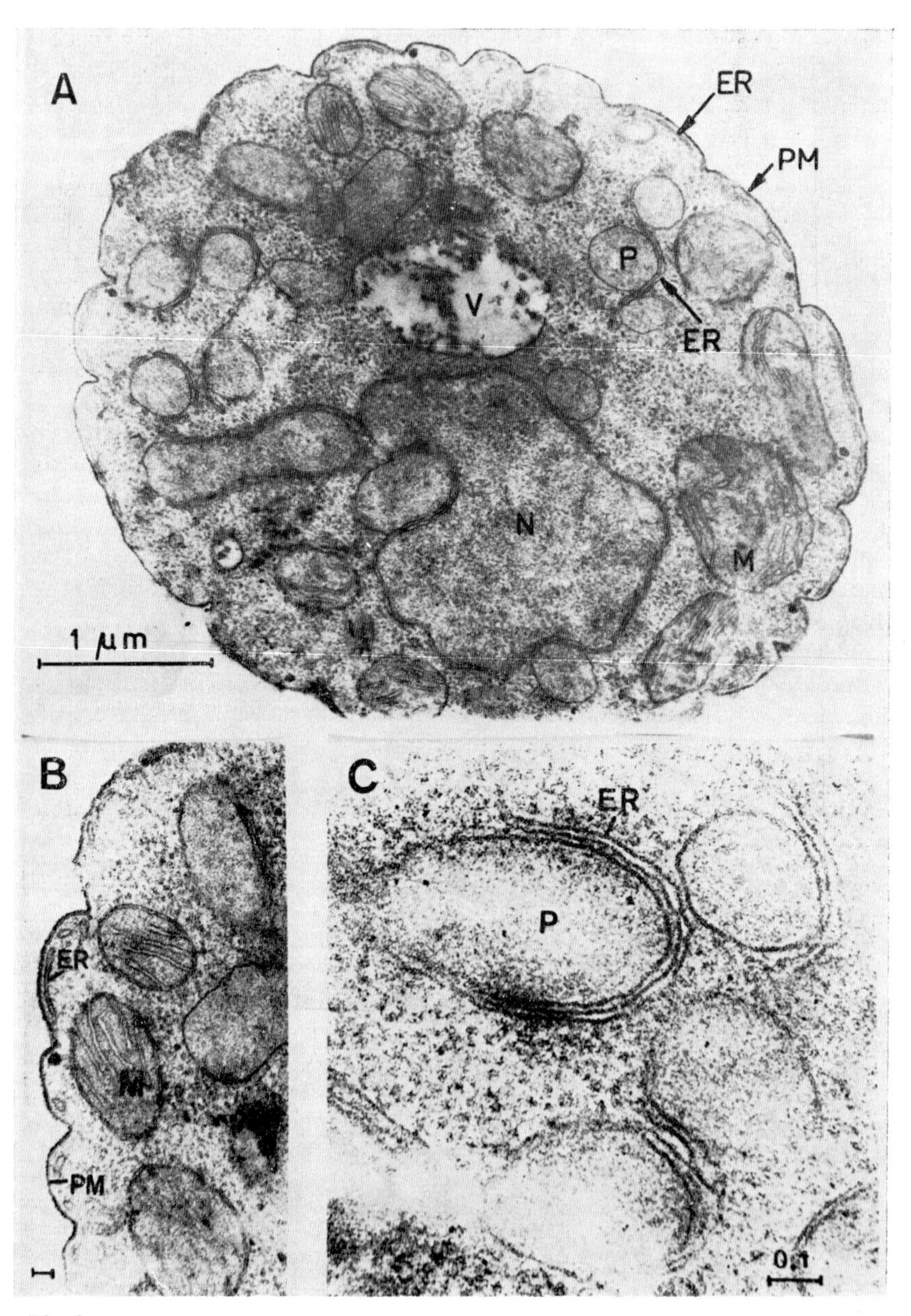

Fig. 3.

ω-hydroxylase catalyzed (RUETTINGER et al., 1977) alkane hydroxylation has been demonstrated. Further mechanisms of alkane functionalization different from monooxygenase hydroxylation which were repeatedly proposed still remain to be proved (for a summary see BÜHLER and SCHINDLER, 1984).

Already in the first studies on the alkane-hydroxylating cytochrome P-450 of yeasts in the early 70s its occurrence in the endoplasmic reticulum was suggested by comparison of the yeast with the well-investigated mammalian liver cell. In both cases after cell disruption cytochrome P-450 sediments with the microsomal fraction. As careful investigations have shown, microsomes of mammalian liver cells consist mainly of fragments of the endoplasmic reticulum (70–80% of the protein content) and cytochrome P-450 is localized in this organelle (BEAUFAY et al., 1974) as shown also by immunoelectron microscopy (FOSTER et al., 1986 and references therein).

However, the cytology of yeast cells has not been so thoroughly studied and it differs from the animal cell in some respects so that a simple comparison with the mammalian cell is not reliably evidenced. Using electron microscopic studies two forms of endoplasmic reticulum can be identified in yeast cells (MAUERSBERGER et al., 1987), one in the neighbourhood of the plasma membrane and the other one surrounding peroxisomes (Fig. 3). It is characteristic of yeast cells that after growth on n-alkanes a proliferation of the endoplasmic reticulum adjacent to the peroxisomes has taken place. Furthermore, the number of peroxisomes and of plasma membrane invaginations is increased.

Several efforts in different laboratories have been undertaken in an attempt to answer the question of intracellular cytochrome P-450 localization by cell fractionation in connection with the biochemical characterization of the fractions obtained (DELAISSE et al., 1981; MAUERSBERGER et al., 1984, 1987; KRAUZOVA and SHARYSHEV, 1987). All results correspond to show that both the mitochondrial/peroxisomal fraction and the post-microsomal supernatante can be left out from consideration for cytochrome P-450 and the n-alkane or fatty acid hydroxylating activity, respectively, are strikingly concentrated in the 100,000 × g sediment.

The yeast microsomal fraction obtained after enzymatic cell wall lysis, followed by cautious osmotic cell disruption and by differential centrifugation

Fig. 3. Ultrastructure of a *C. maltosa* spheroplast after cultivation on hexadecane.
A — spheroplast with nucleus (N), plasma membrane (PM), mitochondria (M), vacuole (V), peroisomes (P) and endoplasmic reticulum (ER).
B — peripherical (plasma membrane associated) ER.
C — ER surrounding peroxisomes.
Bar in A represents 1 μm, in B and C 0.1 μm
(Taken from MAUERSBERGER et al., 1987).

consists mainly of vesicles of the plasma membrane and of intracellular membranes. Furthermore, fragments of mitochondria and peroxisomes occur in small quantities. Further separation of microsomes by gradient centrifugation has made it possible to exclude the occurrence of cytochrome P-450 as being likely in the plasma membrane. However, hitherto it has not succeeded in differentiating between both forms of endoplasmic reticulum by cell fractionation.

At present, an electron microscopic investigation applying antibodies against alkane-hydroxylating cytochrome P-450 and the immunogold technique is under way in our laboratory. Preliminary results support the conception derived from cell fractionation studies.

3.3. Further oxidation to fatty acids

The next step of the alkane catabolic pathway is the oxidation of the n-alkan-1-ol. Early studies led to the assumption that the oxidation of fatty alcohols is catalyzed by a NAD-dependent dehydrogenase (LEBEAULT et al., 1970). Since the middle of the 80s a growing body of evidence has indicated that molecular oxygen instead of NAD is the essential electron acceptor of fatty alcohol oxidation and that simultaneously H_2O_2 arises. Consequently, the enzyme responsible for fatty alcohol oxidation is a H_2O_2-forming oxidase. ILCHENKO and TSFASMAN (1988) have isolated this enzyme from alkane-grown *Torulopsis candida*. Studies of other laboratories with *C. maltosa* (KRAUZOVA et al., 1985; BLASIG et al., 1988; MAUERSBERGER et al., 1987) and *C. tropicalis*, (KEMP et al., 1988) support the assumption of a wide distribution of fatty alcohol oxidase in alkane-assimilating yeasts. Thus, many of the early conceptions concerning the biochemistry of fatty alcohol oxidation in yeasts have had to be revised. But, conclusions that only oxidase-type enzymes are involved in fatty alcohol oxidation in this group of eukaryotic microorganisms should be considered with caution. Such a supposition could possibly prove to be an oversimplification.

The subsequent step was clearly demonstrated as a NAD-dependent dehydrogenase reaction by purification of the enzyme from *T. candida* (ILCHENKO and TSFASMAN, 1987). Consuming 1 mole H_2O per mole fatty aldehyde the latter is converted to fatty acid. Studies with subcellular fractions of *Yarrowia lipolytica* (YAMADA et al., 1980; MAUERSBERGER, 1986, unpublished) allow the same conclusion. These first three steps of n-alkane oxidation up to the fatty acid outlined proceed without variation of the chain length.

The intracellular localization of the fatty alcohol and aldehyde oxidation remained unclear for a long time. Approximately 10 years ago FUKUI and TANAKA (1980) proposed a scheme of alkane catabolism in yeasts. Following this model the fatty alcohol as the first intermediate, can be oxidized not only in endoplasmic reticulum but also in mitochondria and peroxisomes. In a recent cell fractionation study of alkane grown *C. maltosa* applying Percoll gradient centrifugation, KRAUSZOVA and SHARISHEV (1987) were able to show the predominant localization of fatty alcohol oxidase and fatty aldehyde dehydrogenase in the peroxisomes accompanied, e. g. by catalase and isocitrate lyase. As is known from the studies of FUKUI's group in alkane-grown cells, this organelle is characterized by a high level of various oxidases.

Considering the different intracellular localization of the three reactions — alkane hydroxylation in the endoplasmic reticulum, alkanol and aldehyde oxidation in peroxisomes — the fatty alcohol must be an essential transport form for long alkyl chains in the alkane-assimilating yeast cell.

3.4. The fate of fatty acids

The arising fatty acids are intermediates with a central position in the metabolism which can be used directly for the biosynthesis of lipids or for obtaining acetyl-CoA. Of course, for lipid biosynthesis desaturation and, when the occasion arises, elongation or chain shortening of the fatty acids originating from n-alkanes are necessary. However, de-novo-synthesis of fatty acids is without importance in cells of *C. maltosa* growing on long chain n-alkanes (BLASIG et al., 1988). Cytochrome P-450 catalyzed fatty acid hydroxylation for initiating diterminal oxidation is not achieved to any remarkable extent in most n-alkane-assimilating yeasts. However, in mutants diterminal oxidation can become the main pathway allowing the industrial production of long chain dicarboxylic acids (UEMURA et al., 1988).

The cellular fatty acid composition of alkane-grown yeasts is quite stable. For *C. maltosa* it was shown that 96 to 98% of the fatty acids have a chain length of 16—18 C atoms and the relation saturated/unsaturated is about 35/65 (BLASIG et al., 1984) independent of the chain length of the n-alkane utilized. The content of odd-numbered fatty acids, mainly C_{17}, is increased in cells grown on odd-numbered n-alkanes or a distillation fraction of long-chain n-alkanes (e. g. $C_{14}-C_{18}$) demonstrating the direct incorporation of fatty acids arising from alkane oxidation into lipids. As experiments with odd-numbered n-alkanes as carbon source show, even a high percentage of fatty acids with this unusual chain length does not impair the yeast cell (BLASIG et al., 1988).

As shown by MISHINA et al. (1978), two different acyl-CoA-synthetases (ACS I and II) introduce the fatty acids into the lipid biosynthesis (normally the minor part) or into the β-oxidation. In alkane utilizing yeasts β-oxidation is inducible and exclusively localized in peroxisomes (FUKUI and TANAKA, 1980), in contrast to animal cells, where both mitochondrial and peroxisomal β-oxidation systems exist (HASHIMOTO, 1982). A marked proliferation of peroxisomes is one of the morphological characteristics of the alkane-utilizing yeast cells (TANAKA et al., 1982).

In most fungi β-oxidation is introduced by fatty acyl-CoA oxidase (Kunau et al., 1988) converting acyl-CoA to the 2-enoyl-CoA. The following reactions are catalyzed by a trifunctional protein displaying the activities of enoyl-CoA hydratase and 3-hydroxyacyl-CoA dehydrogenase and of 3-hydroxyacyl-CoA epimerase. This three-functional protein has been found in all fungi but not in higher eucaryotes and procaryotes (Kunau et al., 1988). In the final reaction, 3-oxoacyl-CoA was cleaved by 2-oxoacyl-CoA thiolase to acetyl-CoA and a saturated acyl-CoA having two carbons less than the original fatty acyl-CoA.

Catabolizing an odd-numbered n-alkane the concluding cycle of β-oxidation supplies propionyl-CoA which is further metabolized via the methyl citrate cycle (Tabuchi and Uchiyama, 1975). The markedly prevailing acetyl-CoA is the starting point for gaining energy and for all anabolic pathways needed in the alkane-utilizing yeast cell.

Thus, n-alkane or fatty acid utilization does not supply C_3-units which are needed, e. g. for the gluconeogenesis and the biosynthesis of the pyruvate family of amino acids. This bottleneck is overcome by the glyoxylate cycle (Tanaka et al., 1982). After the condensation of oxaloacetate with acetyl-CoA to citric acid and its conversion to isocitric acid the latter is cleaved by isocitrate lyase forming succinate and glyoxylate. Malate synthetase catalyzes the condensation of glyoxylate and acetyl-CoA to malate which is dehydrogenated to oxaloacetate. With this, the acceptor compound for acetyl-CoA is recycled. As known, malate and oxaloacetate are sources for pyruvate and phosphoenolpyruvate, the C_3 key intermediates.

4. Properties of the alkane-induced cytochromes P-450

Although cytochrome P-450-catalyzed n-alkane hydroxylation is known from approximately ten yeast species, the enzymes concerned with have been investigated only from *C. maltosa* and to a limited degree from *C. tropicalis*. Already in the early studies (Lebeault et al., 1971; Gallo et al., 1971; Duppel et al., 1973) it was shown that the alkane-hydroxylating enzyme system is NADPH- and O_2-dependent and sensitive to CO. At first, the cytochrome P-450 from *C. maltosa* was obtained in a highly purified state by Riege et al. (1981), the cytochrome P-450 reductase was purified at first from *C. tropicalis* (Bertrand et al., 1980) and later from *C. maltosa* (Honeck et al., 1982). Tables 2 and 3 summarize the most essential properties of the enzymes of *C. maltosa*.

Reconstitution of both highly purified proteins of *C. maltosa* to the catalytically active alkane-hydroxylating enzyme system was successfully carried out by Honeck et al. (1982) providing for the first time the conclusive proof that the cytochrome P-450 system is the "alkane hydroxylase" of yeast.

Table 2. Selected properties of the alkane-hydroxylating cytochrome P-450 of *C. maltosa*[1,2]

Molecular mass[3]	59 705 Da	(apoprotein)
prosthetic group	1 heme/molecule	
e-transfer by	NADPH-P-450 reductase (containing 1 FAD + 1 FMN, M_r 79 kDa in SDS-PAGE)	
Spin state	low	
g-values[1]	2.41; 2.26; 1.93	
Spectral properties	Maxima (nm)	coefficients ($mM^{-1} \times cm^{-1}$)
ferric form	568	11.3
	535	11.6
	417	11.2
	360	
ferrous form	550	
	412	70
ferrous, CO-complex	555	14.6
ferrous, CO difference	447	91

[1] Schunck et al., 1983a, b.
[2] Corresponding results of *C. tropicalis* cytochrome (s) $P\text{-}450_{alk}$ have not yet been published till now.
[3] Schunck et al., 1989.

A second proof was furnished by selective inhibition of the alkane hydroxylation in the living cell by CO resulting in an immediate stop of growth. Growth is continued in the presence of CO, when fatty alcohol, the product of alkane hydroxylation, is added as carbon source (Schunck et al., 1987a, b; Wiedmann et al., 1988a). This result demonstrates that the cytochrome P-450 catalyzed hydroxylation is the only reaction in yeasts for initiating the alkane degradation.

Antibodies against the alkane hydroxylating cytochrome P-450 of *C. maltosa* were used to establish a radio immuno-assay (Kärgel et al., 1984) in order to quantitate the in vitro translation product of cytochrome P-450 specific mRNA (Wiedmann et al., 1986) and to isolate a cytochrome P-450 cDNA clone (Schunck et al., 1988) (see section 7). Further studies applying the RIA showed that the alkane-induced cytochromes P-450 of yeasts are immunologically very similar. However, they differ clearly from other cytochromes P-450. From these findings the conclusion was drawn that the former have resembling structures and little similarities to all others tested cytochromes P-450 of bacteria and mammalian liver microsomes (Kärgel et al., 1985). This suggestion was confirmed later by the elucidation of amino acid sequences (cf. section 8).

Table 3. Amino acid composition of cytochrome P-450_{alk} of *C. maltosa* and of *C. tropicalis* (derived from sequences in Fig. 5)

Amino acid	*C. maltosa*	*C. tropicalis*
Ala	40	32
Arg	26	27
Asn	24	18
Asp	24	26
Cys	5	5
Gln	19	25
Glu	39	40
Gly	31	31
His	12	12
Ile	28	31
Leu	50	68
Lys	37	36
Met	8	8
Phe	36	36
Pro	24	24
Ser	21	27
Thr	34	37
Trp	6	6
Tyr	25	25
Val	32	29

Probably, the same alkane hydroxylating cytochrome P-450 as obtained by Riege et al. (1981) has been purified from n-hexadecane grown *C. maltosa* cells by Avetissova et al. (1985) and Sokolov et al. (1986). These authors got first evidence for the occurrence of an additional cytochrome P-450 form during cultivation on shorter chain n-alkanes, like decane.

Loper et al. (1985) reported the purification of two immunologically related cytochrome P-450 proteins from n-tetradecane-grown *C. tropicalis* cells having apparent Mw of 53 and 54 kDa, resp., in SDS-PAGE. Results about the properties of these cytochromes P-450 of *C. tropicalis* have not been published. Antibodies raised against the 54 kDA protein were used to screen a genomic library (Sanglard et al., 1987). The cloned gene was shown to code for a cytochrome P-450 form which catalyzes fatty acid ω-hydroxylation after heterologous expression in *S. cerevisiae* (Sanglard et al., 1988). Very recently, the orthologous cytochrome P-450 of *C. maltosa* has been detected by cloning the cDNA and its heterologous expression (Schunck et al., 1990).

Experiments using microsomes of alkane-grown *C. maltosa* demonstrated specific activities of 1—5 nmoles converted hexadecane/nmole P-450 $\times$ min and up to 18 nmoles for the reconstituted system, resp. (SCHUNCK and RIEGE, 1983). The addition of a lipid component has no essential influence, however, nonionic detergents are stimulating. Such velocities are insufficient to explain the growth rates on n-alkanes of these yeasts. As in vivo estimations of the n-hexadecane hydroxylation by *C. maltosa* prove (SCHUNCK et al., 1987a, b), the actual catalytic rate in the living cell is 1,5—3 μmoles/nmole P-450 $\times$ m. That means, it is 2—4 orders of magnitude higher than the specific activities of all mammalian cytochrome P-450s investigated up to now. Comparable high activities are known from bacterial cytochromes P-450 (GUNSALUS et al., 1974; NARHI and FULCO, 1986).

Conflicting opinions exist about the substrate specificity of alkaneinducible cytochromes P-450. Influenced by the diversity of reactions being catalyzed by mammalian microsomal P-450s several investigators looked for a conversion of xenobiotics by yeast cells or by microsomes. In many cases the simultaneous rise and decrease of cytochrome P-450 and the ability to convert a compound was accepted as sufficient proof. However, it was neglected that the molar activity with these substrates was always very low ($<$ 1 nmole converted substrate/nmole P-450 $\times$ min) and such a simple control like a CO inhibition test was frequently omitted. Thus, arguments for an involvement of yeast cytochrome P-450 in these reactions are quite weak.

Studies with the reconstituted alkane-hydroxylating enzyme system of *C. maltosa* applying both cytochrome P-450 and the reductase in a highly purified state provided evidence that this system doesn't attack benzphetamine, ethylmorphine, aminophenazone, and benzo(a)pyrene (SCHUNCK and RIEGE, 1983). On the other hand, n-alkanes between 10 and 18 C atoms are converted with very high regioselectivity to the primary alcohols of the same chain length (SCHUNCK et al., 1987a). No other reaction product was detectable.

Experiments with cells revealed that the alkane-induced cytochromes P-450 can hydroxylate n-alkanes from 8 to 40 C atome, the side chain of long chain alkyl benzenes (HUTH et al., unpublished) and very slowly also the isoalkane pristane. Applying microsomes of n-alkane grown *C. maltosa* the ω-hydroxylation of palmitic acid was confirmed (BLASIG et al., 1988). Experiments with two different recombinant cytochromes P-450 both originating from *C. maltosa* show that one form prefers hexadecane and the other one lauric acid as substrate for the terminal hydroxylation (SCHUNCK et al., 1990). Thus, all results testify the specificity for long alkyl chains and the high regioselectivity of the alkane-induced cytochromes P-450 of *C. maltosa*. To our knowledge, hitherto there are no convincing results for the alkane-induced cytochromes P-450 of *C. tropicalis*, too, which evidence other facts.

All cytochromes P-450 are distinguished by peculiar physical properties, which can be studied very elegantly by ESR and optical spectroscopy (see preceeding volumes of this series). The alkane-hydroxylating cytochrome P-450 of *C. maltosa* is no exception (Tab. 2 and Fig. 4). The Soret peak of the ferric form at 417 nm indicates the low spin state (S = 1/2). Hexadecane causes a transition to the high spin state (S = 5/2) which is promoted by a high phosphate concentration. This changes can be monitored by UV spectroscopy (Fig. 5). The arising type I difference spectrum is characteristic for substrate binding. Interestingly, an almost complete reduction of cyt P-450 by its

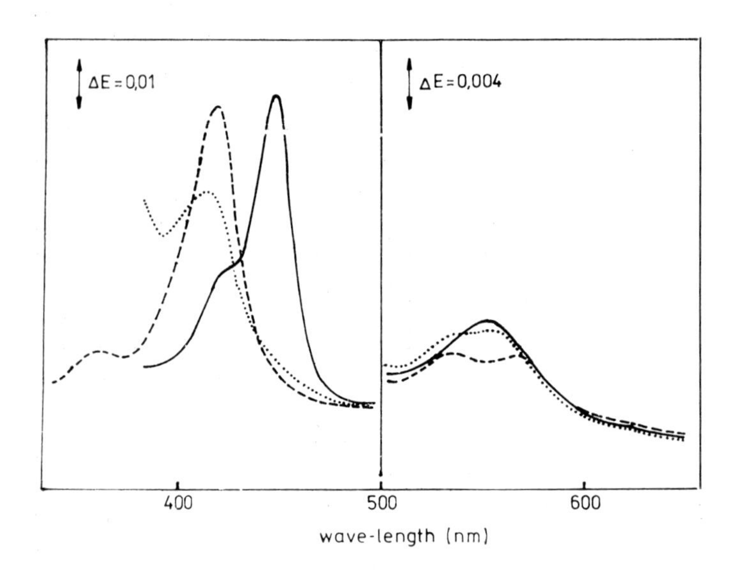

Fig. 4. Spectra of the highly purified alkane-hydroxylating cytochrome P-450 of *C. maltosa*. (—) oxidized, (···) reduced by dithionite, (– – –) reduced by dithionite + CO.

NADPH-dependent reductase was only achieved in the presence of the n-alkane substrate (RIEGE et al., 1981). This finding suggests a significant shift of the redox potential upon substrate binding as studied in detail with the cytochrome P-450_{cam} system (GUNSALUS et al., 1874).

Unlike hexadecane, benzphetamine and imidazole cause inverse type I and type II spectra. (For nomenclature in cytochrome P-450 spectroscopy see JEFCOATE, 1978.) Further details of spectroscopic studies on yeast alkane-hydroxylating cytochromes P-450 were already reviewed earlier (MÜLLER et al., 1984).

Studies about details of the catalytic mechanism of yeast-alkane hydroxylating cytochrome P-450 have hitherto been missing. However, no reason has yet been put forward to assume striking differences to the generally accepted cycle of cytochrome P-450 catalysis (cf. RUCKPAUL et al., 1989; vol. 1 of this series).

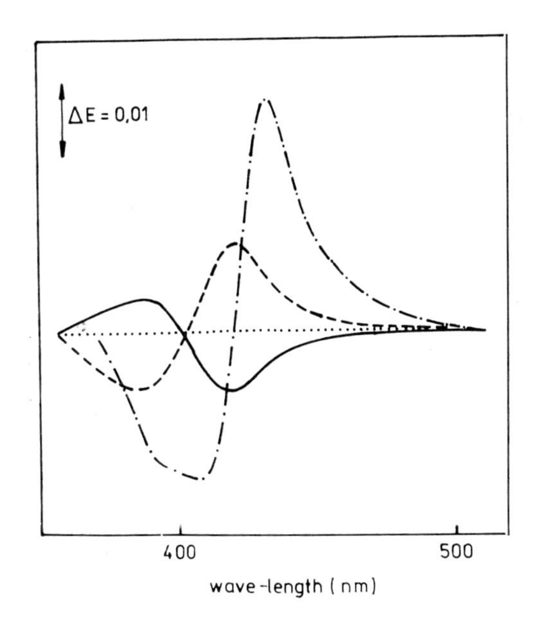

Fig. 5. Difference spectra after binding of (—) hexadecane (- - -), benzphetamine, (— · — · —) imidazole to alkane-hydroxylating cytochrome P-450 (200 mM potassium phosphate, pH 7.25).

5. Cytochrome P-450 and glycolipid biosynthesis

As outlined in section 1, already in 1970 the subterminal hydroxylation of oleic acid in a *Torulopsis* strain producing glycolipids was ascribed to a cytochrome P-450 system, because the reaction was strongly CO-sensitive, NADPH- and O_2-dependent (HEINZ et al., 1970). However, in the following years no report characterizing this cytochrome P-450 was published.

Further indications of a cytochrome P-450 catalyzed fatty acid hydroxylation in connection with the biosynthesis of glycolipids by various yeast species, e. g. *Torulopsis apicola*, *T. dattila* and *T. glabrata* were reported many years later by KLEBER et al. (1988). In addition to sugar moieties, hydroxylated long chain fatty acids are precursors for these compounds. Independently, whether glucose, n-hecadecane or a mixture of both were used as the carbon source, cytochrome P-450 was detected in cells of *T. apicola* in the late stationary growth phase. At this time, secretion of large amounts of glycolipids takes place. This correlation was considered as to indicate the participation of cytochrome P-450 in glycolipid biosynthesis or more exactly as evidence for the cytochrome P-450 type of the fatty acid hydroxylating enzyme (KLEBER et al., 1988). However, experimental confirmation is still required.

6. Cytochrome P-450$_{14DM}$ in alkane-assimilating yeasts

Pioneering work on the cytochrome P-450$_{14DM}$ was done with *S. cerevisiae* by YOSHIDA's group and the properties and physiological importance of this form are reviewed separately (YOSHIDA and AOYAMA, this volume). As a consequence of the importance of the lanosterol-demethylating cytochrome P-450 for ergosterol biosynthesis (cf. section 2) it should be widely distributed among yeasts (Tab. 1) including alkane-utilizers.

Several reports in recent years confirm this idea and, for the sake of completeness, the most important results are mentioned here, without anticipating the detailed discussion in the overview of YOSHIDA and AOYAMA (this volume).

First of all, LOPER et al., (1985) isolated a cytochrome P-450 form from glucose-grown cells of *C. tropicalis* which was immunologically related to cytochrome P-450$_{14DM}$ of *S. cerevisiae*. The same group was successful in cloning and sequencing the gene (CHEN et al., 1988). Simultaneously, SANGLARD et al. (1986) demonstrated that partially purified cytochrome P-450 from glucose-grown *C. tropicalis* cells catalyzes the 14α-demethylation of lanosterol. Furthermore, KÄRGEL et al. (1990) were successful in the immunological detection of this cytochrome P-450 form in *C. maltosa*. All these results support strongly the idea of a widespread occurrence of cytochrome P-450$_{14DM}$.

7. Regulation of cytochrome P-450 biosynthesis

Dithionite-recuded CO-difference spectra have been widely used to characterize cytochrome P-450 formation in yeast cells under the influence of such extracellular factors as carbon source and oxygen. This method yields a reliable quantitation of the total cytochrome P-450 content, provided that there is no severe spectral interference with mitochondrial cytochromes.

Suitable procedures were published to prepare yeast microsomes of sufficient purity for clear spectral determinations (KÄPPELI et al., 1982; MAUERSBERGER et al., 1984). Moreover, a simple method was found to improve already the estimation of cytochrome P-450 in whole cells (SHARISHEV et al., 1983; SCHUNCK et al., 1987b). It involves omitting dithionite and adding antimycin A or CN- in order to prevent reduction of cytochrome oxidase.

As outlined in section 2 multiple cytochrome P-450 forms may occur in yeasts (cf. Table 1) which differ in respect of their physiological functions and regulation of expression. Therefore, the unequivocal identification of the induced cytochrome P-450 form(s) and the availability of specific probes have become the most crucial points of regulation studies.

Based on specific antibodies and recent progress in DNA cloning, the first details have been reported concerning regulation mechanisms of alkane inducible cytochrome P-450 forms and of the lanosterol 14α-demethylating cytochrome P-450.

On the other hand, a number of earlier studies cannot be clearly interpreted with respect to the cytochrome P-450 form(s) actually occurring under the applied cultivation conditions. However, they will be included in this review because they might stimulate future experiments to elucidate both the real degree of cytochrome P-450 multiplicity and the regulation of individual cytochrome P-450 forms in yeasts.

7.1. Induction by long-chain n-alkanes

Following the initial reports of LEBEAULT et al. (1971) and GALLO et al. (1971), induction of cytochrome P-450 by long-chain n-alkanes has been reported for a large number of yeast species being able to utilize these unconventional substrates (cf. Tab. 1). Moreover, various compounds having chemical structures related to n-alkanes were found to induce cytochrome P-450 formation (GILEWICZ et al., 1977; MAUERSBERGER et al., 1981). Among these, long chain n-alkanes and some phenyl alkanes turned out to be the most potent inducers. A remarkable induction effect has been also described for long-chain secondary alcohols. A 3–4 times lower cytochrome P-450 content was detected in fatty alcohol and fatty acid grown cells when compared with the values reached during cultivation on n-alkanes (ILCHENKO et al., 1980; MAUERSBERGER et al., 1981).

n-Tetradecane and n-hexadecane mediated cytochrome P-450 induction can be almost completely repressed by the simultaneous addition of glucose to the culture medium in concentrations above 0,5% (TAKAGI et al., 1980; MAUERSBERGER et al., 1981). Glycerol and galactose were found to be non or less repressive and are therefore suitable substrates when the induction potential of non-degradable hydrocarbons is to be tested (GILEWICZ et al., 1979; MAUERSBERGER et al., 1981).

Alkane-mediated P-450 induction requires protein- and lipid de-novo synthesis as suggested by inhibition with cycloheximide and cerulenin (TAKAGI et al., 1980; MAUERSBERGER et al., 1981).

It is now well established that the spectral cytochrome P-450 content of n-alkane grown cells can be almost completely attributed to cytochrome P-450 forms involved in terminal alkane and fatty acid hydroxylation. However, the possibility that several alkane-inducible cytochrome P-450 forms may exist has not been thoroughly considered for many years (see section 4). Moreover, it remains unclear whether identical cytochrome P-450 forms are induced using the different alkane-related hydrocarbons listed above. Interestingly, fatty acid ω-hydroxylation was not detectable in microsomes of *C. tropicalis* cells after induction with long-chain secondary alcohols (GILEWICZ et al., 1979).

Using a radioimmunoassay based on antibodies against the highly purified alkane hydroxylating cytochrome P-450, KÄRGEL et al. (1984) demonstrated that microsomes of n-alkane grown *C. maltosa* cells have a 100–300 fold higher specific content of the corresponding protein than detectable after cultivation on glucose. The relative amounts of cytochrome P-450 specific mRNA were found to parallel the changes in the spectrally and immunologically determined cytochrome P-450 content (WIEDMANN et al., 1986).

SANGLARD et al. (1987) used a restriction fragment of a cloned cytochrome P-450 gene of *C. tropicalis*, to probe blots of mRNA from glucose and n-tetradecane-grown cells of this yeast. A 500–1000 fold elevation of the cytochrome P-450 specific mRNA was found in cells cultivated on the alkane substrate. The promotor region was found to contain a 17bp sequence with dyad symmetry and two 12 bp perfect inverted repeats. As proposed by SANGLARD and LOPER (1989) these elements may be involved in control of induction.

The results indicate that the formation of the alkane inducible cytochrome P-450 forms is mainly regulated at the pretranslational, most probably transcriptional, level. Cells cultivated on glucose and n-alkanes can be considered to represent the repressed and induced state, respectively. A derepressed state has not yet been defined. Microsomes isolated from glycerol grown *C. maltosa* cells are active in alkane hydroxylation (MAUERSBERGER et al., 1984) and contain significant amounts of cytochrome P-450, which is immunologically indistinguishable from that of alkane grown cells (KÄRGEL et al., 1984). Moreover, the occurrence of cytochrome P-450 during cultivation on glucose upon transition to the stationary phase may indicate derepression of alkane-inducible cytochrome P-450 forms (ILCHENKO et al., 1980).

7.2. Influence of oxygen on alkane-mediated cytochrome P-450 induction

During batch cultivation of *C. maltosa* on n-hexadecane, cytochrome P-450 formation and growth course correlate in a characteristic manner (MAUERSBERGER and MATYASHOVA, 1980): the cellular cytochrome P-450 content is strongly increased in the lag-phase of growth upon transition from glucose to n-alkane utilization, reaches a nearly constant level in the exponential phase and declines rapidly after comsumption of the carbon source in the stationary phase.

In a feed-controlled continuous cultivation of *C. tropicalis* on n-hexadecane, the cytochrome P-450 content was found to be linearly related to the substrate uptake rate (GMÜNDER et al., 1981).

These results underline the key position of cytochrome P-450 in alkane degradation (see section 3) and suggest that the n-alkane substrate itself acts as cytochrome P-450 inducer. In particular, under oxygen-limitation conditions the cytochrome P-450 catalyzed alkane hydroxylation appears to represent the rate-limiting step of the whole pathway (GMÜNDER et al., 1981; SCHUNCK et al., 1987a). In fact, significant differences were found in the pO_2 dependence of utilization of n-alkanes, fatty alcohols, fatty acids and glucose. It was concluded that the cytochrome P-450 system is distinguished by the lowest oxygen affinity among the oxygen activating enzymes involved in alkane assimilation (SCHUNCK et al., 1987a).

As studied in detail with *C. maltosa* (MAUERSBERGER et al., 1980, 1984;

Schunck et al., 1987b) and with *C. tropicalis* (Gmünder et al., 1981; Sanglard et al., 1984), the cellular cytochrome P-450 content is strongly increased by transition to oxygen-limited growth during cultivation on long-chain n-alkanes. Enhanced cytochrome P-450 formation was also observed maintaining oxygen saturation but inhibiting more selectively the alkane hydroxylation reaction by introducing low concentrations of carbon monoxide into the culture medium (Schunck et al., 1987a, b). However, this effect did not occur when, instead of n-alkanes, fatty alcohols were used as the carbon source. In conclusion, enhanced cytochrome P-450 formation appears to depend on two factors:

- a decrease of the enzymatic activity of the alkane hydroxylating cytocrome P-450 system (O_2-deficiency or CO-inhibition) and
- the presence of n-alkanes, which act both as substrate and as inducer of cytochrome P-450.

The most simple interpretation of these findings may be that the cytochromo P-450 formation is mainly regulated by the intracellular inducer concentration which depends on the relative rates of alkane transport into the cell and the actual alkane hydroxylating activity of the enzyme system.

According to this scheme oxygen limitation results in enhanced substrate induction. Immunological studies revealed that it is indeed the alkane hydroxylating cytochrome P-450 which increases during oxygen limitation. The effect is not caused by a shift of the apo/holo enzyme ratio as shown by cycloheximide inhibition and immunological quantitation of the cytochrome P-450 protein (Schunck et al., 1987b). In fact, Wiedmann et al. (1988a) found a parallel increase of the cytochrome P-450 encoding mRNA which was estimated by means of in vitro translation experiments using antibodies against the alkane hydroxylating cytochrome P-450.

However, to qualify the simple scheme of inducer accumulation, it must be considered that oxygen limitation does not only affect the synthesis of the alkane inducible cytochrome P-450. Thus, the cytochrome $P\text{-}450_{14DM}$ underlies a similar regulation as demonstrated during growth of *S. cerevisiae* (Ishidate et al., 1969; Trinn et al., 1982) and *C. tropicalis* (Sanglard et al., 1984) on glucose. Moreover, in different degrees, cytochrome b_5, catalase, cytochrome oxidase and other hemoproteins respond to oxygen limitation (Gmünder et al., 1981; Mauersberger et al., 1984; Sanglard et al., 1984). Therefore, it may be assumed that derepression of heme biosynthesis plays an important role (Sanglard et al., 1984). It remains for further studies on promotor functions to distinguish between enhanced substrate induction, heme control and the not completely excluded possibility of a more direct oxygen effect. Interestingly, a heme control region has been proposed for the *S. cerevisiae* catalase *T* gene (Spevak et al., 1986) but it could not be identified in the promotor of the peroxisomal catalase gene from *C. tropicalis* (Okada et al., 1987).

7.3. Regulation of cytochrome P-450$_{14DM}$

Probably independently of the yeast species, the lanosterol demethylating cytochrome P-450$_{14DM}$ is the prevailing cytochrome P-450 form during cultivation on glucose. The cytochrome P-450$_{14DM}$ has been most extensively studied from *S. cerevisiae* (see Yoshida and Aoyama, this volume). More recently, *C. tropicalis* was shown to produce an immunologically related cytochrome P-450 form (Loper et al., 1985). Both the cytochrome P-450$_{14DM}$ genes of *S. cerevisiae* (Kalb et al., 1986, 1987) and of *C. tropicalis* (Chen et al., 1987, 1988) have been isolated and sequenced.

The cytochrome P-450 content of yeasts growing on glucose ranges according to different reports from the nearly undetectable (Gilewicz et al., 1979; Mauersberger et al., 1981, 1984) to 200 pmoles per mg microsomal protein (Sanglard et al., 1984). In general, semianaerobic growth was found to support a high cytochrome P-450 content. The required conditions were studied in detail using continuous culture techniques. Accordingly, the extent of cytochrome P-450 formation is related to the type of glucose metabolism (fermentative, respirative and stages in between) which differs depending on the cultivation conditions and the yeast strain used (for a review see Käppeli et al., 1986).

Due to its essential physiological function, the cytochrome P-450$_{14DM}$ should also be present under various other growth conditions, perhaps only with the exception of strictly anaerobic growth where yeasts become ergosterol dependent. Lanosterol demethylating activity was detected in microsomes of alkane-grown *C. tropicalis* cells, however, being about 50-fold lower than after cultivation on glucose (Sanglard et al., 1986). The presence of a low amount of cytochrome P-450$_{14DM}$ was also demonstrated in alkane-grown *C. maltosa* cells using antibodies against the related cyt P-450$_{14DM}$ from *S. cerevisiae* (Kärgel et al., 1990). Moreover, Northern analysis revealed the presence of a common transcript when mRNA's from glucose and n-tetradecane grown *C. tropicalis* cells were probed with cytochrome P-450$_{14DM}$-specific DNA (Chen et al., 1988).

In a physiological context, the fact remains difficult to explain that a fermentative glucose metabolism is required to bring about a high P-450$_{14DM}$ content. As discussed by Käppeli (1986), enhanced ergosterol biosynthesis could be required to adapt in some way the composition of cellular membranes to ethanol exposure.

8. Primary structures of the alkane-inducible cytochromes P-450

8.1. Comparison with structures of members of the cytochrome P-450 superfamily

Within the last two years, the first complete primary structures of alkane-inducible yeast cytochrome P-450 forms became available after cloning and sequencing the respective genes or cDNA's (Fig. 6).

SANGLARD and LOPER (1989) succeeded in the isolation of P-450 gene from a genomic library of *C. tropicalis*. Its expression in *S. cerevisiae* yielded a P-450 protein (designated as P-450 LIIA1) active in lauric acid ω-hydroxylation. Later on, SANGLARD and FIECHTER (1989) provided evidence for the existence of additional related genes in *C. tropicalis* by Southern blot analysis. Moreover, a gene fragment was isolated which represented the C-terminal half of a putative second alkane-inducible P-450 form in *C. tropicalis* (SANGLARD and FIECHTER, 1989).

As a first step in the elucidation of alkane-inducible P-450 genes from *C. maltosa*, SCHUNCK et al. (1989) reported the isolation of a cDNA and later on TAKAGI et al. (1989) of a genomic clone. The deduced primary structures of the encoded proteins are almost identical and correspond to the major cyt P-450 form after cultivation of *C. maltosa* on long-chain n-alkanes, which was purified at first by RIEGE et al. (1981) (designated in the following as cyt P-450_{Cm-1}). Alignment analysis revealed a similarity of 57,6% with the above mentioned *C. tropicalis* P-450 LIIA1 (Fig. 6).

Expression of the P-450_{Cm-1} cDNA in *S. cerevisiae* resulted in a P-450 system, which hydroxylated n-hexadecane with significantly higher activity than lauric acid (SCHUNCK et al., 1990). More recently, the cDNA for a second alkane-inducible P-450 form of *C. maltosa* (cyt P-450_{Cm-2}) was isolated and expressed in *S. cerevisiae* (SCHUNCK et al., 1990). It shows a different substrate specificity in being much more active with respect to lauric acid hydroxylation. As shown in Figure 6 this P-450_{Cm-2} is very closely related to the *C. tropicalis* P-450 LIIA1 (76,8% similarity) suggesting that the respective genes are orthologous. Both alkane-induced P-450s of *C. maltosa* exhibit a similarity of 54%.

Table 4 shows the results of pairwise comparisons with representative members of 13 reported P-450 gene families. The data provide evidence that the alkane-inducible yeast P-450s belong to a new gene family as first suggested by SANGLARD et al. (1989). According to the recently introduced nomenclature, CYP 52 and P-450 LII should be used to designate the gene locus and the protein, respectively (NEBERT et al. 1989).

The comparison of the alkane-induced yeast P-450s with members of other families shows that a distant relationship appears to exist among the III (steroid-inducible, biological function unclear), IV (mammalian lauric acid hydroxylating), LII (yeast alkane-inducible) and CII (*B. megaterium* fatty acid hydroxylating BM-3) families (Table 4 and NEBERT, personal communication). The similarity with rat P-450 pcn 1 (III) is significant, however, only in the C-terminal parts (SANGLARD et al., 1988).

The alkane-inducible and lanosterol demethylating yeast P-450s clearly belong to different gene families (Tab. 4 and 5, Fig. 7). A similarity of about 65% was determined for the P-450_{14DM} of *S. cerevisiae* and *C. tropicalis*, whereas the alkane-inducible P-450 and the P-450_{14DM} of *C. tropicalis* share

1	—	—	—	—	—	—	—	—	—	—	—	—	M	A	I	E	Q	I	I	E
																				:
1′	M	S	V	S	—	—	F	V	H	N	V	L	E	V	V	T	P	Y	V	E
	:	:		:								:				:	:	:	:	:
1″	M	S	S	S	P	S	I	A	Q	E	F	L	A	T	I	T	P	Y	V	E
29	F	L	S	I	A	L	R	N	K	F	Y	E	Y	K	L	K	C	E	N	—
						:						:	:	:						
39′	P	F	L	S	I	L	H	T	K	Y	L	E	Y	K	F	N	A	K	P	L
				:		:	:	:	:	:	:	:		:	:		:	:	:	:
41″	N	L	I	S	M	L	H	T	K	Y	L	E	R	K	F	K	A	K	P	L
67	K	V	R	K	A	G	Q	L	A	D	Y	T	D	T	T	F	D	K	Y	P
	:																			
79′	K	W	H	G	T	V	M	E	F	A	C	—	N	V	—	W	N	N	K	F
	:			:	:	:	:		:	:	:					:			:	
81″	K	S	K	G	T	V	M	Q	F	A	C	—	D	L	—	W	D	K	K	L
103	—	T	V	D	P	E	N	I	K	A	V	L	A	T	Q	F	N	D	F	A
		:		:	:	:	:		:	:		:	:	:	:	:	:	:	:	
117′	E	T	T	D	P	E	N	V	K	A	I	L	A	T	Q	F	N	D	F	S
	:	:		:	:	:	:	:	:	:	:	:	:	:	:	:	:	:	:	:
119″	E	T	K	D	P	E	N	V	K	A	I	L	A	T	Q	F	N	D	F	S
142	G	E	G	W	K	H	S	R	A	M	L	R	P	Q	F	A	R	E	Q	I
	:		:	:	:	:	:	:	:	:	:	:	:	:	:	:	:	:	:	
157′	G	A	G	W	K	H	S	R	A	M	L	R	P	Q	F	A	R	E	Q	V
	:	:	:	:	:	:	:	:		:	:	:	:	:	:	:	:	:	:	:
159″	G	A	G	W	K	H	S	R	T	M	L	R	P	Q	F	A	R	E	Q	V
182	K	G	K	T	F	D	L	Q	E	L	F	F	R	F	T	V	D	T	A	T
		:	:	:	:	:		:	:	:	:	:	:		:	:	:			:
197′	Q	G	K	T	F	D	I	Q	E	L	F	F	R	L	T	V	D	S	S	T
		:		:	:	:	:	:	:	:	:	:	:	:	:	:	:	:		:
199″	H	G	Q	T	F	D	I	Q	E	L	F	F	R	L	T	V	D	S	A	T
221	A	P	N	D	I	P	G	R	E	N	F	A	E	A	F	N	T	S	Q	H
					:		:	:	:		:	:		:	:	:		:	:	
237′	S	T	K	N	I	A	G	R	E	E	F	A	D	A	F	N	Y	S	Q	D
		:	:				:	:		:	:	:	:	:	:	:	:	:	:	
239″	T	T	K	D	F	D	G	R	N	E	F	A	D	A	F	N	Y	S	Q	T
261	R	D	C	N	A	K	V	H	K	L	A	Q	Y	F	V	N	T	A	L	N
							:	:	:		:				:			:	:	
277′	N	K	S	I	K	T	V	H	K	F	A	D	F	Y	V	Q	K	A	L	S
		:	:	:			:	:	:	:	:	:		:	:	:	:	:	:	
279″	R	K	S	I	A	I	V	H	K	F	A	D	H	Y	V	Q	K	A	L	E
301	V	K	Q	T	R	D	P	K	V	L	Q	D	Q	L	L	N	I	M	V	A
		:	:	:	:	:	:	:	:	:		:	:	:	:	:	:		:	:
315	A	K	Q	T	R	D	P	K	V	L	R	D	Q	L	L	N	I	L	V	A
	:	:	:	:	:	:	:	:	:	:	:	:	:	:	:	:	:	:	:	:
317″	A	K	Q	T	R	D	P	K	V	L	R	D	Q	L	L	N	I	L	V	A

Fig. 6 (continued see page 113—115).

```
E V L P Y L T K W Y     T I I F G A A V T Y
          : : : : :         :           :
Y Y Q E N L T K W Y     I L I P T I L L T L
:   : : :   : : : :         : :       :   :
Y C Q E N Y T K W Y     Y F I P L V I L S L
P V Y F Q D A G L F     G I – P A L I D I I
        : :               :   :
T N F A Q D Y S F G     V I T P L M L M Y F
        : : :   :         : : : :   :   :
A V Y V Q D Y T F C     L I T P L V L I Y Y
N L S S Y M T V A G     V L K I V – – – F –
                          :   :
L V L N G N V R T V     G L R I M G L N I I
  :             :       : :   :   :       :
I V S D P K A K T I     G L K I L G I P L I
L G A R H A H F D P     L L G D G I F T L D
; ;   : .               : : : : . . . . : :
L G T R H D F L Y S     L L G D G I F T L D
: : : : : : : : : :     : : : : : : : : : :
L G T R H D F L Y S     L L G D G I F T L D
A H V K A L E P H V     Q I L A K Q I K L N
: : : :   : : : : :     :   :   :
A H V K L L E P H V     Q V L F K H V R K S
  : : : : : : : :       : : : : : :   :
S H V K L L E P H M     Q V L F K H I R K H
E F L F G E S V H S     L Y D E K L G I – P
: : : : :   : :   :     :   :       :     :
E F L F G G S V E S     L R D A S I G M V P
: : :   :   :   : :     : : :   :   :
E F L L G E S A E S     L R D E S V G L T P
Y L A T R T Y S Q I     F Y W L T N P K E F
:   :   :       :       : : :     :       :
Y N A Y R F L L Q Q     F Y W I L N G S K F
    : : : : : : : :       : : : : : : :   :
N Q A Y R F L L Q Q     M Y W I L N G S E F
A T E K E V E E K S     K G G Y V F L Y E L
  :         :   :           : : : : : : : :
L T D D D L E – K –     Q E G Y V F L F E L
: : :   : : :   :         : : : : : :   : :
L T D E D L E – K –     K E G Y V F L F E L
G R D T T A G L L S     F A M F E L A R N P
: : : : : : : : : :     :     : : :   : : :
G R D T T A G L L S     F L F F E L S R N P
: : : : : : : : : :     : : : : : : : : : :
G R D T T A G L L S     F L F F E L S R N P
```

341	K	I	W	N	K	L	R	E	E	V	E	V	N	F	G	L	G	D	E	A
					:	:		:	:	:				:	:					:
355′	T	V	F	E	K	L	K	E	E	I	H	N	R	F	G	A	K	E	D	A
			:		:	:		:	:	:		:		:	:				:	:
357″	E	I	F	A	K	L	R	E	E	I	E	N	K	F	G	L	G	Q	D	A
381	N	E	T	L	R	M	Y	P	S	V	P	I	N	F	R	T	A	T	R	D
	:	:		:	:		:	:	:	:	:		:	:	:		:	:.	:	
395′	N	E	A	L	R	V	Y	P	S	V	P	H	N	F	R	V	A	T	R	N
	:	:		:	:		:	:	:	:	:	:	:	:	:	:	:	:	:	:
397″	N	E	T	L	R	I	Y	P	S	V	P	H	N	F	R	V	A	T	R	N
421	S	S	V	V	Y	S	V	Y	K	T	H	R	L	K	Q	F	Y	G	E	D
			:		:					:	:	:					:	:	:	:
435′	Q	N	V	M	Y	T	I	S	A	T	H	R	D	P	S	I	Y	G	E	D
	:		:	:	:	:	:		:	:	:	:	:			:	:	:	:	:
437″	Q	V	V	M	Y	T	I	L	A	T	H	R	D	K	D	I	Y	G	E	D
461	Y	L	P	F	N	G	G	P	R	I	C	L	G	Q	Q	F	A	L	T	E
	:		:	:	:	:	:	:	:	:	:	:	:	:	:	:	:	:	:	:
475′	Y	V	P	F	N	G	G	P	R	I	C	L	G	Q	Q	F	A	L	T	E
	:	:	:	:	:	:	:	:	:	:	:	:	:	:	:	:	:	:	:	:
477″	Y	V	P	F	N	G	G	P	R	I	C	L	G	Q	Q	F	A	L	T	E
500	E	T	Y	P	P	N	K	C	I	H	L	T	M	N	H	N	E	G	V	F
			:	:	:						:	:						:		
515′	T	R	Y	P	P	R	L	Q	N	S	L	T	V	S	L	C	D	G	A	N
	:		:	:	:		:	:	:		:	:		:	:			:	:	
517″	T	E	Y	P	P	K	L	Q	N	T	L	T	L	S	L	F	E	G	A	E

Fig. 6. Comparison of the deduced amino acid sequences of three alkane-inducible yeast line cyt P-450_{Cm-2} (Schunck et al., 1990), the lower line cyt P-450_{alk} of *C. tropicalis* We point to a correction of the sequence published by Schunck et al. (1989) between the A consequent application of the nomenclature guidelines proposed by Nebert et al. lard and Loper (1989) LIIA1 of C. t., cyt P-450_{Cm-1} of *C. maltosa* sequenced by LIIA1 of C. m.
A pre-alignment was performed using the program ALIGN in the PRONUC software alignment was modified by hand to minimize the number of gaps. Two hydrophobic positions, regardless of the coincidence of the amino residues. The symbol (:) indicates.

only 10% identica lamino acids (Chen et al., 1988). Even in the C-terminal parts, which usually are more similar than the N-terminal ones, comparing P-450 sequences (Gotoh and Fujii-Kuriyama, 1989), there is a distinct difference in the distance between two conserved regions (Fig. 7). Likewise, the differences of P-450_{14DM} structures to the mammalian microsomal cytochromes P-450 are more distinct (Tab. 5). It has been calculated that both yeast P-450 gene families split from one another approximately 1,300 million years ago (Chen et al., 1988).

The similarity with members of other P-450 families is mainly restricted to some positions (Tab. 5), known to be highly conserved among the whole

```
R V D E I S F E T L     K K C E Y L K A V L
: :   : :   : :   :     :   : : : : : :
R V E E I T F E S L     K L C E Y L K A C V
: : : : :   : :   :     :   : : : : : :
R V E E I S F E T L     K S C E Y L K A V I

T T L P R G G G K D     G N S P I F V P K G
: : : : : : : : : :     :   : : :       : :
T T L P R G G G K D     G M S P I A I K K G
: : : : : : : :         :   : : : : : : : :
T T L P R G G G E G     G L S P I A I K K G

A Y E F R P E R W F     E P S T R K L G W A
:     : : : : : : :     : :   : : : : : : :
A N V F R P E R W F     E P E T R K L G W A
:   : : : : : : : :     : : : : : : : : : :
A Y V F R P E R W F     E P E T R K L G W A

A S Y V|I A R L A Q     M F E H L E S K – D
: : : :     : :   :       :     :
A S Y V|T V R L L Q     E F H T L T Q D A N
: : : :|: : : : : :     : :     :   : :   :
A S Y V|T V R L L Q     E F G N L K Q D P N

I S A K – – – *         523
:
I Q M Y – – – *         538
  : : :
V Q M Y L I L *         543
```

cyt P-450s. The upper line exhibits cyt P-450$_{Cm-1}$ (Schunck et al., 1989), the middle (Sanglard and Loper, 1989). The residue number of each protein is shown on the left side. positions 221 and 232, which causes a shift of the numbering of cyt P-450$_{Cm-1}$. (1989) leads to following designations: cyt P-450$_{alk}$ of *C. tropicalis* sequenced by Sang- Schunck et al. (1989) LIIA2, and cyt P-450$_{Cm-2}$ sequenced by Schunck et al. (1990)

package (Computer Methods and Programs in Biomedicine **24** (1987) 27–36). This segments in the N-terminal region (underlined sequences) were placed at corresponding identical amino acids. The HR-2 region is boxed.

superfamily (for reviews see: Black and Coon, 1987; Gotoh and Fujii-Kuriyama, 1989).

As shown in Figures 6 and 7, the alkane-inducible cytochromes P-450 contain the typical HR 2 region near the C-terminus which includes an invariant cysteine functioning as the 5th ligand to the heme iron, invariant Phe and Gly and several other conserved positions (Table 5). To the latter it belongs a basic residue interacting with heme propionate. The conserved regions named as helix I and helix K, resp., in P-450$_{cam}$ with invariant Gly and Thr the former and invariant Glu and Arg the latter are present, too. The aromatic region which is the forth domain highly conserved in mammalian cytochromes P-450,

Table 4. Optimized similarity scores of cyt P-450's from different families obtained The parameter K tup was set at 2. The source for each cyt P-450 is: c: rat liver (SOGAWA et al., 1985); La ω: rat liver (HARDWICK et al., 1987); scc: bovine adrenal cortex (MORO adrenal cortex (ZUBER et al., 1986); C 21: bovine, adrenal cortex (YOSHIOKA et al., Cm-1: *C. maltosa* (SCHUNCK et al., 1989); Cm-2: *C. maltosa* (SCHUNCK et al., 1990); cam: al., 1989)

	rat c (IA1)	rat b (IIB1)	rat pcnl (IIIA1)	rat La ω (IVA1)	bovine scc (XIA1)	bovine 11 β (XIB1)
rat c (IA1)	2694					
rat b (IIB1)	713	2529				
rat pcn 1 (IIIA1)	328	385	2635			
rat La ω (IVA1)	236	292	470	2673		
bovine scc (XIA1)	216	208	315	283	2743	
bovine 11 β (XIB1)	235	229	302	212	1116	2598
bovine 17 α (XVIIA1)	688	622	431	299	250	218
bovine C 21 (XXIA1)	532	548	340	252	278	252
S. c. 14 DM (LIA)	90	70	101	110	220	159
C. t. alk (LIIA)	125	209	375	300	175	206
C. m., Cm-2 (LII)	123	215	386	321	58	171
C. m., Cm-1 (LII)	144	234	330	233	234	228
P. p. cam (CIA1)	83	63	(34)	64	49	71
B. m., Bm-3 (CII)	245	267	429	452	204	200

however, is distinctly different in the part proximal to the N-terminus (Tab. 5).

The similarity between the three cytochromes P-450 sequenced is underlined by the fact that 39 out of 41 positions suggested as conserved (Tab. 5) are identical in all three proteins. As the only striking one the exchange Pro 418 → Lys 432/Lys 434 can be noted. P-450_{alk} of *C. tropicalis* and P-450_{Cm-2} are identical in all 41 amino acid residues ascertained as conserved. In four of these positions (403, 460, 463 and 491 of the multiple alignment of GOTOH and FUJII-KURIYAMA, 1989) there are distinctions among mammalian microsomal cyt P-450s and yeast cyt P-450_{alk}.

Summarizing, we conclude that the basic principles of heme-binding and of those structural elements which could be mainly responsible for the stabilization of the three-dimensional structure of cytochromes P-450 are also conserved in the alkane-inducible yeast cytochromes P-450. Thus, spatial structure should resemble that of other microsomal P-450 forms. The same counts for P-450_{14DM}.

The most pronounced differences appear to exist in the central region of the primary structure which probably contains sequences involved in the forma-

using the algorithm FASTP (LIPMAN and PEARSON, 1985)
et al., 1984); b: rat liver (FUJII-KURIYAMA et al., 1982); pcn 1: rat liver (GONZALEZ
HASHI et al., 1984); 11 β: bovine adrenal cortex (MOROHASHI et al., 1987); 17 α: bovine
1986); 14 DM: *S. cerevisiae* (KALB et al., 1987); alk: *C. tropicalis* (SANGLARD et al., 1989);
Pseudomonas putida (UNGER et al., 1986); Bm-3: *Bacillus megaterium* (RUETTINGER et

bovine 17α (XVIIA1)	**bovine C 21** (XXIA1)	**S. c. 14 DM** (LIA)	**C. t. alk** (LIIA)	**C. m. Cm-2** (LIIA)	**C. m. Cm-1** (LIIA)	**P. p. cam** (CIA1)	**B. m. Bm-3** (CII)
2528							
739	2623						
146	178	2754					
169	190	156	2805				
182	158	139	2291	2761			
141	205	1751	1502	1534	2677		
82	78	57	60	49	63	2084	
269	302	86	203	259	204	39	2349

tion of the substrate binding site. For a detailed discussion of the complexity and evolution of the P-450 gene superfamily the reader is refered to the reviews of NEBERT et al. (1987, 1989) and GOTOH and FUJII-KURIYAMA (1989).

8.2. Structure and function of the N-terminus

The alkane-inducible cytochromes P-450 are integral membrane proteins localized in the endoplasmic reticulum of the yeast cell (see section 3). Their N-terminal regions contain two long stretches of neutral and hydrophobic amino acids (Fig. 5). These may serve as transmembrane segments anchoring the protein to the lipid bilayer similar to that proposed in a recent model for the membrane topology of vertebrate microsomal cytochromes P-450 (NELSON and STROBEL, 1988). The sequence preceding the first putative transmembrane segment is considerably longer in yeast cytochromes P-450 (22–30 amino acids) than in their mammalian counterparts (2–8 amino acids) (NELSON and STROBEL, 1988). This applies both to the alkane-inducible cytochromes P-450

Table 5. Comparison of four yeast cyt P-450 forms with conserved positions in 33 mammalian microsomal cyt P-450 forms

Belongs to conserved	Positon in the multiple alignment and conserved amino acid(s)[1]	Positon in the sequence of cyt P-450			
		C. maltosa Cm-1	*C. tropicalis*[2] alk	*C. tropicalis* 14 DM	*S. cerevisiae* 14 DM
	150 Trp, Ala	Trp 145	Trp 162	?	?
	154 Arg, Lys[4]	Arg 149	Arg 166	Arg 138	Arg 146
	182 Ile, Val, Met	Leu 167	Leu 184	Val 159	Val 167
	223 Phe+	Phe 203	Phe 220	?	?
helix I	351 Ala, Gly, Glu	Ala 320	Ala 336	Gly 307	Gly 314
helix I	352 Gly+	Gly 321	Gly 337	Gly 308	Gly 315
helix I	354 Glu, Asp	Asp 323	Asp 339	His 310	His 317
helix I	355 Thr *	Thr 324	Thr 340	Thr 311	Thr 318
helix I	360 Leu, Ile, Val	Leu 329	Leu 345	Ser 316	Ser 323
helix I	362 Tyr, Trp, Phe	Phe 321	Phe 347	Trp 318	Trp 325
	367 Leu, Met	Leu 336	Leu 352	Leu 323	Leu 330
	377 Lys, Arg	Lys 345	Lys 361	Asp 332	Glu 339
	381 Glu, Asp	Glu 349	Glu 365	Glu 336	Glu 343
	382 Ile, Leu, Val	Val 350	Ile 366	Leu 337	Gln 344
	403 Leu, Met, Ile	Cys 373	Cys 389	Leu 359	Met 363
helix K	405 Tyr, Leu	Tyr 375	Tyr 391	Leu 361	Leu 365
helix K	408 Ala, Met	Ala 378	Ala 394	Asn 364	Gln 368
helix K	412 Glu*	Glu 382	Glu 398	Glu 368	Glu 372
helix K	415 Arg*	Arg 385	Arg 401	Arg 371	Arg 375
	426 His, Arg[4]	Arg 395	Arg 411	Arg 381	Arg 385
	434 Leu, Phe, Ile	Leu 403	Leu 419	Leu 387	Met 391
	441 Pro, Asp	Pro 418	Lys 434	Pro 397	Pro 401
	442 Lys, Glu	Lys 419	Lys 435	Lys 398	Ala 402
	446 Val, Ile, Leu	Val 423	Val 439	Val 402	Val 406
aromatic region	460 Phe, Trp	Gly 438[6]	Gly 454[6]	Phe 416	Phe 420
aromatic region	463 Pro+	Ala 441	Ala 457	Pro 419	Ala 423
aromatic region	466 Phe+	Phe 444	Phe 460	Phe 422	Phe 426
aromatic region	468 Pro+	Pro 446	Pro 462	Pro 424	Ile 428
aromatic region	470 Arg, His, Asn	Arg 448	Arg 464	Arg 426	Arg 430
aromatic region	471 Phe+[3]	Trp 449	Trp 465	Trp 427	Trp 431
HR 2 region	490 Phe *	Phe 464	Phe 480	Phe 463	Phe 463
HR 2 region	491 Gly, Ser	Asn 465	Asn 481	Gly 464	Gly 464
HR 2 region	493 Gly*	Gly 467	Gly 483	Gly 466	Gly 466
HR 2 region	495 Arg+[4]	Arg 469	Arg 485	His 468	His 468
HR 2 region	497 Cys*[5]	Cys 471	Cys 487	Cys 470	Cys 470
HR 2 region	498 Ile, Leu, Val, Ala	Leu 472	Leu 488	Ile 471	Ile 471
HR 2 region	499 Gly, Ala, Asp	Gly 473	Gly 489	Gly 472	Gly 472
HR 2 region	503 Ala, Gly	Ala 477	Ala 493	Ala 476	Ala 476
HR 2 region	506 Glu, Asn, Gln	Glu 480	Glu 496	Gln 479	Gln 479
HR 2 region	509 Leu, Val	Tyr 483	Tyr 499	Thr 482	Val 482
	515 Leu+	Leu 488	Leu 504	Ile 488	Ile 488

(Fig. 6) and the cytochromes 450_{14DM} of *S. cerevisiae* (KALB et al., 1987) and of *C. tropicalis* (CHEN et al., 1988).

In mammalian cytochromes P-450, the immediate N-terminal sequence including the first hydrophobic segment is required for targeting the protein to the endoplasmic reticulum (SAKAGUCHI et al., 1987; MONIER et al., 1988). It functions as a combined insertion-halt-transfer signal and determines in a way so far not known the orientation of the cytochrome P-450 molecule in the membrane. SZCZESNA-SKORUPA et al. (1988) have demonstrated that the halt-transfer function can be removed by introduction of positively charged amino acids in front of the first hydrophobic segment. Instead, the mutagenized cytochrome P-450 sequence was found to direct translocation of a fused protein across the microsomal membrane. The insertion of the *C. maltosa* cytochrome P-450 has been studied using different in-vitro translation systems (WIEDMANN et al., 1988b). Signal recognition particles (SRP, see WALTER and BLOBEL, 1981) were found to arrest cytochrome P-450 translation on free ribosomes in a wheat germ cell-free system programmed with mRNA from alkane-induced *C. maltosa* cells. Membrane insertion occurred strictly co-translational and required SRP when dog pancreatic microsomal membranes were added. These results are in agreement with those of SAKAGUCHI et al. (1984, 1987) for a rabbit liver microsomal cytochrome P-450 form. The *C. maltosa* cytochrome P-450 should therefore contain an uncleavable signal sequence as

---

Legend Table 5

[1] Numbers and conserved amino acids according the multiple alignment of GOTOH and FUJII-KURIYAMA (1989) considering 33 mammalian microsomal cyt P-450s, 3 mammalian mitochondrial cyt P-450s, 1 chicken cyt P-450, and the bacterial cyt $P\text{-}450_{cam}$.
[2] In the 41 positions ascertained as conserved cyt $P\text{-}450_{alk}$ of *C. tropicalis* and cyt P-450 Cm-1 are identical.
* Invariant according the multiple alignment of 38 cyt P-450s and the four yeast cyt P-450s.
+ Invariant in 33 mammalian microsomal cyt P-450s.
[3] Trp in three mammalian mitochondrial cyt P-450s.
[4] Contacts heme propionate group.
[5] Heme ligand.
[6] 1 residue before that is Tyr (see Fig. 5).

Comment: We considered the conserved positions mentioned by GOTOH and FUJII-KURIYAMA (1989) and some additional ones derived from their alignment. To the latter counts the conserved aromatic residue (alignment position 362), which is in a fixed distance to the invariant Thr in the helix I region. Comparing the positions listed alkane-inducible cyt P-450s are nearly identical (39 out of 41). The consensus with the mammalian microsomal cyt P-450s is high, except the aromatic region. Both cyt $P\text{-}450s_{14DM}$ show mutually some small differences and vary more distinctly from mammalian microsomal cyt P-450s in some positions conserved. Especially, the aromatic region in cyt $P\text{-}450_{14DM}$ of *S. cerevisiae* includes deviations.

```
437  Y G E D A Y E F R P E R W - - - - - - - - - - - - - - - - -
453  Y G E D A Y V F R P E R W - - - - - - - - - - - - - - - - -
415  W F E H P E H F N P R R W E S D D T K A S A V S F N S E D T
419  Y F P N A H Q F N I H R W - - - - N K D S A S S Y S V G E E

450  - - - - F E P S T R K L G W A Y L P F N G G P R I C L G Q Q
466  - - - - F E P E T R K L G W A Y V P F N G G P R I C L G Q Q
445  V D Y G F G K I S K G V S S P Y L P F G G G R H R C I G E Q
445  V D Y G F G A I S K G V S S P Y L P F G G G R H R C I G E H

476  F A L T E A S Y V     cyt P-450Cm-1   of C. maltosa
492  F A L T E A S Y V     cyt P-450alk    of C. tropicalis
475  F A Y V Q L G T I     cyt P-450 14DM  of C. tropicalis
475  F A Y C Q L G V L     cyt P-450 14DM  of S. cerevisiae
```

Fig. 7. Sequence alignment of a part from the C-terminal portion of yeast cyt P-450s cloned as example for the large differences between LI and LII. Two highly conserved domains, the aromatic and the HR2 region are underlined or boxed, resp. The spacing between both domains varies between 11 and 18 residues in mammalian cyt P-450s (GOTOH and FUJI-KUNIYAMA, 1989) but amounts 14 residues in alkane-inducible cyt P-450 and 31–35 residues in both cyt P-450_{14DM}. The source of each sequence is: cyt P-450_{Cm-1} of C.m. (SCHUNCK et al. 1989) and cyt P-450_{alk} of C.t. (SANGLARD and LOPER, 1989), cyt P-450_{14DM} of C.t. (CHEN et al., 1988) and of S.c. (KALB et al., 1987).

generally found in proteins destined for the endoplasmic reticulum (for a review see RAPOPORT, 1986).

However, it is still largely unknown how the targeting information of the signal sequence is actually decoded in yeast. In marked contrast to the dog pancreatic membranes, *C. maltosa* microsomes can carry out significant post-translational insertion of cytochrome P-450 as well as post-translational translocation of prepro-α-factor (WIEDMANN et al., 1988b). The same property was-reported for *S. cerevisiae* microsomes, where the post-translational translocation of the prepro-α-factor has recently been studied in detail (HANSEN et al., 1986; FECYCZ and BLOBEL, 1987; ROTHBLATT et al., 1987). According to these authors, in addition to unknown peculiarities of the yeast microsomes a cytosolic factor and ATP hydrolysis are required to support this process. Correct targeting of the post-translationally transported proteins may arise from a specific interaction of their signal sequences with a membrane bound signal sequence receptor (WIEDMANN et al., 1988b). In this scheme, the efficiency of membrane insertion would depend on the accessibility of the signal sequence of the protein after its synthesis.

9. Outlook

As outlined above alkane-assimilating yeasts contain a highly active alkane hydroxylating cytochrome P-450-system. Its molecular organization is very similar to the cytochrome P-450 systems of mammalian cell endoplasmic reticulum. Since the early days of cytochrome P-450 research (GUNSALUS et al., 1974) it has been evident that the specific activities of various cytochrome P-450 forms can differ several orders of magnitude. Very high catalytic rates have been demonstrated for several bacterial cytochromes P-450 known as soluble enzymes (GUNSALUS et al., 1974; NARHI and FULCO, 1986) and for yeast alkane-hydroxylating cytochromes P-450 which are membrane integrated like all mammalian cytochromes P-450. As already mentioned, their specific activities exceed those of mammalian cytochromes P-450 by 2–3 orders of magnitude. However, the alkane-hydroxylating enzyme system reaches this high rate only in the intact cell. Considering the great similarity of the molecular organization of yeast and mammalian microsomal cytochrome P-450 systems, it is a fascinating task for an enzymologist to elucidate the reasons for the large differences in catalytic activities. Both structural peculiarities of the enzymes in connection with the substrate structure and the efficient transport of substrate and product in yeast cells optimized in the course of phylogenetic development could be responsible.

Based on the recent cloning of alkane-inducible cytochrome P-450 forms from *C. tropicalis* and *C. maltosa*, progress in the elucidation of structure-function relationships determining their catalytic activity and substrate specificity can be expected.

Moreover, further studies are necessary to elucidate the real degree of cytochrome P-450 multiplicity in yeasts. A particular problem is the actual number and function of cytochrome P-450 forms induced by long-chain n-alkanes and related compounds. Probably, only progress in cloning and heterologous expression of the respective genes will provide unequivocal results.

In addition to the investigation of the cytochromes P-450, the regulation of their genes is a further attractive problem. Up to now results show that carbon sources (n-alkanes, glucose) have a decisive influence on the biosynthesis of yeast cytochromes P-450. However, we are just beginning to understand details of regulation mechanisms.

Hitherto, for the heterologous expression of mammalian cytochromes P-450, *S. cerevisiae* has been used as host organism. Possibly, alkane assimilating yeasts are very suitable for this aim. Their ability to uptake lipophilic compounds very effectively could turn out to be an advantage for the development of efficient biotransformation systems. The first steps to establish host-vector systems have already been taken, using *C. maltosa* (Kunze et al., 1985; Takagi et al., 1986), *Y. lipolytica* (Davidov et al., 1985; Gaillardin et al., 1985), and *C. albicans* (Kurtz et al., 1987).

10. References

Aoyama, Y. and Yoshida, Y., (1978), Biochem. Biophys. Res. Commun. **85**, 28–34.

Asperger, O., Naumann, A., and Kleber, H.-P., (1984), Appl. Microbiol. Biotechnol. **19**, 398–403.

Avetisova, S. M., Sokolov, Y. I., Kozlov, V. I., Davydov, R. M., and Davidov, E. R., (1985), in: Cytochrom P-450-Biochemistry, Biophysics and Induction, (L. Vereczky and K. Magyar, eds.), Akademiai Kiado, Budapest.

Azari, M. R. and Wiseman, A., (1982), Anal. Biochem. **122**, 129–138.

Bassel, J. B. and Mortimer, R. K., (1982), Curr. Genet. **5**, 77–88.

Bassel, J. B. and Mortimer, R. K., (1985), Curr. Genet. **9**, 579–586.

Beaufay, H., Amar-Costenec, A., Feytmans, E., Thines-Sempoux, D., Wibo, M., Robbi, M., and Berthet, J., (1974), Cell Biol. **61**, 188–200.

Bertrand, J. C., Gilewicz, M., Bazin, H., and Azoulay, E., (1980), Biochem. Biophys. Res. Commun. **94**, 880–887.

Black, S. D. and Coon, M. J., (1987), Adv. Enzymology **60**, 35–87.

Blasig, R., Schunck, W.-H., Jockisch, W., Franke, P., and Müller, H.-G., (1984), Appl. Microbiol. Biotechnol. **19**, 241–246.

Blasig, R., Mauersberger, S., Riege, P., Schunck, W.-H., Jockisch, W., Franke, P., and Müller, H.-G., (1988), Appl. Microbiol. Biotechnol. **28**, 589–597.

Bühler, M. and Schindler, J., (1984), Aliphatic hydrocarbons, in: Biotechnology, Vol. 6a, 329–385, (Rehm, H. J. and Reed, G., eds.), Verlag Chemie, Weinheim, BRD.

Cardini, G. and Jurtschuk, P., (1970), J. Biol. Chem. **245** 2789–2796.

Chen, C., Turi, T., Sanglard, D., and Loper, J. C., (1987), Biochem. Biophys. Res. Comm. **146**, 1311–1317.

CHEN, C., KALB, V. F., TURI, T. G., and LOPER, J. C., (1988), DNA **7**, 617–626.

DELAISSE, J. M., MARTIN, P., VERHUYEN-BOUVY, M. F., and NYNS, E. J., (1981), Biochim. Biophys. Acta **676**, 77–90.

DUPPEL, W., LEBEAULT, J. M., and COON, M. J., (1973), Eur. J. Biochem. **36**, 583–592

FECYCZ, J. T. and BLOBEL, G., (1987), Proc. Natl. Acad. Sci. USA **84**, 3723–3727.

FISCHER, W., BRÜCKNER, B., and MEYER, H. W., (1982), Z. Allg. Mikrobiol. **22**, 227–236.

FOSTER, J. R., ELCOMBE-CLIFFARD, R., BOOBIS, A. R., DAVIES, D. S., SESARDIC, D., MCQUADE, J., ROBSON, R. T., HAYWARD, C., and LOCK, E. A., (1986), Biochem. Pharmacol. **35**, 4543–4554.

FUJII-KURIYAMA, Y., MIZUKAMI, Y., KAWAJIRI, K., SOGAWA, K., and MURAMATSU, M., (1982), Proc. Natl. Acad. Sci. USA **79**, 2793–2797.

FUKUI, S. and TANAKA, A., (1980), Adv. Biochem. Eng. **19**, 216–237.

FULCO, A. J. and RUETTINGER, R. T., (1987), Life Sciences **40**, 1769–1775.

GALLO, M., BERTRAND, J. C., ROCHE, B., and AZOULAY, E., (1971), FEBS Lett. **19**, 45–49.

GALLO, M., BERTRAND, J. C., ROCHE, B., and AZOULAY, E., (1973), Biochim. Biophys. Acta **296**, 624–638.

GALLO, M., ROCHE, B., and AZOULAY, E., (1976), Biochim. Biophys. Acta **419**, 425–434.

GILEWICZ, M., ZACEK, M., BERTRAND, J. C., and AZOULAY, E., (1979), Can. J. Microbiol. **25**, 201–206.

GMÜNDER, F. K., KÄPPELI, O., and FIECHTER, A., (1981), Eur. J. Appl. Microbiol. Biotechnol. **12**, 135–142.

GONZALEZ, F. J., NEBERT, D. W., HARDWICK, J. P., and KASPAR, C. B., (1985), J. Biol. Chem. **260**, 7435–7441.

GOTOH, O. and FUJII-KURIYAMA, Y., (1989), in: Frontiers in Biotransformation, (K. RUCKPAUL and H. REIN, eds.), Akademie-Verlag Berlin, Vol. **1**, 195–243.

GUNSALUS, I. C., MERCKS, J. R., LIPSCOMB, J. D., DEBNENNER, P. G., and MÜNCH, E., (1974), in: Molecular Mechanisms of Oxygen Activation, (O. HAYAISHI ed.), Academic Press, New York, 559–613.

HANSEN, W., GARCIA, P. D., and WALTER, P., (1986), Cell **45**, 397–406.

HARDWICK, J. P., SONG, B.-J., HUTERMAN, E., and GONZALEZ, F. J., (1987), J. Biol. Chem. **262**, 801–810.

HASHIMOTO, T., (1982), Ann. N. Y. Acad. Sci. **386**, 5–12.

HATA, S., NISHINO, T., KATZUKI, H., AOYAMA, H., and YOSHIDA, Y., (1983), Biochem. Biophys. Res. Commun. **116**, 162–166.

HEINZ, E., TULLOCH, A. O., and SPENCER, J. F. T., (1970), Biochim. Biophys. Acta **202**, 49–55.

HILL, D. E., BOULAY, R., and ROGERS, D., (1988), Nucl. Acid. Res. **16**, 365–366.

HONECK, H., SCHUNCK, W.-H., RIEGE, P., and MÜLLER, H.-G., (1982), Biochem. Biophys. Res. Commun. **106**, 1318–1324.

ILCHENKO, A. P., MAUERSBERGER, S., MATYASHOVA, R. N., and LOZINOV, A. B., (1980), Mikrobiologija (Moscow) **49**, 452–458.

ILCHENKO, A. P. and TSFASMAN, I. M., (1987), Biochimija (Moscow) **52**, 58–65.

ILCHENKO, A. P. and TSFASMAN, I. M., (1988), Biochimija (Moscow) **53**, 263–271.

ISHIDATE, K., KAWAGUCHI, K., and TAKAWA, K., (1969), J. Biochem. **65**, 385–392.

JEFCOATE, C. R., (1978), Methods in Enzymology **52**, 258–279.

KALB, V. F., WOODS, C. W., TUR, T. G., DEY, C. R., SUTTER, T. S., and LOPER, J. C., (1987), DNA **6**, 529–537.

KÄPPELI, O., MÜLLER, M., and FIECHTER, A., (1978), J. Bacteriol. **133**, 952–958.

KÄPPELI, O., SAUER, M., and FIECHTER, A., (1982), Anal. Biochem. **126**, 179–182.

KÄPPELI, O., WALTHER, P., MUELLER, M., and FIECHTER, A., (1984), Arch. Microbiol. **138**, 279–282.

Käppeli, O., Sanglard, D., and Laurila, H. O., (1985), in: Cytochrome P-450-Biochemistry, Biophysics and Induction, (L. Vereczkey and K. Magyar eds.), Akademiai Kiado, Budapest, 443—446.

Käppeli, O., (1986), Microbiol. Reviews **50**, 244—258.

Kärenlampi, S. O., Marin, E., and Hanninen, O., (1980), J. Gen. Microbiol. **120**, 529—533.

Kärgel, E., Schmidt, H. E., Schunck, W.-H., Riege, P., Mauersberger, S., and Müller, H.-G., (1984), Anal. Lett. **17** ⟨B18⟩ 2011—2024.

Kärgel, E., Schunck, W.-H., Riege, P., Honeck, E., Claus, R., Kleber, H.-P., and Müller, H.-G., (1985), Biochem. Biophys. Res. Commun. **128**, 1261—1267.

Kärgel, E., Aoyama, Y., Schunck, W.-H., Müller, H.-G., and Yoshida, Y., (1990), Yeast **6**, 61—67.

Kemp, G. D., Dickinson, F. M., and Ratledge, C., (1988), Appl. Microbiol. Biotechnol. **29**, 370—374.

King, D. J., Azari, M. R., and Wiseman, A., (1984), Xenobiotica **14**, 187—206.

King, D. J. and Wiseman, A., (1987), in: Enzyme induction, mutagen activation and carcinogen testing in yeast, (A. Wiseman ed.), John Wiley & Sons New York.

Kleber, H.-P., Asperger, O., Stüwer, O., Stüwer, B., and Hommel, R., (1988), in: Cytochrome P-450, Biochemistry and Biophysics, (Schuster, I., ed.), Taylor & Francis London, 169—172.

Krauzova, V. I., Ilchenko, A. P., Sharyshev, A. A., and Lozinov, A. B., (1985), Biochimija (Moscow) **50**, 726—732.

Krauzova, V. J. and Kuvichkina, T. N., (1986), Biochimija **51**, 23—27 (in Russian).

Krauzova, V. J. and Sharyshev, A. A., (1987), Biochimija **52**, 599—606 (in Russian).

Kunau, W.-H., Bühne, S., de la Garza, M., Kionka, C., Mateblowski, M., Schultz-Borchard, U., and Thieringer, R., (1988), Biochem. Soc. Transact **16**, 418—420.

Laurila, H. O., Käppeli, O., and Fiechter, A., (1984), Arch. Microbiol. **140**, 257 to 259.

Lebeault, J. M., Roche, B., Duvnjak, Z., and Azoulay, E., (1970), Biochim. Biophys. Acta **220**, 373—385.

Lebeault, J. M., Lode, E. T., and Coon, M. J., (1971), Biochem. Biophys. Res. Commun. **42**, 413—419.

Lindenmayer, A., and Smith, L., (1964), Biochim. Biophys. Acta **93**, 445—461.

Lipman, D. J. and Pearson, W. R., (1985), Science **227**, 1435—1441.

Loper, J. C., Chen, C., and Dey, C. R., (1985), Hazardous waste and hazardous materials **2**, 131—141, Mary Ann Liebert, Inc. Publishers.

Marchal, R., Metche, M., and van Decasteele, (1982), J. Gen. Microbiol. **128**, 1125—1132.

Mauersberger, S., Matyashova, R. N., Müller, H.-G., and Losinov, A. B., (1980), Eur. J. Appl. Microbiol. Biotechnol. **9**, 285—294.

Mauersberger, S., Schunck, W.-H., and Müller, H.-G., (1981), Z. Allg. Mikrobiol. **21**, 313—321.

Mauersberger, S., Schunck, W.-H., and Müller, H.-G., (1984), Appl. Microbiol. Biotechnol. **19**, 29—35.

Mauersberger, S., Kärgel, E., Matyashova, R. N., and Müller, H.-G., (1987), J. Basic Microbiol. **27**, 565—582.

Meissel, M. N., Medvedova, G. A., and Kozlova, T. M., (1976), Mikrobiologija **45**, 844—852.

Mishina, M., Kamiriyo, T., Tashiro, S., and Numan, S., (1978), Eur. J. Biochem. **82**, 347—354.

Monier, S., Van Luc, P., Kreibich, G., Sabatini, D. D., and Adesnik, M., (1988), J. Cell. Biol. **107**, 457—470.

Morohashi, K., Fujii-Kuriyama, Y., Okada, Y., Sogawa, K., Hirose, T., Inayama, S., and Omura, T., (1984), Proc. Natl. Acad. Sci. USA **81**, 4647—4651.

Morohashi, K., Yoshioka, H., Gotoh, O., Okada, Y., Yamamoto, K., Miyata, T., Sogawa, K., Fujii-Kuriyama, Y., and Omura, T., (1987), J. Biochem. **102**, 559—568.

Müller, H. G., Schunck, W.-H., Riege, P., and Honeck, H., (1984), in: Cytochrome P-450, (K. Ruckpaul and H. Rein, ed.), Akademie Verlag Berlin.

Narhi, L. O. and Fulco, A. J., (1986), J. Biol. Chem. **261**, 7160—7169.

Nebert, D. W. and Gonzalez, F. J., (1987), Ann. Rev. Biochem. **56**, 945—993.

Nebert, D., Nelson, D. R., Adesnik, M., Coon, M. J., Estabrook, R. W., Gonzalez, F. J., Guengerich, F. P., Gunsalus, I. C., Johnson, E. F., Kemper, B., Levin, W., Phillips, I. R., Sato, R., and Waterman, M. R., (1989), DNA **8**, 1—13.

Nelson, D. R. and Strobel, H. W., (1988), J. Biol. Chem. **263**, 6038—6050.

Okada, H., Ueda, M., Sugaya, T., Atomi, H., Mozaffar, S., Hishida, T., Teranishi, Y., Okazaki, K., Takechi, T., Kamiryo, T., and Tanaka, A., (1987), Eur. J. Biochem. **170**, 105—110.

Osumi, M., Fukuzumi, F., Teranishi, Y., Tanaka, A., and Fukui, S., (1975a), Arch. Microbiol. **103**, 1—11.

Osumi, M., Fukuzumi, F., Yamada, N., Nagatami, T., Teranishi, Y., Tanaka, A., and Fukui, S., (1975b), J. Ferment. Technol. **53**, 244—248.

Poole, R. K., Loyd, D., and Chance, B., (1974), Biochem. J. **138**, 201—210.

Rapoport, T. A., (1986), CRC Crit. Rev. Biochem. **20**, 73—137.

Rehm, H. J. and Reiff, I., (1981), Adv. Biochem. Eng. **19**, 175—215.

Riege, P., Schunck, W.-H., Honeck, H., and Müller, H.-G., (1981), Biochem. Biophys. Res. Commun. **98**, 527—534.

Röber, B. and Reuter, G., (1984), Z. Allgem. Mikrobiol. **24**, 317—328.

Rothblatt, J. A., Webb, J. R., Ammerer, G., and Meyer, D. J., (1987), EMBO J., 3455—3463.

Ruckpaul, K., Rein, H., and Blanck, J., (1989), Frontiers in Biotransformations **1**, 1—63.

Ruettinger, R. T., Griffith, G. R., and Coon, M. J., (1977), Arch. Biochem. Biophys. **183**, 528—553.

Ruettinger, R. T., Wen, C.-P., and Fulco, A. J., (1989), J. Biol. Chem. **264**, 10987 bis 10995.

Sakaguchi, M., Mihara, K., and Sato, R., (1984), Proc. Natl. Acad. Sci. USA **81**, 3361—3364.

Sakaguchi, M., Mihara, K., and Sato, R., (1987); EMBO J. **6**, 2425—2431.

Sanglard, D., Käppeli, O., and Fiechter, A., (1984), J. Bacteriol. **157**, 297—302.

Sanglard, D., Käppeli, O., and Fiechter, A., (1986), Arch. Biochem. Biophys. **251**, 276—286.

Sanglard, D., Chen, C., and Loper, J. C., (1987), Biochem. Biophys. Res. Commun. **144**, 251—257.

Sanglard, D. and Fiechter, A., (1989), FEBS-Lett. **256**, 128—134.

Sanglard, D. and Loper, J. C., (1989), Gene, **76**, 121—136.

Schauer, F. and Schauer, M., (1986), Wiss. Z. Univ. Greifswald **35**, (4) 14—23.

Schunck, W.-H., Riege, P., and Müller, H.-G., (1983a), Biochimija (Moscow, in Russian.) **48**, 518—526.

Schunck, W.-H., Riege, P., Honeck, H., and Müller, H.-G., (1983b), Z. Allgem. Mikrobiol. **23**, 653—660.

Schunck, W. H. and Riege, P., (1983), Ph. D. Theses Academy of Sciences of GDR, Berlin.

Schunck, W.-H., Mauersberger, S., Huth, J., Riege, P., and Müller, H.-G., (1987a), Arch. Microbiol. **147**, 240—244.

SCHUNCK, W.-H., MAUERSBERGER, S., KÄRGEL, E., HUTH, J., and MÜLLER, H.-G., (1987b), Arch. Microbiol. **147**, 245–248.

SCHUNCK, W.-H., KIESSLING, U., STRAUSS, M., KÄRGEL, E., WIEDMANN, B., MAUERSBERGER, S., GAESTEL, M., GROSS, B., and MÜLLER, H.-G., (1989), in: Cytochrome P-450, Biochemistry and Biophysics, (SCHUSTER, I., ed.), Taylor & Francis London, 656–659.

SCHUNCK, W.-H., KÄRGEL, E., GROSS, B., WIEDMANN, B., MAUERSBERGER, S., KÖPKE, K., KIESSLING, U., STRAUSS, M., GAESTEL, M., and MÜLLER, H.-G., (1989), Biochem. Biophys. Res. Commun. **161**, 843–850.

SHARYSHEV, A. A., MATYASHOVA, R. N., KOMAROVA, G. N., (1983), FEMS Symp. Environ. Reg. Microbial Metabolism, Pushchino, Abstract p33.

SLIGAR, S. and MURRAY, R. I., (1986), in: Cytochrome P-450 Structure, Mechanism and Biochemistry, (ORTIZ DE MONTELLANO, P. R., ed.), Plenum Press New York.

SOGAWA, K., GOTOH, O., KAWAJIRI, K., and FUJII-KURIYAMA, Y., (1984), Proc. Natl. Acad. Sci. USA **81**, 5066–5070.

SOKOLOV, JU. I., AVETISOVA, S. M., DAVYDOV, P. M., and DAVIDOV, E. P., (1986), Dokl. Akad. Sci. USSR **286**, 1509–1511.

SPEVAK, W., HARTIG, A., MEINDL, P., and RUIS, H., (1986), Mol. Gen. Genet. **203**, 73–78.

SZCZESNA-SKORUPA, E., BROWNE, N., MEAD, D., and KEMPER, B., (1988), Proc. Natl. Acad. Sci. USA **85**, 738–742.

TABUCHI, I. and UCHIYAMA, H., (1975), Agric. Biol. Chem. **39**, 2035–2042.

TAKAGI, M., MORIYA, K., and YANO, K., (1980), Cell. Mol. Biol. **25**, 363–369 and 371–375.

TAKAGI, M., OHKUMA, M., KOBAYASHI, N., WATANABE, M., and YANO, K., (1989), Agric. Biol. Chem. **53**, 2217–2226.

TANAKA, A., OSUMI, M., and FUKUI, S., (1982), Ann. N. Y. Acad. Sci. **386**, 183–199.

TANAKA, A., UEDA, M., OKADA, H., and FUKUI, S., (1988), Ann. N. Y. Acad. Sci. **501**, 449–453.

TITTELBACH, M., ROHDE, H.-G., and WEIDE, H., (1976), Z. Allg. Mikrobiol. **16**, 155–156.

TRINN, M., KÄPPELI, O., and FIECHTER, A., (1982), Eur. J. Appl. Microbiol. Biotechnol. **15**, 64–68.

UEMURA, N., TAOKA, A., and TAKAGI, M., (1988), in: Proceedings of the World Conference on Biotechnology for the Fats and Oils industry, (APPLWHITE, T. H. ed.), American Oil Chemist's Society.

UNGER, B. P., GUNSALUS, I. C., and SLIGAR, S. G., (1986), J. Biol. Chem. **261**, 1158–1163.

WALTER, P. and BLOBEL, G., (1981), J. Cell Biol. **91**, 557–561.

WIEDMANN, B., WIEDMANN, M., KÄRGEL, E., SCHUNCK, W.-H., and MÜLLER, H.-G., (1986), Biochem. Biophys. Res. Commun. **136**, 1148–1154.

WIEDMANN, B., WIEDMANN, M., MAUERSBERGER, S., SCHUNCK, W. H., and MÜLLER, H. G., (1988a), Biochem. Biophys. Res. Commun **150**, 859–865.

WIEDMANN, M., WIEDMANN, B., VOIGT, S., WACHTER, E., MÜLLER, H.-G., and RAPOPORT, T., (1988), EMBO J. **7**, 1763–1768.

WOODS, L. F. J. and WISEMAN, A., (1980), Biochim. Biophys. Acta **613**, 52–61.

YAMADA, T., NOUVA, H., KAWAMOTO, S., TANAKA, A., and FUKUI, S., (1980), Arch. Microbiol. **128**, 145–151.

YANG, S. K. and BAO, Z., (1987), Molec. Pharmacol. **32**, 73–80.

YOSHIDA, Y. and AOYAMA, Y., (1989), Frontiers in Biotransformation **4** (this Volume).

YOSHIOKA, H., MOROHASHI, K., SOGAWA, K., YAMANE, M., KOMINAMI, S., TAKEMORI, S., OKADA, Y., OMURA, T., and FUJII-KURIYAMA, Y., (1986), J. Biol. Chem. **261**, 4106–4109.

ZUBER, M. X., JOHN, M. E., OKAMURA, T., SIMPSON, E. R., and WATERMAN, M. R., (1986), J. Biol. Chem. **261**, 2475–2482.

Chapter 4
Cytochromes P-450 in the Ergosterol Biosynthesis

Y. Yoshida and Y. Aoyama

1. Introduction: The biosynthetic pathway and related enzymes

Sterols are synthesized from acetic acid via mevalonic acid and squalene as key intermediates by many eukaryotes. Conversion of acetic acid to squalene proceeds through the well-known metabolic pathway named "mevalonic acid pathway", the trunk of terpene biosynthesis which is found in almost all

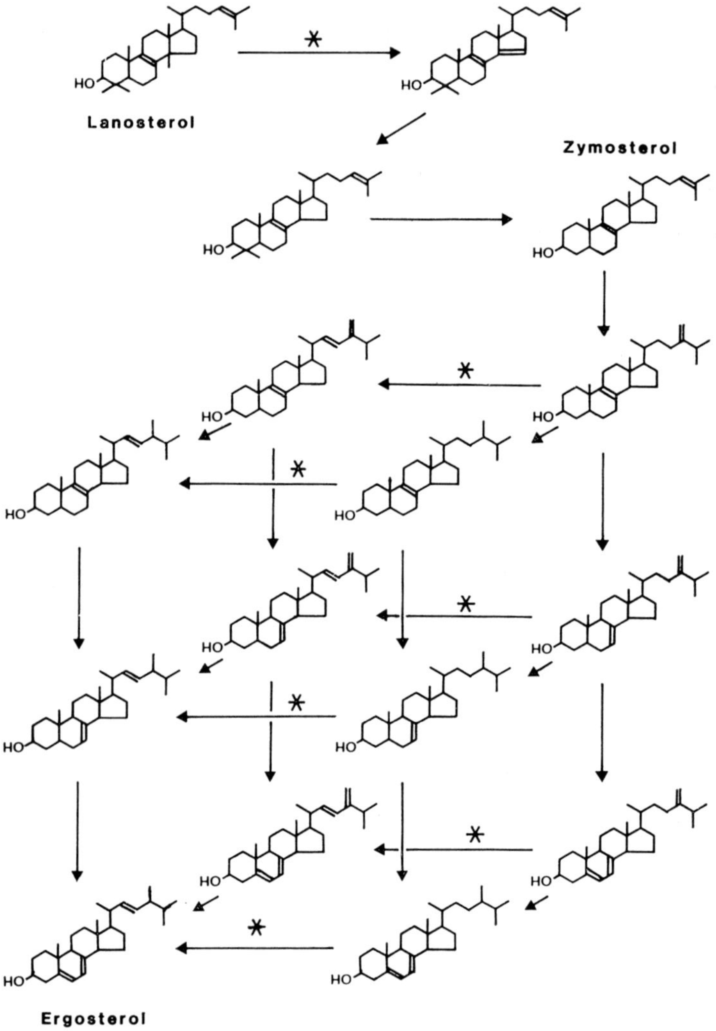

Fig. 1. Biosynthetic pathway of ergosterol in yeast. Steps marked with * are mediated by cytochrome P-450.

organisms from primitive prokaryotes to higher eukaryotes. The pathway proper to sterol biosynthesis branches at squalene and is found almost exclusively in eukaryotes.

Ergosterol (ergosta-5,7,22-trien-3β-ol) is the principal sterol of yeast and fungi which is synthesized from squalene via lanosterol (lanosta-8,24-dien-3β-ol). Figure 1 represents the biosynthetic pathway of ergosterol by yeast (Fryberg et al. 1973). Lanosterol is converted to zymosterol (cholesta-8,24-dien-3β-ol) by removal of the three methyl groups (C-30, C-31, C-32) and then transmethylated at C-24 to fecosterol (ergosta-8,24(28)-dien-3β-ol). The metabolic pathway from lanosterol to fecosterol is different for yeast and filamentous fungi. It is considered that in filamentous fungi lanosterol is first converted to 24-methylene-24,25-dihydrolanosterol and then undergoes the three-step demethylations to give fecosterol, because filamentous fungi treated with azole antifungals do not accumulate lanosterol but 24-methylene-24,25-dihydrolanosterol (Kato and Kawase, 1976; Henry and Sisler, 1979; Baldwin and Wiggins, 1984; Vanden Bossche, 1985). Fecosterol is converted to ergosterol by isomerization from Δ^8 to Δ^7, Δ^5-desaturation, reduction of $\Delta^{24(28)}$, and Δ^{22}-desaturation. In any case, removal of the three methyl groups (C-30, C-31 and C-32) is the essential process in ergosterol biosynthesis. It is noteworthy that removal of the three methyl groups of lanosterol is also essential in cholesterol biosynthesis in mammals.

Most reactions included in the biotransformation of squalene to ergosterol by yeast are mediated by enzymes in the endoplasmic reticulum and the microsomal electron transport system is responsible for several reactions in the bioconversion (Fig. 2). The reactions catalyzed by the microsomal electron transport system are squalene epoxidation, removal of C-30, C-31 and C-32, and Δ^5- and Δ^{22}-desaturation. These reactions are essentially monooxygenations requiring both NAD(P)H and O_2.

The 14α-demethylation (removal of C-32) and the Δ^{22}-desaturation are mediated by cytochrome P-450 and are described in section 2. The 4-demethylation

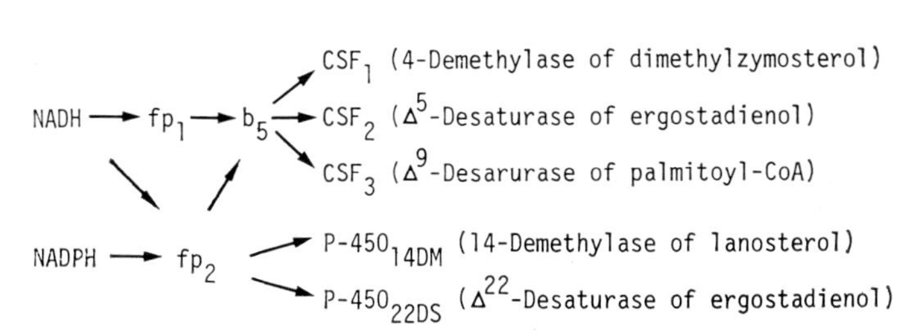

Fig. 2. Construction and function of the microsomal electron-transport system of yeast. The abbreviations fp_1, fp_2, b_5 and CSF represent NADH-cytochrome b_5 reductase, NADPH-cytochrome P-450 reductase, cytochrome b_5 and cyanide-sensitive terminal monooxygenase, respectively.

(removal of C-30 and C-31) (AOYAMA et al. 1981b) and the Δ^5-desaturation (OSUMI et al. 1979) are catalyzed by another cyanide-sensitive monooxygenase (CSF). Electrons necessary for these reactions are provided by NADH to CSF through the electron-transport system consisting of NADH-cytochrome b_5 reductase and cytochrome b_5 (AOYAMA et al., 1981b; OSUMI et al., 1979). By analogy with the mammalian system which is functional in the desaturation of long-chain fatty acyl-CoAs (STRITTMATTER et al., 1974), CSF may be an iron-sulfur protein. However, CSF has not yet been isolated from yeast.

Epoxidation of squalene is catalyzed by squalene epoxidase (RYDER et al., 1984) which may be a flavin-dependent monooxygenase like the corresponding enzyme of mammals (ONO et al., 1980). Electrons necessary for this reaction are supplied from NADPH via NADPH-cytochrome P-450 reductase (RYDER et al., 1985; ONO et al., 1980). Squalene epoxidase has not yet been isolated from yeast.

2. Role of cytochrome P-450 in sterol biosynthesis

2.1. Lanosterol 14α-demethylation

Lanosterol is converted to 4,4-dimethylzymosterol upon incubation with the microsomes from *Saccharomyces cerevisiae* in the presence of NADPH and O_2 (OHBA et al., 1978; AOYAMA et al., 1981a). This reaction, the removal of C-32, is the initial step in the ergosterol biosynthesis from lanosterol (Fig. 1), and is known to be a cytochrome P-450-dependent reaction (MITROPOULOS et al., 1976; OHBA et al., 1978; AOYAMA and YOSHIDA, 1978a, b; AOYAMA et al., 1981a). The 14α-demethylation of lanosterol is also included in the cholesterol biosynthesis in mammals and the mechanism of this reaction (AKHTAR et al., 1969, 1977; ALEXANDER et al., 1972; WATKINSON et al., 1969) is the same as that of the demethylation catalyzed by the yeast enzyme described below.

The cytochrome P-450 responsible for lanosterol 14α-demethylation has been isolated from *S. cerevisiae* and termed as cytochrome $P\text{-}450_{14DM}$ (see section 3.). A reconstituted system consisting of cytochrome $P\text{-}450_{14DM}$ and NADPH-cytochrome P-450 reductase (AOYAMA et al., 1978), both purified from *S. cerevisiae* microsomes, metabolizes lanosterol not to 4,4-dimethylzymosterol but to 4,4-dimethyl-5α-cholesta-8,14,24-trien-3β-ol which is the 14-desaturated derivative of the former compound (AOYAMA and YOSHIDA, 1978a; AOYAMA et al., 1984). This fact suggests that the 14α-demethylation of lanosterol by yeast microsomes is not only completed by the cytochrome $P\text{-}450_{14DM}$-mediated reaction. Actually, we found that the dimethylcholestatrienol formed from lanosterol by cytochrome $P\text{-}450_{14DM}$ was reduced to 4,4-dimethylzymosterol by an NADPH-dependent 14-reductase in microsomes (AOYAMA and YOSHIDA, 1986). Thus, the 14α-demethylation of lanosterol is considered to proceed as in Figure 3 and the cytochrome P-450_{14DM}-containing electron-transport system catalyzes the former step of this bioconversion.

The 14α-demethylation of lanosterol by cytochrome P-450_{14DM} is accompanied by the introduction of a double bond at C-14 (AOYAMA and YOSHIDA, 1978a; AOYAMA et al., 1984), and the methyl group is removed as formic acid (MITROPOULOS et al., 1976). This fact suggests that the demethylation

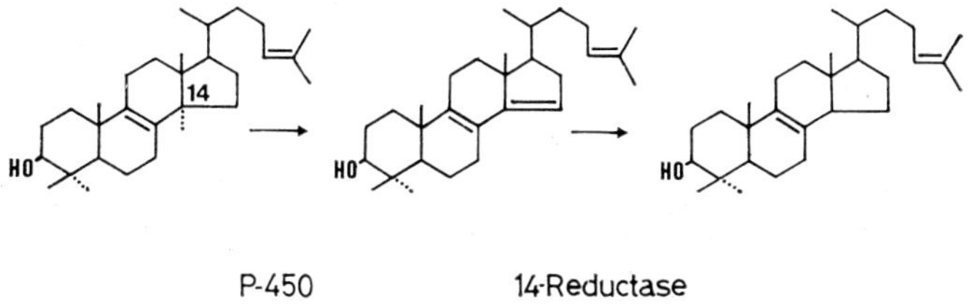

P-450 14-Reductase

Fig. 3. Reaction process of the 14α-demethylation of lanosterol. The conversion of lanosterol to 4,4-dimethylcholestatrienol is catalyzed by cytochrome P-450_{14DM}, and then the latter compound is reduced to 4,4-dimethylzymosterol by a NADPH-requiring reductase.

consists of three monooxygenations as shown in Figure 4. First, lanosterol is hydroxylated at C-32 to 32-hydroxylanosterol. Then, the second monooxygenation occurs at the same carbon and the 32-hydroxy-intermediate is converted to the 32-gem-diol derivative which may be in equilibrium with 32-oxolanosta-8,24-dien-3β-ol. C-32 of this second intermediate is further attacked by the third oxygen activated by the cytochrome. At the same time, the C-C bond between C-14 and C-32 is cleaved, C-32 is eliminated together with hydrogen at C-15 as formic acid, and the double bond is formed between C-14 and C-15 (AOYAMA et al., 1987a; STEVENSON et al., 1988). 4,4-Dimethyl-5α-cholesta-8,14,24-trien-3β-ol is the only product formed from lanosterol by the reconstituted system and no detectable accumulation of the intermediates occurs during the incubation under usual assay conditions (AOYAMA et al., 1984, 1987a). Accordingly, cytochrome P-450_{14DM} must mediate these three

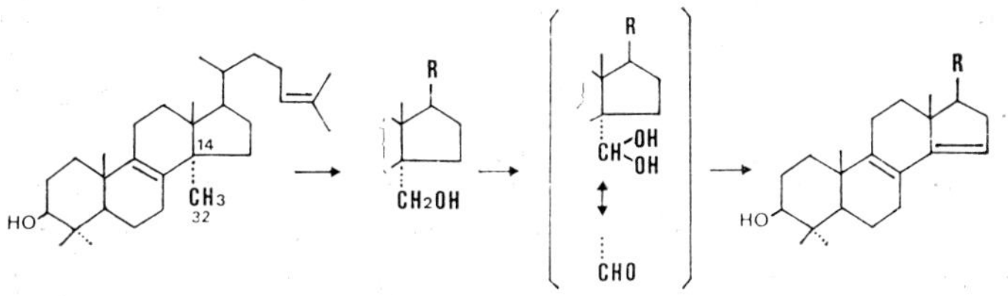

Fig. 4. A postulated reaction process of lanosterol 14α-demethylation catalyzed by cytochrome P-450_{14DM}.

successive monooxygenations without release of any intermediate. In fact, we confirmed that 32-hydroxy-24,25-dihydrolanosterol, which is the first intermediate of the 14α-demethylation of 24,25-dihydrolanosterol, shows considerably higher affinity to the cytochrome than 24,25-dihydrolanosterol (AOYAMA et al., 1987a). We also found that the conversion of 32-hydroxy-24,25-dihydrolanosterol to 4,4-dimethyl-5α-cholesta-8,14-dien-3β-ol by the reconstituted system consumed 2 mol of NADPH per mol of the product formation (AOYAMA et al., 1987a). Furthermore, we recently found that 32-oxolanost-8-en-3β-ol was also converted to the same demethylated product by cytochrome $P\text{-}450_{14DM}$ (AOYAMA et al., 1989). These lines of evidence firmly support the postulated mechanism that cytochrome $P\text{-}450_{14DM}$ catalyzes the whole process of lanosterol 14α-demethylation consisting of three monooxygenations (Fig. 4) without releasing the intermediates and converts the sterol to its 14-demethylated and 14,15-unsaturated derivative.

2.2. Δ^{22}-Desaturation of 22,23-dihydroergosterol

The final step of ergosterol biosynthesis in yeast is the Δ^{22}-desaturation of 22,23-dihydroergosterol (ergosta-5,7-dien-3β-ol) (Fig. 1). It was reported by HATA et al. (1983a, 1987) that this reaction was inhibited by CO, SKF 525-A and metyrapon and required NADPH and molecular oxygen. These findings suggest the contribution of a cytochrome P-450-containing monooxygenase system to this reaction. The cytochrome P-450 responsible for this reaction was tentatively named cytochrome $P\text{-}450_{22DS}$ (HATA et al., 1983a). Purification of cytochrome $P\text{-}450_{22DS}$ has not yet been achieved and little is known of the properties of this cytochrome.

The Δ^{22}-desaturation is inhibited neither by the antibodies to cytochrome $P\text{-}450_{14DM}$ nor by buthiobate, a specific inhibitor for the cytochrome (HATA et al., 1983a, 1987). A mutant yeast, *S. cerevisiae* N-22, which is defective in Δ^{22}-desaturation showed considerable lanosterol 14α-demethylase activity, whereas *S. cerevisiae* SGl which is a lanosterol 14α-demethylase defective mutant (see section 3.3.) showed high Δ^{22}-desaturase activity (HATA et al., 1983a, 1987). In addition, the Δ^{22}-desaturase activity was higher in the microsomes from aerated cells of *S. cerevisiae* but high 14α-demethylase activity was found in the microsomes from semianaerobically grown cells of the same yeast (HATA et al., 1983a, 1987). These lines of evidence indicate that cytochrome $P\text{-}450_{22DS}$ is a different enzyme from cytochrome $P\text{-}450_{14DM}$.

The Δ^{22}-desaturation of dihydroergosterol is the first example of a cytochrome P-450-dependent desaturation reaction. Other known desaturation reactions occurring in yeast microsomes such as the Δ^{9}-desaturation of palmitoyl CoA (TAMURA et al., 1976) and the Δ^{5}-desaturation of 5,6-dihydroergosterol (ergosta-7,22-dien-3β-ol) (OSUMI et al., 1979) are catalyzed by another type of monooxygenase, CSF, and it is supposed that these reactions consist of an enzyme-catalyzed hydroxylation and the spontaneous dehydration of the hydroxylated intermediates. The Δ^{22}-desaturation may also consist of hydroxylation at either C-22 or C-23 and dehydration. In this case, however, the dehydration may not be a spontaneous reaction because 23-hydroxydihydroergosterol which was found by HATA et al. (1983b) in yeast cells, is a relatively stable compound. Accordingly, the dehydration may be an enzymecatalyzed reaction, but it is not clear whether cytochrome P-450_{22DS} catalyzes both of these reactions.

3. Properties of cytochrome P-450_{14DM}

3.1. Molecular properties of cytochrome P-450_{14DM}

The content of cytochrome P-450_{14DM} in microsomes of *S. cerevisiae* reaches its maximum when the yeast is grown semi-anaerobically in a relatively high glucose medium in the late logarithmic phase (ISHIDATE et al., 1969; YOSHIDA et al., 1974; KOERENLAMPI et al., 1981). Cytochrome P-450_{14DM} has been purified from the semi-anaerobically grown cells of *S. cerevisiae* (YOSHIDA et al., 1977). The cytochrome is solubilized with sodium cholate and purified by successive chromatography with AH-Sepharose 4B (Pharmacia), hydroxylapatite (Bio-Gel HT, Bio-Rad) and CM-Sephadex C-50 (Pharmacia). The cytochrome P-450_{14DM} preparation thus obtained is gel electrophoretically homogeneous with a specific content of about 15 nmol of cytochrome P-450/mg protein, and the overall yield of cytochrome P-450 from the microsomes is usually 25—30% (YOSHIDA et al., 1977; YOSHIDA and AOYAMA, 1984).

An apparent monomeric molecular weight of cytochrome P-450_{14DM} was assumed to be 58,000 by a polyacrylamide gel electrophoresis (YOSHIDA and AOYAMA, 1984), while the molecular weight was calculated to be 60,700 from its amino acid sequence (KALB et al., 1987; ISHIDA et al., 1988). One mole of the cytochrome contained one mole of protoporphyrin IX as prosthetic group (YOSHIDA and AOYAMA, 1984).

Recently, the primary structures of cytochromes P-450_{14DM} of *S. cerevisiae* (KALB et al., 1987; ISHIDA et al., 1988) and *C. tropicalis* (CHEN et al., 1988) were determined (primary structure of the *S. cerevisiae* cytochrome is shown in Fig. 12). Cytochrome P-450_{14DM} of *S. cerevisiae* and of *C. tropicalis* consist of 530 and 528 amino acid residues, respectively. The primary structures of these two cytochromes are highly conserved; i. e., they share 66.5% identical and 23.1% conservatively replaced amino acids in 516-amino acid alignment (CHEN et al., 1988). According to their alignment, occurrence of three or four relatively long identical segments can be pointed out (CHEN et al., 1988). One of them around Cys-470 of the *S. cerevisiae* cytochrome (see Fig. 12) is assigned as the heme-binding domain and others may be the domain responsible for interaction with the substrate and with the reductase. In this respect, it is noteworthy that the longest one from Asn-290 to Ala-320 of the *S. cerevisiae* protein (CHEN et al., 1988; see Fig. 12) is considered to include the "distal helix" (ISHIDA et al., 1988; see Figs. 13 and 14), because the "distal helix" is considered to locate over the oxygen activating site of the cytochrome based on the crystallographic analysis of cytochrome P-450_{cam} (POULOS et al., 1985).

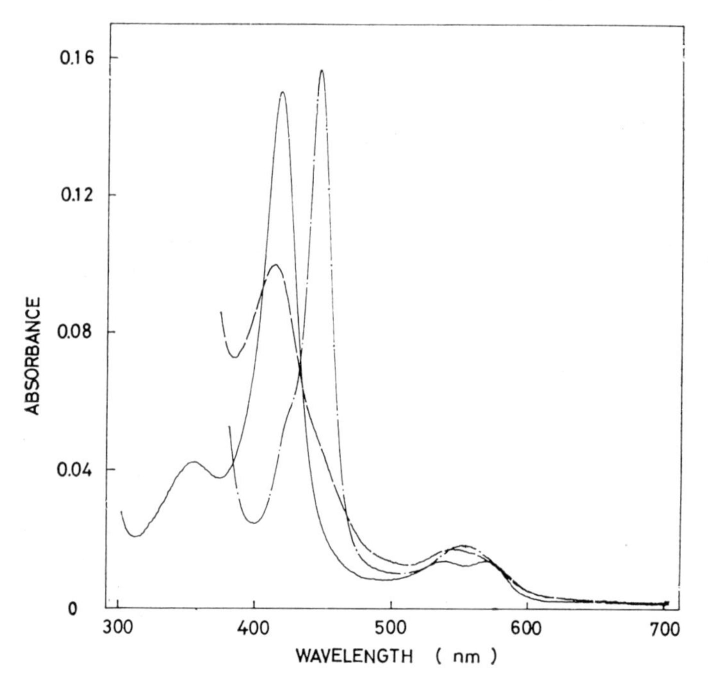

Fig. 5. Absorption spectra of cytochrome P-450_{14DM} purified from *S. cerevisiae*. ——, — — and —·— represent oxidized, reduced and reduced CO-bound forms, respectively.

In addition, CHEN et al. (1988) also pointed out that less conserved regions are found in more hydrophilic parts of the protein. This fact seems to suggest the possibility that the molecular surfaces of cytochromes P-450_{14DM} of *S. cerevisiae* and *C. tropicalis* are significantly different, and this is consistent with our immunological finding that cytochrome P-450_{14DM} of *Candida* sp. formed no precipitation line with the antibodies to cytochrome P-450_{14DM} of *S. cerevisiae* (KÄRGEL et al., 1990).

Figure 5 represents the absorption spectra of cytochrome P-450_{14DM} purified from *S. cerevisiae* microsomes (YOSHIDA & AOYAMA, 1984). Oxidized form shows the Soret band at 417 nm and the broad α and β bands at 570 and

540 nm, respectively, but no absorption band is observed at around 650 nm. These spectral characteristics suggest that the oxidized form is in a low-spin state and this is confirmed by the electron paramagnetic resonance spectrum showing three characteristic signals at $g = 1.92$, 2.27, and 2.45 (Yoshida and Aoyama, 1984). The Soret band of the reduced CO-complex appears at 447 nm (Yoshida et al., 1977; Yoshida and Aoyama, 1984).

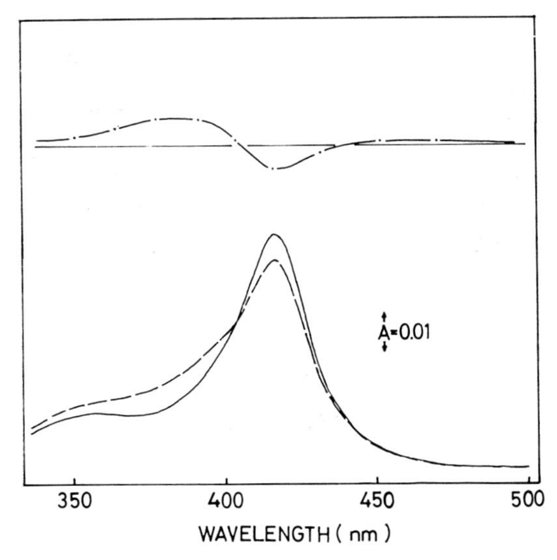

Fig. 6. Spectral change of the ferric cytochrome P-450_{14DM} upon binding with lanosterol. The absorption spectrum of the ferric substrate-free form (——) was converted to the substrate-bound form (— —) by the addition of excess lanosterol. This spectral change is due to the high to low spin state conversion of the cytochrome as shown by the difference spectrum (—·—). Note that the spin-state change is not complete even in the presence of excess substrate.

The addition of lanosterol to the ferric resting form of cytochrome P-450_{14DM} induces Type I spectral change (Fig. 6), and lanosterol was the only compound which induces Type I spectral change in this cytochrome (Aoyama and Yoshida, 1978b; Yoshida and Aoyama, 1984). This fact suggests that lanosterol interacts with the cytochrome at a site near the heme and this site must be the substrate-binding site. The spectral change caused by lanosterol is incomplete (Yoshida and Aoyama, 1984). As can be seen in Figure 6, the Soret peak is remained at 416–417 nm and the absorption band at around 390 nm which is assigned as the Soret band of the high-spin form is observed only as a weak shoulder even in the presence of excess lanosterol. This fact suggests that a considerable portion of the cytochrome remains in the low-spin state even in the presence of excess substrate. In other words, cytochrome P-450_{14DM} is in an equilibrium between high- and low-spin states and the equilibrium is partially shifted to the high-spin form upon interaction with lanosterol. The spectrophotometrically-determined dissociation constant of lanosterol is 7 μM (YOSHIDA and AOYAMA, 1984) and this value is comparable to the Km of lanosterol (see section 3.2.).

Basic amines induce Type II spectral change on cytochrome P-450$_{14DM}$ (YOSHIDA and KUMAOKA, 1975). They include primary amines and imidazole and pyridine derivatives. Azole antifungal agents, potent inhibitors for the demethylase, also induce Type II spectral change, as discussed in section 4.

Cytochrome P-450$_{14DM}$ is relatively stable. The purified preparation can be stored for more than one year at −70 to −80 °C without loss of activity. The cytochrome is resistant to perturbations that denature cytochrome P-450 to "P-420" form. For example, complete conversion of cytochrome P-450$_{14DM}$ to the P-420 form in the presence of 1 M KSCN took about 150 minutes at room temperature (YOSHIDA et al., 1977; YOSHIDA and AOYAMA, 1984).

3.2. Electron transport system constituting the lanosterol 14α-demethylase and catalytic properties of cytochrome P-450$_{14DM}$

Lanosterol 14α-demethylase consists of cytochrome P-450$_{14DM}$ and a flavoprotein, NADPH-cytochrome P-450 reductase (AOYAMA and YOSHIDA, 1978a; AOYAMA et al., 1984). NADPH-cytochrome P-450 reductase has been purified from the microsomes of semi-anaerobically grown cells of *S. cerevisiae* (AOYAMA et al., 1978). This enzyme shows an apparent monomeric molecular weight of 83,000 upon analysis with polyacrylamide gel electrophoresis and contains one molecule each of FAD and FMN as the prosthetic groups. Cytochrome *c* can serve as an electron acceptor for this enzyme and this enzyme acts as NADPH-oxidase in the presence of menadion (AOYAMA et al., 1978) as in the case of the corresponding mammalian enzyme (NISHIBAYASHI-YAMASHITA and SATO, 1970).

A reconstituted system consisting of purified preparations of cytochrome P-450$_{14DM}$ and the reductase and dilauroylphosphatidylcholine catalyzes the conversion of lanosterol to 4,4-dimethyl-5α-cholesta-8,14,24-trien-3β-ol (AOYAMA and YOSHIDA, 1978a; AOYAMA et al., 1984). Km of the reconstituted system for lanosterol is 6 μM and turnover number of the lanosterol demethylation is 10−15 nmol/min × nmol cytochrome P-450$_{14DM}$ under standard assay conditions (AOYAMA et al., 1984). The reconstituted system also catalyzes the 14α-demethylation of 24,25-dihydrolanosterol. However, the Km for this sterol is 20 μM (AOYAMA et al., 1987a), indicating that the cytochrome shows a higher affinity to lanosterol than to 24,25-dihydrolanosterol.

The reconstituted system readily converted 32-hydroxy-24,25-dihydrolanosterol, a postulated intermediate of the 14α-demethylation of 24,25-dihydrolanosterol (see Fig. 4), to 4,4-dimethyl-5α-cholesta-8,14-dien-3β-ol (AOYAMA et al., 1987a). Affinity of the 32-hydroxysterol for the reconstituted system

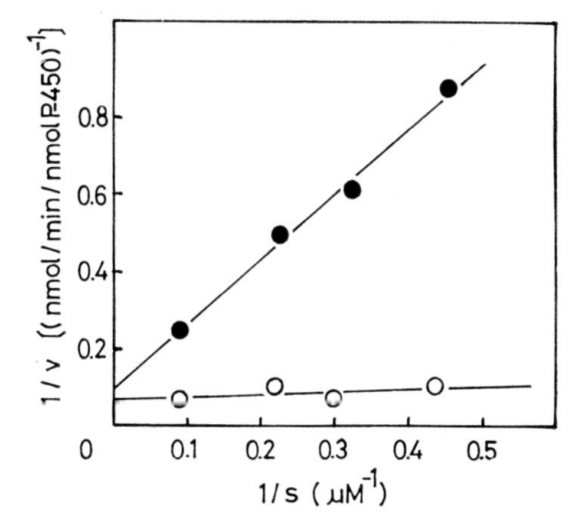

Fig. 7. LINEWEAVER-BURK plot of the metabolism of 24,25-dihydrolanosterol and 32-hydroxy-24,25-dihydrolanosterol by the reconstituted lanosterol 14α-demethylase system. ● and ○ represent 24,25-dihydrolanosterol and 32-hydroxy-24,25-dihydrolanosterol, respectively. This plot indicates high affinity of the 32-hydroxy-intermediate for cytochrome P-450_{14DM}.

is considerably high when compared with 24,25-dihydrolanosterol with a Km of 1 μM for this compound (Fig. 7) (AOYAMA et al., 1987a). Recently, we obtained lines of evidence indicating that the reconstituted system also converted 32-oxodihydrolanosterol to 4,4-dimethyl-5α-cholesta-8,14-dien-3β-ol (AOYAMA

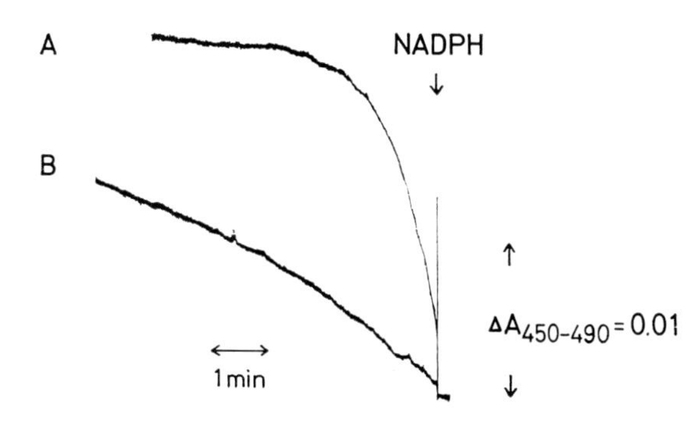

Fig. 8. Acceleration of the NADPH-dependent enzymatic reduction of cytochrome P-450_{14DM} by lanosterol. Reduction of cytochrome P-450 by NADPH in the reconstituted system was determined in the presence (A) and absence (B) of lanosterol.

et al., 1989). These facts indicate that the cytochrome P-450_{14DM} catalyzes the whole process of lanosterol 14α-demethylation without releasing the intermediates (see Fig. 4).

Cytochrome P-450_{14DM} is rapidly reduced in the reconstituted system by NADPH in the presence of substrate (Fig. 8) (AOYAMA and YOSHIDA, 1978b;

AOYAMA et al., 1984). This fact suggests that the cytochrome has a control mechanism to prevent idling of the catalytic cycle occurring in the absence of substrate.

3.3. An altered molecule of cytochrome P-450$_{14DM}$ found in a mutant yeast which is defective in lanosterol 14α-demethylation

In 1983, AOYAMA et al. (1983a) found that a mutant of *S. cerevisiae* SGl which is defective in lanosterol 14α-demethylation (TROCHA et al., 1977) contained a protein which reacted immunochemically like cytochrome P-450$_{14DM}$. This protein has been isolated from the microsomes of the mutant yeast according to

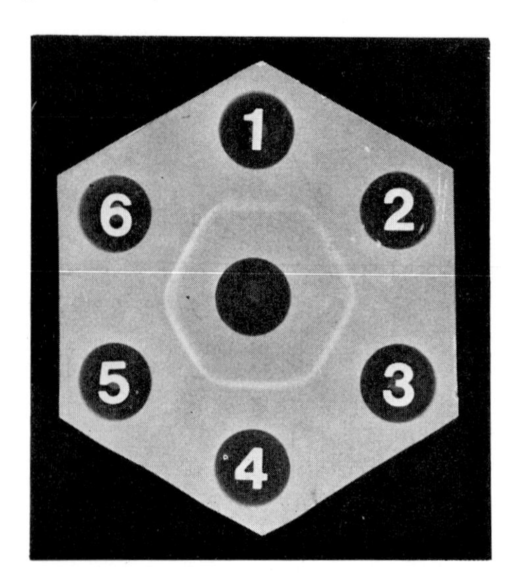

Fig. 9. Immunological identity of cytochrome P-450$_{SG1}$ with cytochrome P-450$_{14DM}$. Rabbit antibodies to cytochrome P-450$_{14DM}$ of *S. cerevisiae* were placed in the center well. Wells 1 and 4 contained cytochrome P-450$_{14DM}$ and wells 2, 3, 5 and 6 contained cytochrome P-450$_{SG1}$.

essentially the same procedure as that developed for the purification of cytochrome P-450$_{14DM}$ (AOYAMA et al., 1987b). The isolated protein was identified as a cytochrome P-450 and tentatively named cytochrome P-450$_{SG1}$ (AOYAMA et al., 1987b). Cytochrome P-450$_{SG1}$ formed a single precipitation line with the rabbit antibodies to cytochrome P-450$_{14DM}$ and the precipitation line was completely fused with that formed between the antibodies and cytochrome P-450$_{14DM}$ (Fig. 9) (AOYAMA et al., 1987b). Furthermore, the peptide maps of chymotrypsin- or *Staphylococcus aureus* V_8 proteinase-digested P-450$_{SG1}$ were superimposable on those of cytochrome P-450$_{14DM}$ (AOYAMA et al., 1987b). These facts indicate that cytochrome P-450$_{SG1}$ is essentially the same protein as cytochrome P-450$_{14DM}$. However, cytochrome P-450$_{SG1}$ shows no lanosterol 14α-demethylase activity and is reduced only slowly by NADPH in the reconstituted system even in the presence of sufficient lanosterol (AOYAMA et al., 1983b, 1987b). These lines of evidence suggest that cytochrome P-450_{SG1} is an altered form of cytochrome P-450_{14DM}.

Spectral properties of cytochrome P-450_{SG1} are different from those of cytochrome P-450_{14DM} and are unusual (Fig. 10) (YOSHIDA et al., 1985). Ferric cytochrome P-450_{SG1} shows its Soret peak at 422 nm and α band was observed as a weak shoulder on the broad β peak, while other known low-spin cytochromes

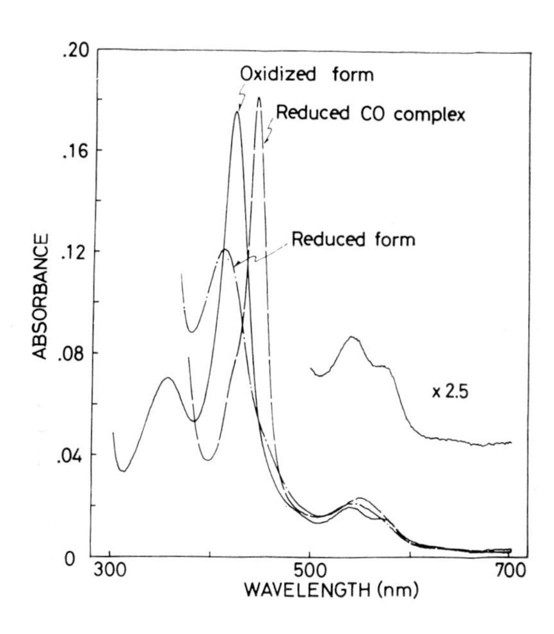

Fig. 10. Absorption spectra of cytochrome P-450_{SG1} purified from the mutant yeast, *S. cerevisiae* SG1. Note that the α-band of the oxidized form at 577 nm is observed as a shoulder, while that of normal cytochrome P-450_{14DM} is observed as a distinct peak (see Figs. 5 and 11).

P-450 including cytochrome P-450_{14DM} show their Soret peaks at 416–418 nm and have slightly higher α than β peaks (Fig. 10) (AOYAMA et al., 1987b). The δ band of ferric cytochrome P-450_{SG1} is observed as a distinct peak at 355 nm, while that of other cytochromes P-450 is observed as a weaker shoulder (Fig. 10) (YOSHIDA et al., 1985; AOYAMA et al., 1987b).

In general the slightly higher α than β peak of ferric low-spin cytochrome P-450 is considered to be due to the unique ligand structure of the cytochrome, H_2O-Fe^{3+}-S^- (YOSHIDA et al., 1982) where S^- represents the fifth thiolate ligand derived from a cysteine residue of the apoprotein. When the native 6th ligand trans to S^- is replaced by a nitrogen containing one such as pyridine, imidazole or their derivatives, the above spectral characteristics of the cytochrome are lost with marked hypochromicity and slight red-shift of the α peak and a few nm red-shift of the Soret band (YOSHIDA et al., 1982). In addition, marked enhancement of the δ band is also observed upon the ligand displacement (Yoshida et al., 1982). The spectral characteristics of ferric cytochrome P-450_{SG1} are coincident with those of a nitrogenous-ligand derivative of ferric cytochrome P-450_{14DM}, especially of the 1-methylimidazole complex (Fig. 11) (Yoshida et al., 1985; Aoyama et al., 1987b). The same conclusion is also derived from the comparison of the EPR spectrum of cytochrome P-450_{SG1} with those of cytochrome P-450_{14DM} and its 1-methylimidazole complex

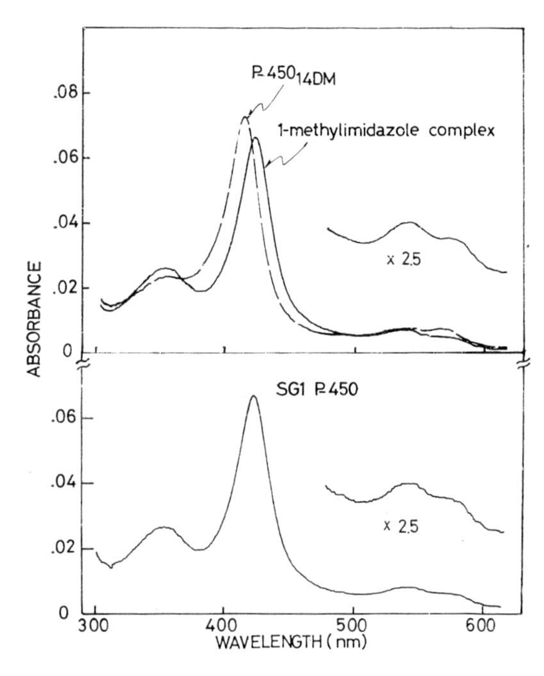

Fig. 11. Spectral change of cytochrome P-450_{14DM} upon formation of the 1-methylimidazole complex. Note that the absorption spectrum of the 1-methylimidazole complex is superimposable on that of the ferric cytochrome P-450_{SG1} (Fig. 10).

(Aoyama et al., 1987b). These facts strongly suggest that the 6th coordination position of ferric cytochrome P-450_{SG1} is occupied by an imidazole of histidine residue in the apoprotein (Aoyama et al., 1987b).

Based on these facts, it can be concluded that cytochrome P-450_{SG1} is a modified cytochrome P-450_{14DM}. The alteration must occur in the heme environments and the sixth ligand of the ferric resting form, H_2O, must exchange with histidine. Cytochrome P-450_{SG1} loses the catalytic activity as the lanosterol demethylase, though it retains the spectral nature of cytochrome P-450. P-450_{SG1} may thus be an interesting object for studying the structure of the active site of cytochrome P-450.

Recently, we isolated the genes of cytochromes P-450_{14DM} and P-450_{SG1} and determined their primary structures (Ishida et al., 1988). Both of these cytochromes consisted of 530 amino acid residues and only one amino acid exchange (Gly to Asp) was found at the 310th residue (Fig. 12). Accordingly, cytochrome P-450$_{SG1}$ was formed by the one-point mutation occurring in the cytochrome P-450$_{14DM}$ gene. This finding was surprising because such an apparently minor amino acid exchange caused significant conformational

1 Met Ser Ala Thr Lys Ser Ile Val Gly Glu Ala Leu Glu Thr Val Asn Ile Gly Leu Ser
21 His Phe Leu Ala Leu Pro Leu Ala Gln Arg Ile Ser Leu Ile Ile Ile Ile Pro Phe Ile
41 Tyr Asn Ile Val Trp Gln Leu Leu Tyr Ser Leu Arg Lys Asp Arg Pro Pro Leu Val Phe
61 Tyr Trp Ile Pro Trp Val Gly Ser Ala Val Val Tyr Gly Met Lys Pro Tyr Glu Phe Phe
81 Glu Glu Cys Gln Lys Lys Tyr Gly Asp Ile Phe Ser Phe Val Leu Leu Gly Arg Val Met
101 Thr Val Tyr Leu Gly Pro Lys Gly His Glu Phe Val Phe Asn Ala Lys Leu Ala Asp Val
121 Ser Ala Glu Ala Ala Tyr Ala His Leu Thr Thr Pro Val Phe Gly Lys Gly Val Ile Tyr
141 Asp Cys Pro Asn Ser Arg Leu Met Glu Gln Lys Lys Phe Val Lys Gly Ala Leu Thr Lys
161 Glu Ala Phe Lys Ser Tyr Val Pro Leu Ile Ala Glu Glu Val Tyr Lys Tyr Phe Arg Asp
181 Ser Lys Asn Phe Arg Leu Asn Glu Arg Thr Thr Gly Thr Ile Asp Val Met Val Thr Gln
201 Pro Glu Met Thr Ile Phe Thr Ala Ser Arg Ser Leu Leu Gly Lys Glu Met Arg Ala Lys
221 Leu Asp Thr Asp Phe Ala Tyr Leu Tyr Ser Asp Leu Asp Lys Gly Phe Thr Pro Ile Asn
241 Phe Val Phe Pro Asn Leu Pro Leu Glu His Tyr Arg Lys Arg Asp His Ala Gln Lys Ala
261 Ile Ser Gly Thr Tyr Met Ser Leu Ile Lys Glu Arg Arg Lys Asn Asn Asp Ile Gln Asp
281 Arg Asp Leu Ile Asp Ser Leu Met Lys Asn Ser Thr Tyr Lys Asp Gly Val Lys Met Thr
301 Asp Gln Glu Ile Ala Asn Leu Leu Ile Gly Val Leu Met Gly Gly Gln His Thr Ser Ala
* * * * * * * * * Asp * * * * * * * * * *
321 Ala Thr Ser Ala Trp Ile Leu Leu His Leu Ala Glu Arg Pro Asp Val Gln Gln Glu Leu
341 Tyr Glu Glu Gln Met Arg Val Leu Asp Gly Gly Lys Lys Glu Leu Thr Tyr Asp Leu Leu
361 Gln Glu Met Pro Leu Leu Asn Gln Thr Ile Lys Glu Thr Leu Arg Met His His Pro Leu
381 His Ser Leu Phe Arg Lys Val Met Lys Asp Met His Val Pro Asn Thr Ser Tyr Val Ile
401 Pro Ala Gly Tyr His Val Leu Val Ser Pro Gly Tyr Thr His Leu Arg Asp Glu Tyr Phe
421 Pro Asn Ala His Gln Phe Asn Ile His Arg Trp Asn Asn Asp Ser Ala Ser Ser Tyr Ser
441 Val Gly Glu Glu Val Asp Tyr Gly Phe Gly Ala Ile Ser Lys Gly Val Ser Ser Pro Tyr
461 Leu Pro Phe Gly Gly Gly Arg His Arg Cys Ile Gly Glu His Phe Ala Tyr Cys Gln Leu
481 Gly Val Leu Met Ser Ile Phe Ile Arg Thr Leu Lys Trp His Tyr Pro Glu Gly Lys Thr
501 Val Pro Pro Pro Asp Phe Thr Ser Met Val Thr Leu Pro Thr Gly Pro Ala Lys Ile Ile
521 Trp Glu Lys Arg Asn Pro Glu Gln Lys Ile ---

Fig. 12. The amino acid sequence of cytochrome P-450$_{14DM}$ deduced from the nucleotide sequence of its cloned DNA. The 310th residue, Gly, is replaced with Asp in cytochrome P-450$_{SG1}$. The membrane-binding domain and the heme-binding fragment are started at Ile-31 and Phe-463, respectively, as indicated by underlining. The segment from Ile-309 to Trp-325 is assigned as the "distal helix" by the alignment described in Figure 13.

changes which were accompanied by the binding of a histidine residue to the heme iron and the complete loss of the catalytic activity.

Alignment of the primary structure of cytochrome P-450$_{14DM}$ with several known cytochrome P-450 sequences (Fig. 13) (Ishida te al., 1988) suggests that the Gly-310 may locate in the so-called "distal helix" or "helix I" which crosses over the distal side of the heme of cytochrome P-450cam (Poulos et al., 1985). Consequently, if conformation of cytochrome P-450 is fundamentally conserved, the amino acid exchange of cytochrome P-450_{SG1} may occur in the apoprotein locating over the distal side of the heme. Interestingly, one histidine, H-317, is found in this region near G-310, the mutation point (Fig. 13). Therefore, it can be assumed that this histidine comes to interact with the heme iron and that interaction may be caused by the conformational change induced

			310							317									
14DM	309	I	G	V	L	M	G	G	Q	H	T	S	A	A	T	S	A	W	325
SG1	309	I	D	V	L	M	G	G	Q	H	T	S	A	A	T	S	A	W	325
CAM	243	G	L	L	L	V	G	G	L	D	T	V	V	N	F	L	S	F	259
SCC	320	T	E	M	L	A	G	G	V	N	T	T	S	M	T	L	Q	W	336
P450b	293	L	S	L	F	F	A	G	T	E	T	T	S	S	T	T	L	R	309
P450c	316	F	D	L	F	G	A	G	F	D	T	I	T	T	A	I	S	W	332

Fig. 13. Alignment of amino acid sequences assignable to distal helices of some cytochromes P-450.

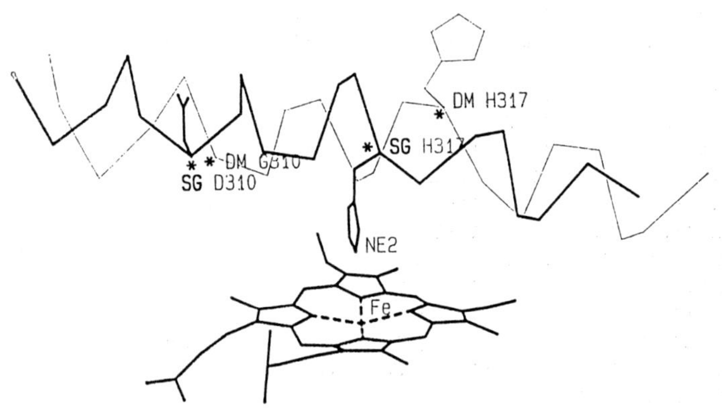

Fig. 14. A computer-aided structural model showing correlation between the distal helices and heme of cytochromes P-450_{14DM} and P-450_{SG1}. The α-carbon backbones of cytochromes P-450_{14DM} and P-450_{SG1} are indicated with thin and thick lines, respectively. In this modeling, displacement of Gly-310 of cytochrome P-450_{14DM} (DM G310) with Asp in cytochrome P-450_{SG1} (SG D310) causes the shift of the helix accompanied by the coordination of His-317 (SG H317) to the heme iron.

by the G to D replacement at 310th residue. Figure 14 represents a possible illustration of such conformational change depicted by a computer simulation of the conformation of the distal helix of cytochromes P-450_{14DM} and P-450_{SG1}. The most important driving force of the conformational change postulated in this simulation is the movement of the polar side chain of D-310 from a hydrophobic heme environment to a more hydrophilic outer environment (ISHIDA et al., 1988).

To obtain more confirmative and detailed information on this structural change, site-directed mutagenesis in this region is now in progress. Precise comparison of the structural differences between cytochromes P-450_{14DM} and P-450_{SG1} may provide important information for understanding the active site structure of cytochrome P-450_{14DM}.

4. Cytochrome P-450_{14DM} as the target of azole antifungal agents

4.1. Effects of azole antifungal agents on sterol biosynthesis

In recent years, numbers of azole compounds have been developed as potent antifungal agents for medicinal and plant protecting purposes. These compounds are known to disturb membrane structure and membrane-bound enzyme systems and inhibit the growth of fungi (for a review, see VANDEN BOSSCHE, 1985). Such disturbance results from the depletion of ergosterol and accumulation of 14-methylsterols such as lanosterol, 24-methylene-24,25-dihydrolanosterol, obtusifoliol and 14-methylfecosterol in the membrane (KATO and KAWASE, 1976; HENRY and SISLER, 1979; BALDWIN and WIGGINS, 1984). As described in section 1., these 14-methylsterols are the intermediates in the biosynthetic pathway of ergosterol. This fact suggests that azole antifungal agents inhibit the 14α-demethylation of methylsterols during the ergosterol biosynthesis.

4.2. Inhibition of lanosterol 14α-demethylase by azole antifungal agents

Azole antifungal agents inhibited the lanosterol demethylation by the reconstituted system consisting of cytochrome P-450_{14DM} and cytochrome P-450 reductase (YOSHIDA et al., 1986; YOSHIDA and AOYAMA 1987). These compounds

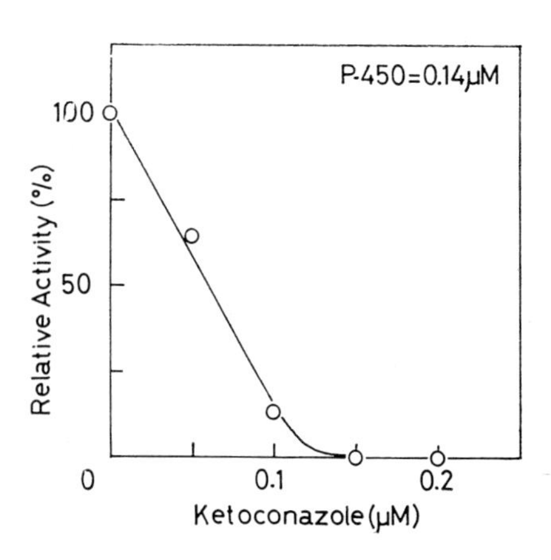

Fig. 15. Inhibition of the reconstituted lanosterol 14α-demethylase activity by ketoconazole.

also inhibited the enzymatic reduction of the cytochrome by NADPH in the presence of NADPH-cytochrome P-450 reductase, while they did not inhibit NADPH-cytochrome *c* reductase activity catalyzed by the reductase itself (Aoyama et al., 1983b; Yoshida and Aoyama, 1987). As described in the following section, azole antifungal agents induce characteristic spectral changes of cytochrome P-450_{14DM}. Consequently, it can be concluded that azole antifungal agents interact with cytochrome P-450_{14DM} and inhibit the enzyme activity.

Figure 15 represents a typical inhibition profile of the lanosterol demethylase activity of the reconstituted system by ketoconazole. The concentration of ketoconazole necessary to inhibit the demethylase activity is very low and complete inhibition is observed when it is added at a concentration equal to cytochrome P-450_{14DM} (Yoshida and Aoyama, 1987). This fact suggests that ketoconazole inhibits the demethylase activity by forming a 1:1 complex due to its extremely high affinity for the cytochrome.

4.3. Spectrophotometric analysis of interaction between antifungal agents and cytochrome P-450_{14DM}

Azole antifungal agents bind to the oxidized form of cytochrome P-450_{14DM} and induce Type II spectral change of the cytochrome. Figure 16 represents a typical example of such spectral change, which is induced by ketoconazole

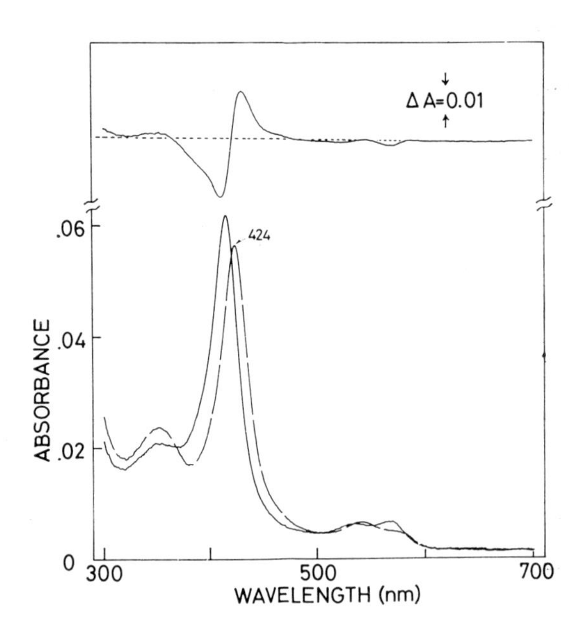

Fig. 16. Spectral change of cytochrome P-450_{14DM} caused by the binding of keteoconazole. —— and — — represent the free and ketoconazole-bound forms, respectively, of ferric cytochrome P-450_{14DM}. The binding of ketoconazole can be determined by using the difference spectrum shown in the upper half of the figure.

(YOSHIDA and AOYAMA, 1987). Azole antifungal agent complexes of the oxidized cytochrome P-450_{14DM} show their Soret peaks at 420−424 nm, suggesting interaction between the heme iron and azole nitrogens. Affinity of azole antifungal agents for cytochrome P-450_{14DM} is very high; for example, binding of ketoconazole to the cytochrome is quantitative, as shown in Figure 17, and an apparent Kd is assumed to be less than 0.01 μM (YOSHIDA and AOYAMA, 1987). This observation agrees well with the inhibitory effect of the antifungal agent described in section 4.2. (see Fig. 15).

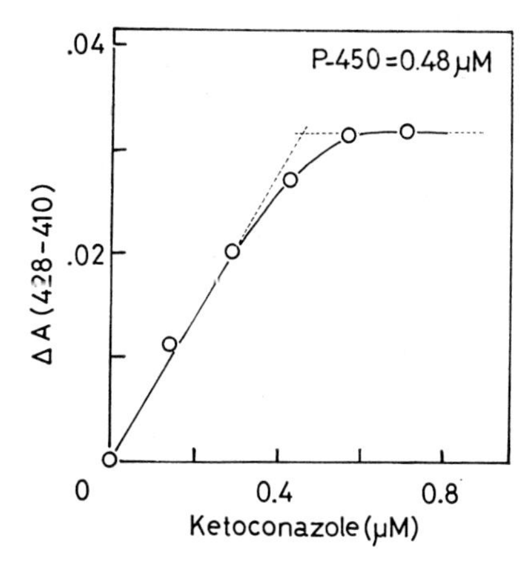

Fig. 17. Spectral titration of cytochrome P-450_{14DM} with ketoconazole. Ketoconazole can bind to the cytochrome quantitatively.

Azole antifungal agents also interact with the reduced form of cytochrome P-450_{14DM}, and the absorption spectra of these complexes indicate that the azole nitrogens still bind to the reduced heme iron (YOSHIDA et al., 1986; YOSHIDA and AOYAMA, 1987). It is well known that the reduced cytochrome P-450 combines with CO and forms the reduced CO-complex showing the Soret band around 450 nm. VANDEN BOSSCHE and his associates reported that ketoconazole and itraconazole interfered with the appearance of the Soret peak at 448 nm in the reduced CO-difference spectrum of yeast microsomes (VANDEN BOSSCHE et al., 1984a, 1984b). In accordance with their observation, itraconazole and ketoconazole complexes of the reduced form of the purified yeast cytochrome P-450_{14DM} are not easily converted to the reduced CO-complex (YOSHIDA and AOYAMA, 1987). These observations suggest that the interaction of itraconazole and ketoconazole with the reduced enzyme is so strong as to interfere with binding of CO to the reduced enzyme.

These lines of spectral evidence, together with the observations described in section 4.2, clearly indicate that azole antifungal agents avidly interact with cytochrome P-450_{14DM} and inhibit the lanosterol metabolism catalyzed by the enzyme. Since fungicidal activity of azole antifungal agents originates from depression of ergosterol concomitant with accumulation of 14-methylsterols, it can be concluded that cytochrome P-450_{14DM} is the target of azole antifungal agents.

5. Summary and prospects

Bioconversion of lanosterol to ergosterol by yeast includes two cytochrome P-450-mediated reactions, the 14α-demethylation of lanosterol and the 22-desaturation of 22,23-dihydroergosterol. The cytochrome responsible for the former reaction, cytochrome P-450_{14DM}, has been studied in detail, while the other cytochrome catalyzing the latter reaction, cytochrome P-450_{22DM}, has yet to be studied.

14α-Demethylation of lanosterol consists of three-step monooxygenations and the methyl group is eliminated as formic acid accompanied by the introduction of a double bond between C-14 and C-15 of the sterol ring. This reaction is a typical cytochrome P-450-catalyzed C-C cleavage reaction and cytochrome P-450_{14DM} catalyzes the whole process.

Since removal of the 14α-methyl group is an indispensable reaction in sterol biosynthesis of eukaryotes, cytochrome P-450_{14DM} may be distributed widely in eukaryotes. Actually, the lanosterol 14α-demethylation in cholesterol biosynthesis by mammals is known to be a cytochrome P-450-dependent reaction and a cytochrome P-450 which catalyzes this reaction has been isolated from rat liver microsomes. Furthermore, occurrence of a similar cytochrome in various lower eukaryotes is assumed from the effects of azole antifungal agents on these organisms. So far as we know, cytochrome P-450_{14DM} is the only cytochrome P-450 which is distributed widely in eukaryotes with essentially the same function. In other words, cytochrome P-450_{14DM} is considered to be one of the most conserved species of cytochrome P-450. In this respect, cytochrome P-450_{14DM} may be an important object for studying the evolution of cytochrome P-450.

6. References

AKHTAR, M., I. A. WATKINSON, A. D. RAHIMTULA, D. C. WILTON, and K. A. MUNDAY, (1969), Biochem. J. **111**, 757–761.

AKHTAR, M., C. W. FREEMAN, D. C. WILTON, R. B. BOAR, and D. B. COPSEY, (1977), Bioorg. Chem. **6**, 473–481.

ALEXANDER, K., M. AKHTAR, R. B. BOAR, J. F. MCGHIE, and D. H. R. BARTON, (1972), J. Chem. Soc. Chem. Comm. **1972**, 383–385.

AOYAMA, Y., Y. YOSHIDA, S. KUBOTA, H. KUMAOKA, and A. FURUMICHI, (1978), Arch. Biochem. Biophys. **185**, 362–369.

AOYAMA, Y. and Y. YOSHIDA, (1978a), Biochem. Biophys. Res. Commun. **85**, 28–34.

AOYAMA, Y. and Y. YOSHIDA, (1978b), Biochem. Biophys. Res. Commun. **82**, 33–38.

AOYAMA, Y., T. OKIKAWA, and Y. YOSHIDA, (1981a), Biochim. Biophys. Acta **665**, 596—601.

AOYAMA, Y., Y. YOSHIDA, R. SATO, M. SUSANI, and H. RUIS, (1981b), Biochim. Biophys. Acta **663**, 194—202.

AOYAMA, Y., Y. YOSHIDA, S. HATA, T. NISHINO, H. KATSUKI, U. S. MAITRA, V. P. MOHAN, and D. B. SPRINSON, (1983a), J. Biol. Chem. **258**, 9040—9042.

AOYAMA, Y., Y. YOSHIDA, S. HATA, T. NISHINO, and H. KATSUKI, (1983b), Biochem. Biophys. Res. Commun. **115**, 642—647.

AOYAMA, Y., Y. YOSHIDA, and R. SATO, (1984), J. Biol. Chem. **259**, 1661—1666.

AOYAMA, Y., and Y. YOSHIDA, (1986), Biochem. Biophys. Res. Commun. **134**, 659—663.

AOYAMA, Y., Y. YOSHIDA, Y. SONODA, and Y. SATO, (1987a), J. Biol. Chem. **262**, 1239—1243.

AOYAMA, Y., Y. YOSHIDA, T. NISHINO, H. KATSUKI, U. S. MAITRA, V. P. MOHAN, and D. B. SPRINSON, (1987b), J. Biol. Chem. **262**, 14260—14264.

AOYAMA, Y., Y. YOSHIDA, Y. SONODA, and Y. SATO, (1989), J. Biol. Chem. **264**, 18502 to 18505.

BALDWIN, B. C. and T. E. WIGGINS, (1984), Pestic. Sci. **15**, 156—166.

CHEN, C., V. F. KALB, T. G. TURI, and J. C. LOPER, (1988), DNA **7**, 617—626.

FRYBERG, M., A. C. OEHLSCHLAGER, and A. M. UNRAU, (1973), J. Am. Chem. Soc. **95**, 5747—5757.

HATA, S., T. NISHINO, M. KOMORI, and H. KATSUKI, (1981), Biochem. Biophys. Res. Commun. **103**, 272—277.

HATA, S., T. NISHINO, H. KATSUKI, Y. AOYAMA, and Y. YOSHIDA, (1983a), Biochem. Biophys. Res. Commun. **116**, 162—166.

HATA, S., T. NISHINO, Y. ODA, H. KATSUKI, Y. AOYAMA, and Y. YOSHIDA, (1983b), Tetrahedron Lett. **24**, 4729—4730.

HATA, S., T. NISHINO, H. KATSUKI, Y. AOYAMA, and Y. YOSHIDA, (1987), Agric. Biol. Chem. **51**, 1349—1354.

HENRY, M. J. and H. D. SISLER, (1979), Antimicrob. Agents Chemother. **15**, 603—607.

ISHIDA, N., Y. AOYAMA, R. HATANAKA, Y. OYAMA, S. IMAJO, M. ISHIGURO, T. OSHIMA, H. NAKAZATO, T. NOGUCHI, U. S. MAITRA, V. P. MOHAN, D. B. SPRINSON, and Y. YOSHIDA, (1988), Biochem. Biophys. Res. Commun. **156**, 317—323.

ISHIDATE, K., K. KAWAGUCHI, and K. TAGAWA, (1969), J. Biochem. **65**, 385—392.

KÄRGEL, E., Y. AOYAMA, W.-H. SCHUNCK, H.-G. MÜLLER, and Y. YOSHIDA, (1990), Yeast **6**, 61—67.

KALB, V. F., C. W. WOODS, T. G. TURI, C. R. DEY, T. R. SUTTER, and J. C. LOPER, (1987), DNA **6**, 529—537.

KATO, T. and Y. KAWASE, (1976), Agric. Biol. Chem. **40**, 2379—2388.

KOERENLAMPI, S. O., E. MARIN, and O. P. HAENNINEN, (1981), Biochem. J. **194**, 407—413.

KUBOTA, S., Y. YOSHIDA, H. KUMAOKA, and A. FURUMICHI, (1977), J. Biochem. **81**, 197—205.

MITROPOULOS, K. A., G. F. GIBBONS, and B. E. A. REEVES, (1976), Steroids **27**, 821—829.

NISHIBAYASHI-YAMASHITA, H. and R. SATO, (1970), J. Biochem. **67**, 199—210.

OHBA, M., R. SATO, Y. YOSHIDA, T. NISHINO, and H. KATSUKI, (1978), Biochem. Biophys. Res. Commun. **85**, 21—27.

ONO, T., K. TAKAHASHI, S. ODANI, H. KONNO, and Y. IMAI, (1980), Biochem. Biophys. Res. Commun. **96**, 522—528.

OSUMI, T., T. NISHINO, and H. KATSUKI, (1979), J. Biochem. **85**, 819—826.

POULOS, T. L., B. C. FINZEL, I. C. GUNSALUS, G. C. WAGNER, and J. KRAUT, (1985), J. Biol. Chem. **260**, 16122—16130.

Ryder, N. S., M. C. Dupont, (1984), Biochim. Biophys. Acta **794**, 466–471.
Stevenson, D. E., J. N. Wright, and M. Akhtar, (1988), J. Chem. Soc. Parkin Trans. 1988, 2043–2052.
Strittmatter, P., D. Spatz, D. Corcoran, M. J. Rogers, B. Setlow, and R. Redline, (1974), Proc. Nat. Acad. Sci. USA **71**, 4565–4567.
Tamura, Y., Y. Yoshida, R. Sato, and H. Kumaoka, (1976), Arch. Biochem. Biophys. **175**, 284–294.
Trocha, P. J., S. J. Jasne, and D. B. Sprinson, (1977), Biochemistry **16**, 4721–4726.
Vanden Bossche, H., G. Willemsens, P. Marichal, W. Cools, and W. Lauwers, (1984a), in: Mode of Action of Antifungal Agents, (Trinci, A. P. J. and J. F. Ryley, eds.), Cambridge University Press, London, pp. 321–341.
Vanden Bossche, H., W. Lauwers, G. Willemsens, P. Marichal, F. Cornelissen, and W. Cools, (1984b), Pestic. Sci. **15**, 188–198.
Vanden Bossche, H., (1985), in: Current Topics in Medical Mycology vol. 1, (McGinnis, M. R., ed.), Springer-Verlag, New York, pp. 313–351.
Watkinson, I. A. and M. Akhtar, (1969), J. Chem. Soc. Chem. Comm. **1969**, 206.
Yoshida, Y., H. Kumaoka, and R. Sato, (1974), J. Biochem. **75**, 1201–1210.
Yoshida, Y. and H. Kumaoka, (1975), J. Biochem. **78**, 785–794.
Yoshida, Y., Y. Aoyama, H. Kumaoka, and S. Kubota, (1977), Biochem. Biophys. Res. Commun. **78**, 1005–1010.
Yoshida, Y., Y. Imai, and C. Hashimoto-Yutsudo, (1982), J. Biochem. **91**, 1651–1659.
Yoshida, Y. and Y. Aoyama, (1984), J. Biol. Chem. **259**, 1655–1660.
Yoshida, Y., Y. Aoyama, T. Nishino, H. Katsuki, U. S. Maitra, V. P. Mohan, and D. B. Sprinson, (1985), Biochem. Biophys. Res. Commun. **127**, 623–628.
Yoshida, Y., Y. Aoyama, H. Takano, and T. Kato, (1986), Biochem. Biophys. Res. Commun. **137**, 513–519.
Yoshida, Y. and Y. Aoyama, (1987), Biochem. Pharmacol. **36**, 229–235.

Chapter 5

Cytochrome P-450 from *Rhizopus nigricans*

T. Hudnik-Plevnik and K. Breskvar

1. Introduction

Rhizopus nigricans ATCC 6227b is one of the best known biotechnologically important microorganisms, because of its ability to introduce a hydroxyl group into a steroidal ring system (PETERSEN and MURRAY, 1952). In 1949, glucocorticoids were shown to alleviate inflammatory joint diseases like rheumatoid arthritis and since then organic chemists have tried to prepare these compounds in the laboratory. The main difficulty was the introduction of the hydroxyl group into the steroid molecule, since the chemical method of hydroxylation is complex and difficult. The use of microorganisms like *Rhizopus nigricans* as chemical tools reduced the synthesis of glucocorticoids from 37 to only 11 steps: this represented a great improvement economically because it substantially reduced the cost of production. Although progesterone is the best substrate for 11α-hydroxylase of *Rhizopus nigricans* (PETERSEN and MURRAY, 1952), many other steroids can also be hydroxylated at different positions, (CHARNEY and HERZOG, 1967). The possibilities of applying this organism are very wide since the steroidal ring can be attacked by this organism at different positions and different stereochemical variants are possible (JONES, 1973).

The involvement of cytochrome P-450 in hydroxylations performed with the fungus *Rhizopus nigricans* ATCC 6227b was not known until 1977 (BRESKVAR and HUDNIK-PLEVNIK, 1977a). The hydroxylation system was shown to be inducible by the substrate progesterone (BRESKVAR and HUDNIK-PLEVNIK, 1978; 1981) and by some other steroids (HUDNIK-PLEVNIK and ČREŠNAR, in preparation). The resolution and the reconstitution of the hydroxylation system revealed the presence of three components: $P\text{-}450_{R.n.}$ as the terminal oxidase and a ferredoxin (rhizoporedoxin) and its reductase (NADPH-rhizoporedoxin reductase) as electron-carriers (ČREŠNAR et al., 1985; BRESKVAR et al., 1987). The purification of P-450 to electrophoretic homogeneity and its characterization was achieved in 1988 (HUDNIK-PLEVNIK and BRESKVAR, 1988). Partial purification and characterization of rhizoporedoxin was achieved in 1986 (ČREŠNAR, 1986). Rhizoporedoxin-reductase is, according to ČREŠNAR et al. (ČREŠNAR et al., 1985) an FAD-flavoprotein receiving electrons from NADPH.

The optimization of the production of progesterone 11α-hydroxylase by *R. nigricans*, i. e. the influence of O_2 and progesterone as substrates for P-450, was extensively studied by DUNNILL and his coworkers (HANISCH et al., 1980); the same author with coworkers succeeded in preparing functional microsomes on a large scale (TALBOYS and DUNNILL, 1985; 1985a; BROAD et al., 1986; BONNERJEA et al., 1988) and studied the hydroxylation of progesterone using immobilized organism for industrial purposes (MADDOX et al., 1981).

The aim of this chapter is, therefore, to inform the reader of the present status of knowledge about P-$450_{R.n.}$ and about the hydroxylation system in which this cytochrome participates; data evolving from both basic as well as applied research efforts is presented, this being considerably interdependent.

2. Biochemistry of cytochrome P-450 in *Rhizopus nigricans*

2.1. Involvement of cytochrome P-450 in hydroxylations performed by *Rhizopus nigricans*

All cytochrome P-450 species characterized until now are essential constituents of multiprotein complexes involved in a variety of oxidative reactions. Cytochrome P-450 binds substrate and molecular oxygen in a series of events involved in oxidation of the substrate. The involvement of such a system in a biochemical process is usually proved using a specific inhibitor acting on terminal oxidase of this multienzyme complex. One of the most generally used tests for the involvement of cytochrome P-450 in the hydroxylation is specific light reversible carbon monoxide inhibition of the studied reaction, although other specific inhibitors of the system are also known (MONTELLANO and REICH, 1986). Carbon monoxide inhibition was tested with homogenate from induced mycelia of *R. nigricans*. As shown in Table 1 the presence of CO in the gas mixture completely inhibited hydroxylation of progesterone (BRESKVAR and HUDNIK-PLEVNIK, 1977a). The inhibition was reversed to a significant level by irradiation of the reaction mixture with light, thus proving that 11α-hydroxylation of progesterone in *R. nigricans* is a cytochrome P-450 mediated process.

Table 1. Light reversible CO inhibition of progesterone hydroxylation (BRESKVAR and HUDNIK-PLEVNIK, 1977a) (Reproduced with permission from Academic Press Inc., New York)

Gas-phase	The composition of $CHCl_3$ extracts (%)		
	unreacted progesterone	11α-OH progesterone	6β-OH progesterone
Air	19	63	18
CO + air 3:1	100	0	0
N_2 3:1	25	55	20
CO + air 3:1 + light	33	53	14

Besides progesterone, *R. nigricans* hydroxylates several other steroid compounds. Studies concerned with the mechanism of substrate binding to hydroxylase (cf. section 3.) indirectly suggest that other steroid hydroxylations obtained with this organism are catalyzed by the same enzyme system and are, therefore, also cytochrome P-450 mediated.

2.2. Intracellular localization and some spectroscopic characteristics of membrane-bound cytochrome P-450 in *Rhizopus nigricans*

The intracellular localization of steroid hydroxylases in different organisms differs with respect to the organism and the tissue involved. Bacterial steroid hydroxylases are usually present in a soluble form (BERG et al., 1976) whereas in higher organisms, such as yeasts or mammalian steroidogenic organs, steroid hydroxylases are usually bound to the membrane and are present in the mitochondria or in the membranes of endoplasmatic reticulum of the cell (KÄPPELI, 1986).

In *R. nigricans* the intracellular localization of 11α-hydroxylase was studied by BRESKVAR and HUDNIK-PLEVNIK (BRESKVAR and HUDNIK-PLEVNIK, 1977; 1977a) and to some extent by BONNERJEA et al. (BONNERJEA et al., 1988). Subcellular fractions of induced fungus were tested for hydroxylation activity using progesterone as a substrate. Progesterone hydroxylase was located in the postmitochondrial fraction of the fungus; the same subcellular fraction was also rich in cytochrome P-450 and NADPH-cytochrome P-450 reductase (section 2.4.). It will be shown in section 2.4. that cytochrome P-450 and NADPH-cytochrome P-450-reductase were also found in the membrane fraction sedimenting at 105.000 g. In spite of the presence of all apparent components of 11α-hydroxylase it was never possible to detect the hydroxylation activity in this fraction (HUDNIK-PLEVNIK, unpublished data). Applying a different homogenization procedure and using the method of PEG precipitation of fungal microsomes, BONNERJEA et al. (BONNERJEA et al., 1988) succeeded in preparing the microsomal fraction of *R. nigricans* showing a significant hydroxylation activity compared to the total activity present in the cell homogenate.

In higher organisms the fraction sedimenting at 105.000 g is called the microsomal fraction. This subcellular fraction usually represents a mixture of membrane fragments originating from different cellular organelles but is predominantly composed of endoplasmatic reticulum (DEPIERRE and DALLNER, 1975). Microsomal fractions in higher organisms and also in eucaryotic microorganisms, such as yeasts, have been extensively studied (DEPIERRE and DALLNER, 1976; COBON et al., 1974); up to the present there has been no experimental data available on the biochemical properties of the corresponding subcellular fraction in filamentous fungi. In order to characterize the fraction which contained cytochrome P-450 and NADPH-cytochrome P-450 reductase of *R. nigricans* some biochemical properties of the 105.000 g sediment were determined (BRESKVAR and HUDNIK-PLEVNIK, in preparation). The membrane fraction was analysed for the activity of certain marker enzymes of different cell organelles, for the content of polar and nonpolar lipids, proteins and RNA. The results of this study indicate that 105.000 g sediment of *R. nigricans* does

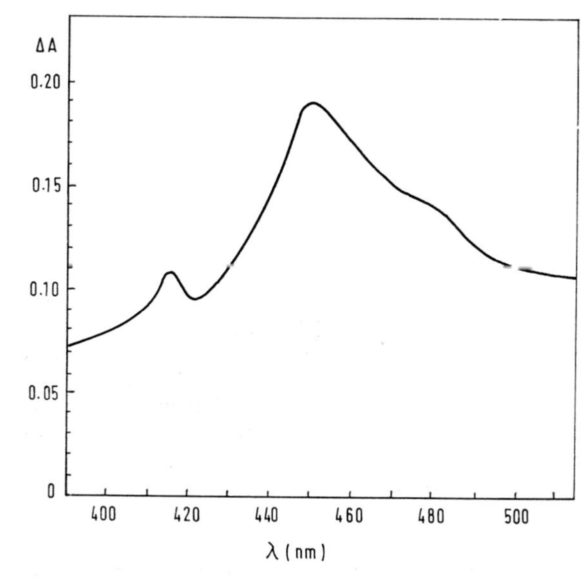

Fig. 1. The reduced CO difference spectrum of the fraction sedimenting at 105.000 g (BRESKVAR and HUDNIK-PLEVNIK, 1977). Reproduced with permission from Academic Press Inc., New York.

not contain nuclear and mitochondrial membranes, whereas the presence of plasma membrane could not be excluded. The membrane fraction shows certain characteristics of smooth endoplasmatic reticulum present in higher organisms; since there is very little data on the existence of endoplasmatic reticulum in this primitive eucaryotic organism the 105.000 g sediment in *R. nigricans* was simply called cytochrome P-450 containing microsomes.

Some spectroscopic properties of cytochrome P-450 bound to the microsomal membranes were determined (BRESKVAR and HUDNIK-PLEVNIK, 1977a). The reduced carbon monoxide difference spectrum with the induced fungal microsomes is presented in Figure 1. The spectrum proves that cytochrome P-450 is present in this subcellular fraction mostly in its active P-450 form. The relative content of its inactive P-420 form differs from one preparation of fungal microsomes to another.

In the presence of progesterone, fungal microsomal preparation exhibited the modified type II difference spectrum (Fig. 2) as defined by SCHENKMAN et al. (SCHENKMAN et al., 1967). The interaction of P-450 with progesterone does not result in a spectral change typical of substrate-enzyme interaction (type I interaction). The phenomenon could be explained by the fact that several agents present during the preparation of microsomal membranes could

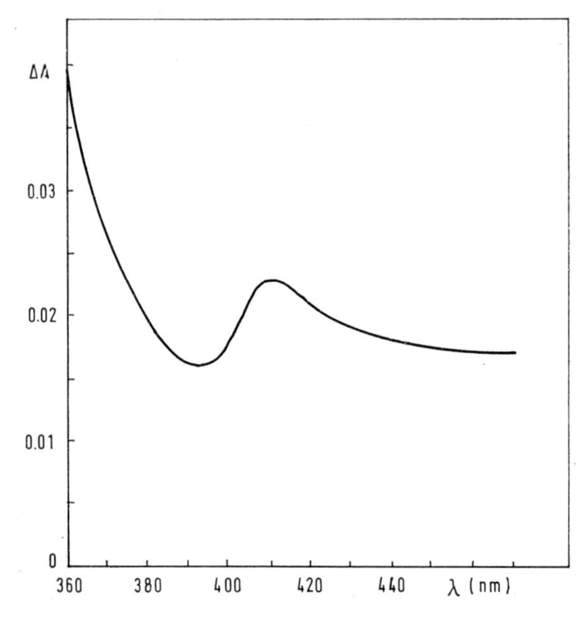

Fig. 2. Progesterone induced spectral difference of the fraction sedimenting at 105.000 g from *Rhizopus nigricans* (Breskvar and Hudnik-Plevnik, 1977).

induce the transition from the low-spin to the high-spin state in the absence of substrate (Ingelman-Sundberg, 1986) and could, therefore, interfere with the measured spectral change.

2.3. Solubilization of cytochrome P-450 from the microsomal membranes

11α-hydroxylase of *R. nigricans* is a very labile enzyme system. Steroid hydroxylase is unstable not only under conditions in which cell integrity is disrupted but also in intact cells either in suspension or immobilized on a support matrix (Broad et al., 1986; Maddox et al., 1981).

In studies performed with the aim of separating and isolating enzyme components of 11α-hydroxylase of *R. nigricans*, attention was paid to certain problems connected with solubilization of enzyme components from the fungal microsomal membranes (Breskvar and Hudnik-Plevnik, 1977a; Breskvar, 1983). Different procedures, each efficient in the solubilization of cytochrome P-450 from the membranes of mammalian tissues (Ryan et al., 1978; Kimura

Table 2. Solubilization of cytochrome P-450 from the microsomal fraction of induced *Rhizopus nigricans* (Breskvar and Hudnik-Plevnik, unpublished data)

Detergent	Solubilized protein [%]	P-450	
		Cyt P-450 solubilized nmol/mg prot	Cyt P-420 solubilized nmol/mg prot
Triton X-100 1%	70	0.04	0.08
Digitonin 1%	52	0.14	0.11
DOC 0.5%	62	0.134	0.05
Cholate 0.5%	50	0.16	0.05

et al., 1978) or from the microsomal membranes of yeasts (Yoshida et al., 1977), were tried for the same purpose with *R. nigricans*. None of these procedures was directly applicable for the solubilization of P-450 of *R. nigricans* in a stable form.

In order to find the best conditions for solubilization of cytochrome P-450 from microsomal membranes in *R. nigricans* a systematic study was undertaken (Breskvar and Hudnik-Plevnik, unpublished data). In the solubilization experiments different detergents were tried in various conditions. The parameters systematically examined were: the concentration of detergent, conditions of homogenization, detergent/protein ratio, duration and temperature of solubilization. A part of this study is presented in Table 2, showing the efficiency of solubilization of microsomal cytochrome P-450 with some detergents under optimal conditions concerning the listed parameters. Solubilization of cytochrome P-450 can be achieved with most of the detergents under optimal conditions. Several problems were solved in this study: i. with all the detergents tested a significant part of the enzyme was converted into its inactive P-420 form, and ii. typical for the above studies were very low yields and poor reproducibility. In these studies several proteolytic activities were detected in all fractions containing cytochrome P-450 (Hudnik-Plevnik and Virant, in preparation). The proteolytic activity found in the microsomal fraction increased several-fold upon solubilization of cytochrome P-450 with detergents. Using different inhibitors of proteinases it was possible to only partly inhibit the proteolytic activity. It was found that microsomal proteinase was successfully inhibited only at pH = 9 and higher. Under these extreme pH conditions it was impossible to solubilize cytochrome P-450 from the fungal microsomes.

2.4. Identification of the functional components by in vitro reconstitution of the system and purification of the enzymes

2.4.1. Reconstitution of the hydroxylation system

The three component nature of steroid 11α-hydroxylase in *R. nigricans* was first suggested by the successful separation of P-$450_{R.n.}$ from its reductase (Breskvar, 1983), followed by the characterization of reductase as a two component system consisting of flavoprotein and iron-sulfur protein (Črešnar et al., 1985).

The direct involvement of all three components, i. e. P-$450_{R.n.}$, rhizoporedoxin and rhizoporedoxin-reductase, in the fungal hydroxylation system was proved by reconstitution of the system on conditions (Breskvar et al., 1987). As illustrated in Figure 3 all three components of steroid hydroxylase were successfully separated from each other in a one step separation procedure by chromatography on DEAE-cellulose. Cytochrome P-450, NADPH-rhizopore-

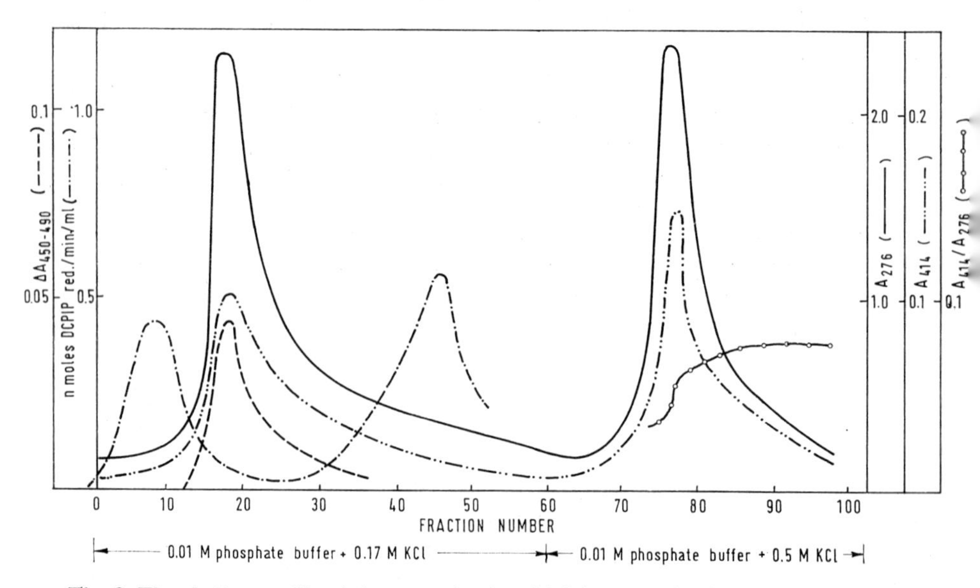

Fig. 3. The elution profile of the postmitochondrial fraction of induced fungus *Rhizopus nigricans* on DEAE-cellulose column (Breskvar et al., 1987). Reproduced with permission from Pergamon Press Ltd., Oxford.

Table 3. The activity of progesterone hydroxylase from *Rhizopus nigricans* in the postmitochondrial supernatant and in the reconstituted system. *One activity unit is defined as the amount of the enzyme catalyzing the reduction of 1 nmole cytochrome c per min by NADPH in the presence of excess rhizoporedoxin (Breskvar et al., 1987) (Reproduces with permission from Pergamon Press Ltd., Oxford)

Components present in incubation medium	Incubation condition		
	1	2	3
Postmitochondrial supernatant (mg prot.)	1.2	—	—
Cytochrome P-450 (nmoles)	—	0.1	0.1
Rhizoporedoxin reductase (units)*	—	8.0	8.0
Rhizoporedoxin (nmoles non-heme Fe)	—	—	0.12
NADPH oxidation (nmoles/min)	18	0	5

doxin-reductase and iron-sulfur protein were eluted from the column in three separated protein fractions suitable for the reconstitution assay. Subsequent reconstitution experiments (shown in Tab. 3) proved that all three components were necessary in order to reconstitute the functional electron transport chain, transferring electrons from NADPH to the substrate (Breskvar et al., 1987).

From the experiments presented above it is apparent that the functional electron transport chain can be reconstituted from three separated components. We still lack final evidence that cytochrome P-450$_{R.n.}$, NADPH-rhizoporedoxin reductase and rhizoporedoxin are also sufficient for the in vitro hydroxylation of progesterone, i. e. reconstitution of the functional hydroxylation system from purified components.

2.4.2. Purification of the components

The problem of solubilization of proteins from microsomal membranes was circumvented by using the postmitochondrial fraction as the starting subcellular extract and by omitting the use of detergents in all purification steps (Hudnik-Plevnik and Breskvar, 1988).

Cytochrome P-450 purification steps involved batch adsorption of postmitochondrial proteins on DEAE-cellulose, batch extraction of proteins with a high salt solution, fractionation of the extracted proteins on DEAE-cellulose and final purification by polyacrylamide slab gel electrophoresis (Hudnik-Plevnik and Breskvar, 1988). The enzyme was purified to a specific content of 8–19 nmoles/mg protein. Electrophoretic homogeneity on SDS-PAGE and Western blot analysis of the purified enzyme indicated the presence of a single molecular species (Breskvar and Hudnik-Plevnik, unpublished results). Other physico-chemical properties of the purified enzyme are described in Section 2.5.

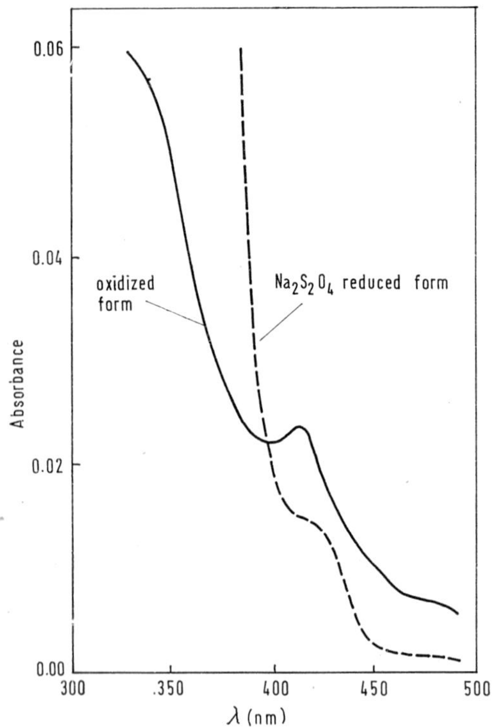

Fig. 4. Absorption spectra of the oxidized and reduced form of partially purified iron-sulfur protein from *Rhizopus nigricans* (Črešnar, 1986). Reproduced with permission from Academic Press Inc., New York.

The iron-sulfur protein was isolated and purified from postmitochondrial supernatant by chromatography on DEAE-cellulose and Sephadex G-100 (Črešnar, 1986). The presence of non-heme iron and acid-labile sulfur determined in the partly purified protein was almost equimolar. Molecular weight of the partly purified protein determined by SDS-PAGE and by chromatography on Sephadex G-100 was 12000 or 12400 daltons, respectively. The spectroscopic properties of partly purified protein are presented in Figure 4; the disappearance of the absorption maximum in the oxidized form at 415 nm upon reduction is typical for iron sulfur proteins. The EPR spectra necessary for the complete characterization of the purified protein are still lacking.

Flavin-containing NADPH-rhizoporedoxin reductase was partly purified from the postmitochondrial supernatant by two fractionation steps on DEAE-cellulose (Črešnar et al., 1985). FAD, as the coenzyme for the reductase, was determined by thin layer chromatography after extraction of the coenzyme from the partly purified enzyme (Črešnar et al., 1985).

2.5. Comparison of $P\text{-}450_{R.n.}$ with other P-450 species

Cytochrome P-450 of *Rhizopus nigricans* has until now been characterized only by its molecular weight, determined by SDS-PAGE, and by some spectroscopic characteristics (Hudnik-Plevnik and Breskvar, 1988; Hudnik-Plevnik and Breskvar, in preparation). The comparison with other P-450 species can, therefore, only be made on the basis of these physico-chemical properties. With the exception of bovine adrenal cortex mitochondria the other sources chosen for comparison of cytochromes P-450 were microorganisms. One can conclude from the data presented in Table 4 that $P\text{-}450_{R.n.}$ has a molecular weight of 46000 daltons, like the P-450 of *Pseudomonas putida* ($P\text{-}450_{cam}$), *Rhizobium japonicum*-c ($P\text{-}450_c$) and of adrenal mitochondria ($P\text{-}450_{11\beta}$). The P-450 of *Saccharomyces cerevisiae* ($P\text{-}450_{S.c.}$) and of *Bacillus megaterium* ($P\text{-}450_{meg}$) have significantly higher M_r compared to $P\text{-}450_{R.n.}$. If absorption maxima of CO-difference spectra of the reduced P-450 are compared, the close similarity of $P\text{-}450_{R.n.}$ to $P\text{-}450_{cam}$ is again apparent, whereas cytochromes from other sources differ from $P\text{-}450_{R.n.}$ to some extent. This observation could, according to Black and Coon (Black and Coon, 1986), indicate that the heme-binding sites of P-450 of *Rhizopus nigricans* and of

Table 4. Comparison of $P\text{-}450_{R.n.}$ with other P-450 species.

P-450 species	$Fe^{II}CO$ λ_{max} (nm)	M_r(av) SDS-PAGE	References
$P\text{-}450_{R.n.}$	446	46000	Hudnik-Plevnik and Breskvar, 1988
$P\text{-}450_{cam}$	446	46000	Dus et al., 1970
$P\text{-}450_c$	447	46000	Appleby, 1978
$P\text{-}450_{11\beta}$	448	46000	Katagiri et al., 1978
$P\text{-}450_{meg}$	450	52000	Berg et al., 1979
$P\text{-}450_{S.c.}$	447	58000	Yoshida and Aoyama, 1984

Table 5. Spectral properties of P-450 of *Rhizopus nigricans* (Hudnik-Plevnik and Breskvar, in preparation)

Form	Absorption peak λ (nm)
ferric	527–529
	408–409
	358
	278
ferrous	548
	520
	412
ferrous, CO bound	446

Pseudomonas putida are similar but differ from the heme-binding sites of the other cytochromes compared.

The absorption characteristics of the oxidized and reduced forms of $P\text{-}450_{R.n.}$ are shown in Table 5. From the comparison of this data with analogous data published for other P-450 species some general conclusions can be drawn. There are no essential differences between P-450 species in the ultraviolet region of the absorption spectrum. We notice, however, a great difference between $P\text{-}450_{R.n.}$ and all other P-450 species with respect to the gamma-Soret band of the oxidized form and, to some extent, to the alpha- and beta-Soret bands which are all shifted to shorter wavelenghts compared to other P-450 cytochromes. No definite explanation for this phenomenon can be given in this respect until the EPR spectrum of $P\text{-}450_{R.n.}$ has been determined.

2.6. Induction of $P\text{-}450_{R.n.}$

The induction of P-450 with progesterone has been described by Breskvar and Hudnik-Plevnik (Breskvar and Hudnik-Plevnik, 1981). As illustrated in Figure 5 the conversion rate of progesterone is closely correlated with the rate of increase of the P-450 content in the fungus. The inducible nature of P-450 in *R. nigricans* was also suggested by the boservation that this cytochrome was never shown to be in microsomes of noninduced mycelia (Breskvar and Hudnik-Plevnik, 1981). In all other studies the induction of P-450 has been followed only by an estimation of the hydroxylation activity, which has been shown to be P-450 dependent (Breskvar and Hudnik-Plevnik, 1977a). Besides progesterone other steroids also act as inducers; this is indicated by two lines of evidence. All bioconversions described until 1977 were performed by adding the steroid to the growing culture and by analysing the transformation products several hours later. Since nowadays it is known that the hydroxylase in *R. nigricans* is an inducible enzyme system (BRESKVAR and HUDNIK-PLEVNIK,

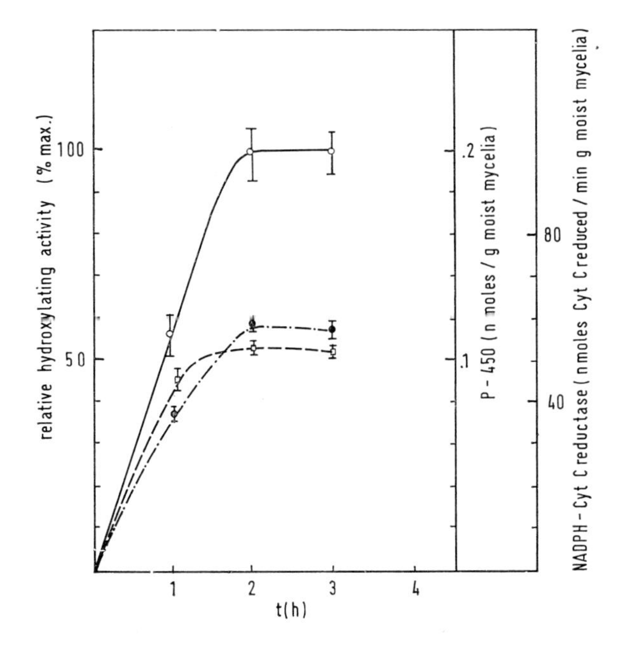

Fig. 5. The kinetics of induction of the 11α-hydroxylase and its enzyme constituents in *Rhizopus nigricans* (BRESKVAR and HUDNIK-PLEVNIK, 1981). Reproduced with permission from Pergamon Press Ltd., Oxford.

1978; HANISCH et al., 1980) one can conclude that all the steroids described until now and which were successfully transformed by *R. nigricans* also acted as inducers to varying extents. A systematic study of inducing steroids was performed by HUDNIK-PLEVNIK and ČREŠNAR (HUDNIK-PLEVNIK and ČREŠNAR, in preparation), who induced the 11α-hydroxylase of progesterone on standard conditions with a series of steroids and followed the rate of transformation of progesterone in the presence of cycloheximide. They showed that deoxycorticosterone and testosterone were better inducers than compared to progesterone, whereas all the other steroids tested were much less efficient (HUDNIK-PLEVNIK and ČREŠNAR, in preparation).

The induction of P-450 in *R. nigricans* requires de novo synthesis of protein. This was shown in experiments in which the induction of the 11α-hydroxylase was inhibited in the presence of cycloheximide and was not significantly affected by chloramphenicol (BRESKVAR and HUDNIK-PLEVNIK, 1978; HANISCH et al., 1980).

Similar as in yeasts (Mauersberger et al., 1980), the induction of P-450 also depends on the concentration of oxygen in the medium (Hanisch et al., 1980). The pattern of dependence of the induction process on pO_2 is, however, very different in the two types of microorganisms. Whereas induction proceeds very efficiently in yeasts, even at extremely low pO_2, for induction of $P\text{-}450_{R.n.}$ a minimal pO_2 of 4% air saturation is obligatory and a clear optimum is observed at about 10% air saturation (Hanisch et al., 1980).

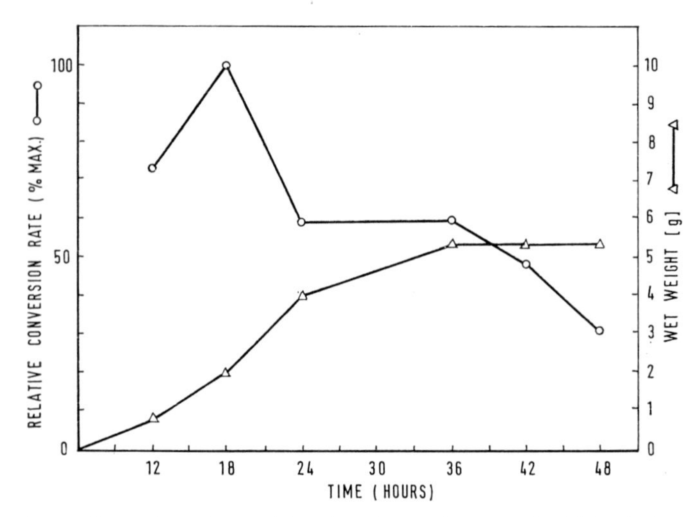

Fig. 6. The effect of age of the fungal mycelium on its ability to transform progesterone. *Rhizopus nigricans* was grown for different periods of time, induced for 2 h with progesterone and assayed for hydroxylation activity (Hudnik-Plevnik and Črešnar, in preparation)

The amount of the induced P-450 is influenced by the concentration of the inducer. If induction was performed with progesterone dissolved in 0.01%. Tween 80 after micronization the induced hydroxylation activity of incubation medium reached a plateau at 0.5 g/l and did not change on increasing further the concentration of progesterone (Hanisch et al., 1980). These experiments were performed with *R. nigricans* grown in the fermentor (Hanisch et al., 1980). When the mycelium was grown in shake flasks and deoxycorticosterone dissolved in dimethylformamide was used for induction, an optimal concentration of the inducer at 100 μg/ml of medium was observed and higher concentrations of the inducer lowered the yield of P-450 (Hudnik-Plevnik and Črešnar, in preparation). The optimal age of mycelium for the induction of P-450 seems to be the late logarithmic phase of growth if the hydroxylation activity is expressed on the basis of wet weight mycelium (Fig. 6) (Hudnik-Plevnik and Črešnar, in preparation).

The biological role of P-450 induction and the mechanism of induction are mentioned only briefly since only a few studies have been made in connection with these problems. As can be concluded from the data presented in Sections 2. and 3., P-450 is involved in the introduction of hydroxyl group(s) into the steroidal ring system, making the molecule more hydrophilic. Experiments in which radioactively labelled progesterone was used as the substrate showed that progesterone was concentrated in different parts of the fungal mycelium (Hudnik-Plevnik, unpublished data). As the induction proceeded during the incubation process the radioactivity was removed from the mycelium; this was not the case if cycloheximide was present in the incubation medium (Hudnik-Plevnik, unpublished data). The explanation for this observation is that the microorganism uses the hydroxylation reaction to remove a foreign steroid which, probably because of its hydrophobicity, binds to cellular membranes and interferes with cellular activities.

There is some indication that in *Rhizopus nigricans*, as in higher organisms, the mechanism of induction operates through specific steroid receptors. Progesterone binding proteins with a $K_d = 3 \times 10^{-9}$ and $n = 100-130$ fmoles/mg protein were found in the cytosolic and microsomal fractions of the fungus (Hudnik-Plevnik and Lenasi, in preparation). The physiological role of these receptors in *Rhizopus nigricans* has yet to be established.

3. Hydroxylation of the steroidal ring system

Steroids which are efficiently hydroxylated by *Rhizopus nigricans* ATCC 6227b were reviewed by Charney and Herzog (Charney and Herzog, 1967); we shall, therefore, concentrate primarily on the information evolving from the studies of the influence of structural factors on the course of steroid hydroxylation by this microorganism.

3.1. Influence of the oxygen functions attached to the ring system

The hydroxylation pattern of the steroid substrate with *Rhizopus nigricans* follows certain rules – as has been shown by Jones and his coworkers (Jones, 1973; Browne et al., 1973; Chambers et al., 1973; Charney et al., 1967; Evans, 1969). The presence of a carbonyl group in the ring system is necessary for the attachment of the substrate to the enzyme since acetals and enol ethers derived from oxoandrostanes are not hydroxylated by the fungus (Evans et al., 1975). Using mono-oxo-5-androstane-3-one and -5-estrane-3-one as substrates, Browne et al. (Browne et al., 1973) discovered that the microorganism very efficiently attacks the 16-position, something which has never been observed with the previously studied substrates bearing 17β-oxygenated side-chains. From mono-oxo-steroids dihydroxy-derivatives are obtained carrying hydroxyl groups at defined positions. 11,16-dihydroxylations occur with 3-ketones and 3,7-dihydroxylation with 17-ketones. The positions preferentially attacked are, therefore, 3 (or 4), 6 (or 7), 11 and 16 (JONES, 1973). On the basis of these findings it was concluded that two modes of hydroxylation are in operation: positions 3, 11 and 16 are involved in the normal mode, and positions 16, 3 and 7 in the reversed mode (JONES, 1973). Since good substrates were 3- and 16-ketones and poor substrates 7- and 11-ketones it was suggested that the binding is less efficient with middle ring ketones (JONES, 1973). With both diketones and keto-alcohols the hydroxylations of the steroidal ring system are highly specific. Substrates with one oxygen substituent in each of the terminal ring are attacked namely at position 11 or 7; those with one keto group in rings B or C are hydroxylated at position 16 if the second group is in ring A and at position 3 if the second group is in ring D (CHAMBERS et al., 1973). The authors speculate that the substrate orients itself so as to maximize hydrophilic binding between its two substituents and two of the enzyme sites; the remaining site then becomes hydroxylated at the nearest C-atom (CHAMBERS et al., 1973). The symmetrical disposition of these sites allows a suitable disubstituted steroid to take up more than one orientation; as mentioned above two such orientations were termed the normal and the reversed mode.

Another important finding was that substitution at position 11 leads exclusively to 11α- products, whereas hydroxylation at other common positions (3,7 and 16) gives α- or β- products, or a mixture of both (CHAMBERS et al., 1973). The site of hydroxylation not only depends on the position of the substrate substitutent but also on the oxidation level: the 17β-hydroxy-3-keto substrate resulted in the 11α-hydroxylated product whereas the 3β-hydroxy-17-keto substrate gave preferentially the 7-hydroxylated product (CHAMBERS et al., 1973).

3.2. Effects of the structure of the side-chain at C-17

Whereas JONES and coworkers (JONES, 1973) studied the bioconversion of steroids with *Rhizopus nigricans* by adding the substrates to the microbial culture and analysing the products several hours later, ŽAKELJ et al. (ŽAKELJ et al., 1986, 1989; ŽAKELJ and BELIČ, 1987) used in their studies fungal mycelia in which the 11α-hydroxylase was induced with progesterone on standard conditions and the bioconversion was performed in the presence of cycloheximide. In this way the possibility of inducing any other P-450 species with the use of different substrates was excluded. This new approach also makes it possible to gain information on the specificity of the induced 11α-hydroxylase.

The structure of the side-chain seems to be very important for the extent of the bioconversion and for the position(s) hydroxylated in the ring system. In all experiments the bioconversion proceeded optimally if progesterone was used as the substrate, suggesting that the enzyme was relatively specific for progesterone (Žakelj and Belič, 1987). A higher polarity of the side chain (17α-21-dihydroxy-progesterone, testosterone) has an inhibitory effect on the total hydroxylation and increases the hydroxylations at 6β- and 7β- relative to 11-α. The presence of the ethynyl-group completely prevents the hydroxylation at C-11 (Žakelj et al., 1986; Žakelj and Belič, 1987). Ethisterone is converted to 7β-hydroxyethisterone, whereas norethisterone gives 10β- and 6β-hydroxy-derivatives (Žakelj et al., 1986).

3-Keto-4-ene steroids with a nonpolar side chain are hydroxylated poorly or not at all (Žakelj et al., 1989), thus suggesting the importance of the oxygen function in the side chain of the steroid substrate. Hydroxylation by the 11α-hydroxylase of *Rhizopus nigricans* is totally prevented if substrates containing more than 2 C-atoms in the side chain are used in the bioconversion. Experiments with purified $P\text{-}450_{R.n.}$ should show if the reason for this phenomenon is interference with the enzyme-substrate interaction or binding takes place normally but hydroxylation is inhibited.

4. Industrial implications

Two main trends are apparent in the biotechnology of 11α-hydroxy-progesterone. One is concerned with the improvement of the existent technological process taking place in large incubation vessels, the other with the introduction of new procedures in which hydroxylation occurs either with immobilized cells or with immobilized microsomes.

4.1. Optimization of the production of 11α-hydroxyprogesterone for the synthesis of glucocorticoids

The yield of biotechnological production of 11α-hydroxyprogesterone is influenced by the induction of the hydroxylating enzyme system and the activity of the synthesized enzymes. Progesterone and oxygen, both substrates for 11α-hydroxylase, were also found to affect the synthesis of hydroxylase (Hanisch et al., 1980). The optimal induction of hydroxylase in the laboratory fermentor was obtained at 0.5 g/l (or higher) of progesterone and at dissolved oxygen tension (DOT) of 10% of air saturation.

The activity of induced hydroxylase was also markedly influenced by oxygen, the optimal DOT for hydroxylation being much higher (about 21%) than the DOT necessary for the induction. By studying the separate effects of oxygen on the synthesis and on the activity of 11α-hydroxylase it was found that the best production of 11α-hydroxyprogesterone in the fermentor was obtained by inducing the enzymes at a low DOT, followed by an increase of DOT after the induction was completed (Hanisch et al., 1980).

4.2. Use of immobilized cells and of microsomes for the production of 11α-hydroxyprogesterone

The use of immobilized cells as an alternative to the use of free cells for 11α-hydroxylation of progesterone was studied by MADDOX et al. (MADDOX et al., 1981). *R. nigricans* cells immobilized in either alginate or agar as support materials were capable of hydroxylation, whereas no activity was detected in the polyacrylamide supported cells. Agar supported cells showed higher hydroxylation rates and higher yields. The main difficulty in the process of immobilization of *R. nigricans* cells was that the hydroxylation activity could not be maintained for longer periods even by changing the medium at regular intervals. One possible explanation for the observed instability of hydroxylase in the immobilized cells is that the cells kept in the stationary phase cannot metabolize normally for indefinite periods (MADDOX et al., 1981).

The other alternative to the cell suspension method in 11α-hydroxyprogesterone production is the use of immobilized microsomes. For this purpose the conventional procedure for microsomal preparation had to be replaced by some other procedure acceptable for a large-scale process. The problem of disruption of fermentor-grown cells of *R. nigricans* was solved by TALBOYS and DUNNILL (TALBOYS and DUNNILL, 1985, 1985a). By using a specially designed concentric cylinder viscometer with a defined gap the above authors succeeded in establishing the conditions for cell homogenization in which the cell wall was disrupted and the activity of the released hydroxylase retained.

The microsomal fraction was isolated from debris-free homogenate with two techniques which could both be used on a process scale: poly(ethylene)glycol (PEG) precipitation and two aqueous phase separation. Both methods resulted in relatively low yields of isolated hydroxylase: 14.4% for PEG and 24% for the two aqueous phase method (BONNERJEA et al., 1988). Although both methods appear promising as an initial purification and concentration step further studies are required in order to optimize the procedure.

5. Conclusions

The purification of cytochrome P-450 from *Rhizopus nigricans*, for which the designation $P\text{-}450_{R.n.}$ is proposed, represents the first successful preparation of this enzyme from a filamentous fungus. The great problems encountered during the purification process, due to the unusual behaviour of the fungal enzyme on the one hand and the presence of a latent protease in the same membrane fraction as P-450 on the other, have been satisfactorily solved. The way is now open for investigations of the $P\text{-}450_{R.n.}$ structure and for studies of physico-chemical and immunological properties. The availability of specific antibodies will facilitate the use of recombinant DNA technology in studies of

P-450 gene structure and its regulation by steroids. Furthermore, the use of genetic engineering in biotechnology of 11α-hydroxylase is now possible.

As for the hydroxylation system in *Rhizopus nigricans,* the complete electron transport chain has been elucidated. Reducing equivalents are provided to P-$450_{R.n.}$ via a membrane bound electron transport system. Electrons from NADPH are transferred to an FAD-containing flavoprotein, for which the name NADPH-rhizoporedoxin reductase has been proposed. This enzyme, in turn, transfers electrons to an iron-sulfur protein (rhizoporedoxin), which then transfers electrons to P-$450_{R.n.}$. The electron transport system of *Rhizopus nigricans* resembles the soluble systems found in the bacteria *Pseudomonas putida* and *Bacillus megaterium* and the membrane bound mitochondrial 11β-hydroxylation system of the mammalian adrenal cortex. Kinetic studies with purified components of the electron transport chain will help to explain the substrate specificity and to study the mechanism of the hydroxylation reaction.

A survey of the recent achievements in P-$450_{R.n.}$research might be advantageous for those who are involved in the exploitation of this organism for biotechnological purposes. In this respect two points should be condidered. Because of the membrane bound protease the use of immobilized microsomes is highly questionable unless an efficient inhibitor is found. A much more promising approach for the future seems to be the use of an immobilized system constructed from the purified components; the purification of P-$450_{R.n.}$ will certainly undergo further simplification. The unsolved problem connected with this approach, however, remains the preparation of a protease-free undissociated P-450-reductase.

6. References

Appleby, C. A., (1978), Methods Enzymol. **52**, 147–166.

Berg, A., J.-A. Gustafsson, and M. Ingelman-Sundberg, (1976), J. Biol. Chem. **251**, 2831–2838.

Berg, A., M. Ingelman-Sundberg, and J. A. Gustafsson, (1979), J. Biol. Chem. **52**, 5264–5271.

Black, S. D. and M. J. Coon, (1986), in: Cytochrome P-450-structure, mechanism and biochemistry, (deMontellano, P. R. O., ed.), Plenum Press, New York. 161–216.

Bonnerjea, J., S. Pontin, M. Hoare, and P. Dunnill, (1988), Appl. Microbiol. Biotechnol. **27**, 362–365.

Breskvar, K. and T. Hudnik-Plevnik, (1977), Croatica Chemica Acta **49**, 207–212.

Breskvar, K. and T. Hudnik-Plevnik, (1977a), Biochem. Biophys. Res. Commun. **74**, 1192–1197.

Breskvar, K. and T. Hudnik-Plevnik, (1978), J. steroid Biochem. **9**, 131–134.

Breskvar, K. and T. Hudnik-Plevnik, (1981), J. steroid Biochem. **14**, 395–399.

Breskvar, K., (1983), J. steroid Biochem. **18**, 51–53.

Breskvar, K., B. Črešnar, and T. Hudnik-Plevnik, (1987), J. steroid Biochem. **26**, 499–501.

Broad, D. F., S. Pontin, and P. Dunnill, (1986), Enzyme Microb. Technol. **9**, 546–548.

BROWNE, J. W., W. A. DENNY, E. R. H. JONES, G. D. MEAKINS, Y. MORISAWA, A. PENDLEBURY, and J. PRAGNELL, (1973), J. Chem. Soc. Perkin Trans I. **104**, 1493 to 1499.

CHAMBERS, V. E. M., W. A. DENNY, J. M. EVANS, E. R. H. JONES, A. KASAL, G. D. MEAKINS, and J. PRAGNELL, (1973), J. Chem. Soc. Perkin Trans. I. **105**, 1500—1511.

CHARNEY, W. and H. L. HERZOG, (1967), Microbiological transformations of steroids, Academic Press, New York.

COBON, G. S., P. D. CROWFOOT, and A. W. LINNANE, (1974), Biochem. J. **144**, 265—275.

ČREŠNAR, B., K. BRESKVAR, and T. HUDNIK-PLEVNIK, (1985), Biochem. Biophys. Res. Commun. **133**, 1057—1063.

ČREŠNAR, B., (1986), PhD. Thesis, University of Ljubljana.

DEMONTELLANO, P. R. O. and N. O. REICH, (1986), in: Cytochrome P-450-structure, mechanism and biochemistry, (DE MONTELLANO, P. R. O., ed.), Plenum Press, New York, 273—314.

DEPIERRE, J. W. and G. DALLNER, (1975), Biochim. Biophys. Acta **415**, 411—472.

DEPIERRE, J. W. and G. DALLNER, (1976), in: Biochemical analysis of membranes, (MADDY, A. H., ed.), Chapman & Hall, London 79—131.

DUS, K. M., M. KATAGIRI, C. A. YU, D. L. ERBES, and I. C. GUNSALUS, (1970), Biochem. Biophys. Res. Commun. **40**, 1423—1430.

EVANS, J. M., R. H. JONES, A. KASAL, V. KUMAR, G. D. MEAKINS, and J. WICHA, (1969), Chem. Comm., 1491.

EVANS, J. M., E. R. H. JONES, G. D. MEAKINS, J. O. MINERS, A. PENDLEBURY, and A. L. WILKINS, (1975), J. Chem. Soc. Perkin Trans. I., 1356—1359.

HANISCH, W. H., P. DUNNILL, and M. D. LILLY, (1980), Biotechnol. Bioeng. **22**, 555 to 570.

HUDNIK-PLEVNIK, T. and K. BRESKVAR, (1988), Cytochrome P-450 in biochemistry and biophysics, (SCHUSTER, I., ed.), Taylor & Francis, Philadelphia, in press.

INGELMAN-SUNDBERG, M., (1986), in: Cytochrome P-450 — structure, mechanism and biochemistry, (DE MONTELLANO P. R. O., ed.), Plenum Press, New York. 119—160.

JONES, E. R. H., (1973), Pure Appl. Chem. **33**, 39—52.

KÄPPELI, O., (1986), Microbiol. Reviews **50**, 244—258.

KATAGIRI, M., S. TAKEMORI, E. ITAGAKI, and K. SUHARA, (1978), Methods Enzymol. **52**, 124—132.

KIMURA, T., J. H. PARCELLS, and H.-P. WANG, (1978), Methods Enzymol. **52**, 132—142.

MADDOX, I. S., P. DUNNILL, and M. D. LILLY, (1981), Biotechnol. Bioeng. **23**, 345—354.

MAUERSBERGER, S., R. N. MATYASHOVA, H.-M. MULLER, and A. B. LOSINOV, (1980), Europ. J. Appl. Microbiol. Biotechnol. **9**, 285—294.

PETERSEN, D. H. and H. C. MURRAY, (1952), J. Amer. Chem. Soc. **74**, 1871.

RYAN, D., A. Y. H. LU, and W. LEVIN, (1978), Methods Enzymol. **52**, 117—123.

SCHENKMAN, J. B., H. REMMER, and R. W. ESTABROOK, (1967), Mol. Pharmacol. **3**, 113—123.

TALBOYS, B. L. and P. DUNNILL, (1985), Biotechnol. Bioeng. **27**, 1726—1729.

TALBOYS, B. L. and P. DUNNILL, (1985a), Biotechnol. Bioeng. **27**, 1730—1734.

YOSHIDA, Y., Y. AOYAMA, H. KUMAOKA, and S. KUBOTA, (1977), Biochem. Biophys. Res. Commun. **78**, 1005—1010.

YOSHIDA, Y., Y. AOYAMA, (1984), J. Biol. Chem. **259**, 1655—1660.

ŽAKELJ-MAVRIČ, M., I. BELIČ, and H. E. GOTTLIEB, (1985), FEMS Microbiol. Letters **33**, 117—120.

ŽAKELJ-MAVRIČ, M. and I. BELIČ, (1987), J. steroid Biochem. **28**, 197—201.

Chapter 6

Genetic Engineering on Cytochrome P-450 Monooxygenases

Y. Yabusaki and H. Ohkawa

1. Introduction

Cytochrome P-450 (P-450) monooxygenases are distributed widely in organisms from prokaryotes to eukaryotes and engaged in the oxidative metabolism of a wide variety of structurally unrelated lipophilic compounds. The enzymes are classified into three types by their functions (Fig. 1). Bacterial enzymes are soluble and responsible for the oxidative catabolism of substrate compounds utilized for bacterial growth. Mitochondrial enzymes in mammalian steroido-

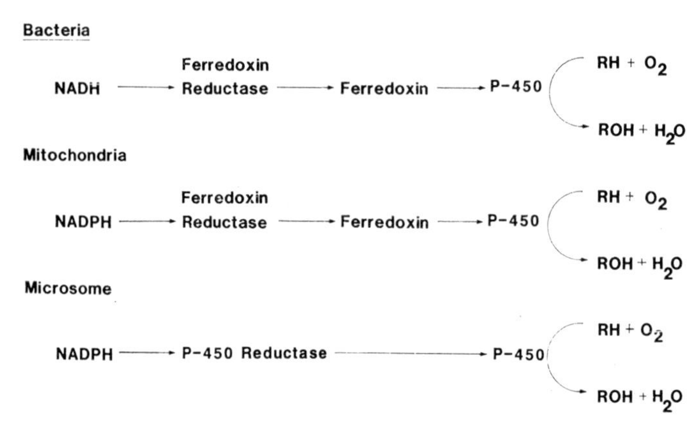

Fig. 1. Three types of P-450 monooxygenases.

genic tissues are membrane-bound and play an important role in the biosynthesis of steroid hormones. These two types of enzymes have their own distinct substrate specificities. On the other hand, microsomal enzymes mostly occurring in mammalian livers metabolize endogenous substrates such as steroids, fatty acids and prostaglandins as well as foreign chemicals such as drugs, carcinogens and environmental contaminants. These three types of P-450 monooxygenases also differ from each other in the components of their electron transport chains. As shown in Figure 1, the electron transport chains in bacteria and mitochondria involve a flavoprotein reductase and an iron-sulfur protein (ferredoxin) for electron transfer from NADH or NADPH to P-450. On the other hand, the microsomal chains perform direct electron transfer from NADPH to P-450 by the catalysis of NADPH-cytochrome P-450 reductase, a flavoprotein containing one molecule each of FAD and FMN. The reaction versatility of the microsomal enzymes responsible for the metabolism of foreign chemicals is due, in part, to the multiplicity of the P-450 molecules with different but overlapping broad substrate specificities. Although P-450 species differ in their catalytic activities, they receive electrons from a common species of either flavoprotein or ferredoxin.

Recent progress in recombinant DNA technology clarified the multiplicity of P-450 at the molecular level. So far, about 70 complete cDNA clones for P-450 molecules have been isolated from various sources and sequenced for the determination of their primary structures. Comparison of the deduced amino acid sequences of a number of P-450 species revealed that P-450 molecules share several common structural features such as a conserved cysteine-containing peptide for heme binding, and that the enzymes consist of at least 10 protein families, based on the sequence similarity, which are derived from a common ancestor (ADESNIK and ATCHINSON, 1986; BLACK and COON, 1986, 1987; NEBERT and GONZALEZ, 1987; GOTOH and FUJII-KURIYAMA, 1988).

Genetic engineering technology also enabled the expression of P-450 genes in heterologous host cells which inherently contain little or no P-450. The functional expression of P-450 enzymes afforded the opportunity to perform identification, determination of enzymatic properties and functional analysis of P-450 monooxygenases. This technology could deal with one P-450 gene product without interference from coexisting homologous enzymes. More sophisticated genetic engineering techniques such as site-directed mutagenesis and construction of chimeric molecules were used for the analysis of structure-function relationships of P-450 monooxygenases. These techniques even clarified the function of one specific amino acid residue in the enzyme molecule, although it was rather difficult to identify the modified residue(s) by chemical modification. In addition, the technology for gene expression made it possible to produce the enzymes in large quantities in heterologous host cells and to apply the enzymes to bioconversion processes of a wide variety of industrially important chemicals. The protein engineering of P-450 monooxygenases is another important approach to analyze structure-function relationships and to improve the enzymes in the reactivity and stability for practical purposes. In this article we review the genetic engineering on P-450 monooxygenases that aids us in a fundamental understanding of P-450 monooxygenases and the possible application of the enzymes to practical bioconversion processes.

2. Expression of P-450 monooxygenase genes in heterologous hosts

2.1. *Escherichia coli*

Escherichia coli is one of the best organized host strains in recombinant DNA technology for the production of heterologous proteins. *E. coli* cells grow fast and are easy to be treated. Bacterial P-450s are soluble enzymes and are produced as enzymatically active ones in *E. coli* by genetic engineering (Tab. 1). The *Pseudomonas putida* P-450cam gene under the control of **lac** promoter, and the *Bacillus megaterium* P-450$_{BM-3}$ gene under its own promoter, were

Table 1. Expression of P-450 monooxygenase genes in *E. coli*

P-450 Form	Source	Vector (Promoter)	Remarks	Reference
P-450cam	*P. putida*	pUC, pEMBL (own promoter)	I, S, N, A(camphor)	UNGER et al. (1986)
P-450$_{BM-3}$	*B. megaterium*	pUC(own promoter)	I, S, N, A(fatty acid)	WEN and FULCO (1987)
P-450c	Rat liver	(lac, tac)	I	NAKAMURA et al. (1984)
P-450 Reductase	Rat liver	pCQV2(λPR)	I, S A(cytochrome c)	PORTER et al. (1987)

I: Immunodetection; S: Spectral properties; N: Amino-terminal amino acid sequence; A: Enzyme activity (substrate in parenthesis).

expressed in *E. coli* cells to produce the corresponding P-450 molecules active in vitro in camphor 5-**exo**-hydroxylation and myristate (ω-2)-hydroxylation, respectively (UNGER et al., 1986; WEN and FULCO, 1987). The recombinant P-450$_{BM-3}$ enzyme was purified from the *E. coli* cells to show the same properties as the enzyme purified from the original strain (NARHI et al., 1988).

Eukaryotic P-450 monooxygenases are associated with microsomal or mitochondrial membranes. While *E. coli* cells contain no intracellular membrane systems, it was attempted to express rat liver microsomal P-450c cDNA in *E. coli* cells directly or as a fused enzyme with β-lactamase. The synthesized protein in the *E. coli* cells contained no heme in the molecule (NAKAMURA et al., 1984). PORTER et al. (1987) also expressed rat microsomal NADPH-cytochrome P-450 reductase cDNA in *E. coli*. Immunoblot analysis with anti-reductase Ig revealed that the recombinant *E. coli* cells produced the reductase protein, 10% of which had the same molecular weight on SDS-polyacrylamide gel electrophoresis as that of the mature enzyme and the rest of them were degraded to smaller peptides, indicating that the reductase produced was fairly unstable in the *E. coli* cells. From these results, the *E. coli* cells appear to be suitable for the expression of bacterial soluble enzymes but not for the mammalian membrane-bound ones.

2.2. Yeast

Yeast contains intracellular organels similar to mammalian cells and can be well used as a simple model of eukaryotic cells. OEDA et al. (1985) first succeeded in the functional expression of a mammalian liver microsomal P-450 cDNA in yeast. Rat P-450c cDNA was placed between alcohol dehydrogenase I (ADH) promoter and terminator regions of the expression vector pAAH5 (AMMERER, 1983) to construct an expression plasmid, which was introduced into *Saccharomyces cerevisiae* AH22 cells by the LiCl method. The LiCl method was recommended for yeast transformation because of its simple procedure and transformation efficiency, similar to that of the protoplast method (ITO et al., 1983). The transformed yeast cells produced P-450c holoenzyme at the level of 4×10^5 molecules per cell as determined by the reduced CO-difference spectrum. Although yeast microsomes possess endogenous P-450 responsible for ergosterol biosynthesis (YOSHIDA and AOYAMA, 1984; AOYAMA et al., 1984; Kalb et al., 1987) and P-450 species inducible under the culture conditions of a high glucose concentration (WISEMAN and WOODS, 1979; WENZEL et al., 1988), these contents were extremely low under the aerobic culture conditions used for the experiment. Actually, no P-450 peak was detected in the spectrum of the whole yeast cells transformed with the vector plasmid carrying no P-450c cDNA. Thus, the P-450 hemoprotein detected in the transformed yeast cells was concluded to be rat P-450c produced therein. The produced rat P-450c was mainly located in the yeast microsomal fraction. Both whole cells and microsomal fraction of the transformed yeast cells showed P-450c-dependent monooxygenase activities toward acetanilide, 7-ethoxycoumarin and benzo(a)pyrene (OEDA et al., 1985; SAKAKI et al., 1986). Therefore, the recombinant P-450c was found to interact with yeast P-450 reductase in the microsomes to exhibit P-450c-dependent monooxygenase activity. The P-450c preparation purified from the recombinant yeast microsomes had the same spectral, immunochemical and enzymatic properties as those of rat liver microsomal enzyme (SAKAKI et al., 1985). Since the amino-terminal amino acid residue of the recombinant P-450c protein started with Pro as with the mature enzyme, posttranslational removal of the initial Met residue similarly occurred in the yeast cells (SAKAKI et al., 1985).

After the functional expression of rat microsomal P-450c cDNA in the yeast cells, a number of P-450 cDNAs were expressed in yeast (Tab. 2). Rat liver microsomal P-450a and $P\text{-}450_{LA\omega}$ cDNAs were expressed by using the ADH promoter, and both enzymes were found to function as testosterone 7α-hydroxylase and laurate ω-hydroxylase, respectively (NAGATA et al., 1987; HARDWICK et al., 1987). HAYASHI et al. (1988b) reported that rat liver microsomal $P\text{-}450_{M-1}$ expressed in the yeast is responsible for testosterone 16α-hydroxylation, but not for cholecalciferol 25-hydroxylation, although the $P\text{-}450_{CC25}$ preparation from rat liver microsomes was reported to show both activities (HAYASHI et al., 1988a). In addition to the ADH promoter, other yeast promoters such as PH05 (acid phosphatase), PGK (phosphoglycerate kinase), GAL7 (galactose-1-phosphate uridyltransferase) and GAL10 (UDP-glucose 4-epimerase) were also utilized for the expression of cloned P-450 cDNAs in yeast (see Tab. 2). These results clearly indicated that the yeast

Table 2. Expression of P-450 monooxygenase cDNAs in yeast

P-450 Form	Source	Vector (Promoter)	Remarks	Reference
P-450c	Rat liver	pAAH5 (ADH1)	I, S, A(benzo(a)pyrene, acetanilide, 7-ethoxycoumarin)	Oeda et al. (1986)
P-450d	Rat liver	pAM82(PH05)	I, S, A(17β-estradiol)	Shimuzu et al. (1986)
P_1-450	Mouse liver	pAAH5(ADH1)	I, A(benzo(a)-pyrene)	Kimura et al. (1987)
P_1-450	Mouse liver	pYeDP1/8-2 (GAL10-CYC1) pYeDP1-10 (PGK)	I, S, A(benzo(a)pyrene, 7-ethoxyresorufin)	Cullin and Pompon (1988)
P-450_{LM6}	Rabbit liver	pYeDP1/8-2 (GAL10-CYC1)	I, S,	Pompon (1988)
P-450_{LM4}		pYeDP1-10 (PGK)	A(benzo(a)pyrene, 7-ethoxyresorufin, 7-ethoxycoumarin, acetanilide)	
P-450a	Rat liver	pAAH5(ADH1)	I, A(testosterone-(7α))	Nagata et al. (1987)
P-$450_{LA\omega}$	Rat liver	pAAH5(ADH1)	I, A(laurate(ω))	Hardwick et al. (1987)
P-450(M-1)	Rat liver	pYcDE2 (ADH1-CYC1)	I, A(testosterone-(16α))	Hayashi et al. (1988b)
P-450e	Rat liver	pDP34(PH05)	I, S, A(?)	Zurbriggen et al. (1989)
P-$450_{16\alpha}$	Rabbit liver	pAAH5(ADH1)	S, A(testosterone-(16α))	Imai (1987)
P-450(ω-1)	Rabbit liver	pAAH5(ADH1)	S, A(laurate(ω-1))	Imai (1988)
P-450human-2	Human liver	pAA7(GAL7)	I, S	Yasumori et al. (1989)
P-$450_{17\alpha}$	Bovine adrenal	pAAH5 (ADH1)	I, S, A(progesterone, pregnenolone)	Sakaki et al. (1989)
$P450_{C21}$	Bovine adrenal	pAAH5(ADH1)	I, A(progesterone, 17α-hydroxyprogesterone)	Sakaki et al. (1990)
P-450 Reductase	Rat liver	pAAH5(ADH1)	I, A(cytochrome **c**)	Murakami et al. (1986)

I: Immunochemical detection using antibody; S: Spectral properties;
A: Enzyme activity (substrate in parenthesis).

system is useful for the functional expression of the microsomal P-450 monooxygenases.

Rat liver NADPH-cytochrome P-450 reductase cDNA was also expressed in yeast to exhibit the enzyme activity (MURAKAMI et al., 1986). The content of the reductase protein in the yeast cells was fairly low as compared with that of P-450c, although both P-450c and rat reductase cDNAs were expressed under the control of the same ADH promoter and terminator. Simultaneous expression of both P-450c and rat reductase cDNAs in the same yeast cells increased the P-450c-dependent monooxygenase activity of the cells two fold as compared with that of the yeast cells expressing P-450c alone. These results also indicated that the recombinant P-450c receives electrons from NADPH by the catalysis of endogenous yeast reductase as well as rat reductase simultaneously expressed, leading to the enhancement of the cellular monooxygenase activity. Recently, yeast NADPH-cytochrome P-450 reductase gene was cloned (YABUSAKI et al., 1988a) and expressed under the control of the ADH or its own promoter in yeast (MURAKAMI et al., 1990). Simultaneous expression of both rat P-450c and yeast reductase resulted in an increase of about eight times higher acetanilide p-hydroxylation activity in the recombinant yeast cells as compared with that of the yeast cells simultaneously expressing both P-450c and rat reductase. Therefore, it is likely that the yeast reductase is more useful for the enhancement of the P-450 monooxygenase activity in the yeast cells than the rat reductase.

2.3. Mammalian cells

Table 3 summarizes the expression of P-450 cDNAs in cultured mammalian cells. ZUBER et al. (1986) first succeeded in the functional expression of steroidogenic P-$450_{17\alpha}$ from bovine adrenal microsomes by using COS1 cells, which were derived from the kidney of African Green monkey (GLUZMAN, 1981). An expression plasmid, in which bovine microsomal P-$450_{17\alpha}$ cDNA was placed under the control of SV40 promoter of pCD vector (OKAYAMA and BERG, 1983), was transfected into COS1 cells. The transformed COS1 cells produced the P-$450_{17\alpha}$ enzyme, which contained a protoheme in the molecule and localized in the microsomes. Of interest were the findings that P-$450_{17\alpha}$ showed in situ both 17α-hydroxylase and $C_{17,20}$-lyase activities toward pregnenolone and progesterone by the interaction with endogenous P-450 reductase present in nonsteroidogenic COS1 cells (ZUBER et al., 1986). However, ESTABROOK et al. (1988) reported that the transformed COS1 cells expressing bovine P-$450_{17\alpha}$ formed no androstenedione from progesterone but formed dehydroepiandrosterone from pregnenolone, indicating the deficiency in the lyase activity toward 17α-hydroxyprogesterone. Similarly, P-450arom in human placental and chicken ovary microsomes catalyzed the several steps in the aromatization

Table 3. Expression of P-450 cDNAs in cultured mammalian cells

P-450 Form	Source	Host/Vector (Promoter)	Remarks	Reference
P_1-450			I, S, A(benzo(a)-pyrene)	
	Mouse liver	NIH3T3/PSC-11		BATTULA et al. (1987)
P_3-450		(Vaccinia Virus)	I, S, A(acetanilide)	
P-450b	Rat liver	V79/pSP64 (SV40)	I, A(7-pentoxyreso-rufin)	DOEHMER et al. (1988)
P-450_{PCN1}	Human liver	COS1/P91023(B) (Adenovirus)	A(nifedipine)	GONZALEZ et al. (1988)
P-$450_{II}/_{E1}$	Human liver	COS1/P91023(B) (Adenovirus)	I, A(N-nitrosodi-methylamine)	UMENO et al. (1988)
P-$450_{17\alpha}$	Bovine adrenal	COS1/pCD (SV40)	I, A(progesterone)	ZUBER et al. (1986)
P-$450_{17\alpha}$	Rat testis	COS1/pSVL (SV40)	A(progesterone)	NAMIKI et al. (1988)
P-450arom	Human placenta	COS1/pCMV20 (Cytomegalo-virus)	I, A(testosterone, androstenedione)	CORBIN et al. (1988)
P-450arom	Chicken ovary	COS1/pCD (SV40)	A(testosterone)	McPHAUL et al. (1988)
P-450scc	Bovine adrenal	COS1/pCD (SV40)	I, A(cholesterol)	ZUBER et al. (1988)
P-$450_{11\beta}$	Bovine adrenal	COS1/pCD (SV40)	I, A(deoxycortico-sterone)	MOROHASHI et al. (1990)
Adrenodoxin	Bovine adrenal	COS1/pCD (SV40)	I, A(enhancement of P-450scc)	ZUBER et al. (1988)

I: Immunochemical detection using antibody; S: Spectral properties;
A: Enzyme activity (substrate in parenthesis).

reaction of testosterone leading to estradiol when the respective cDNAs were expressed in the COS1 cells (Corbin et al., 1988; McPhaul et al., 1988).

Recently, Zuber et al. (1988) reported that bovine adrenal mitochondrial P-450scc cDNA was expressed in the COS1 cells, and that the recombinant P-450scc precursor was processed in the mitochondria to the mature form, which converted 22R-hydroxycholesterol to pregnenolone by coupling with endogenous renoredoxin reductase and renoredoxin in the cells. Double transfection of the COS1 cells with both bovine adrenal P-450scc and adrenodoxin expression plasmids resulted in an increase of pregnenolone production in the cells, indicating that adrenodoxin precursor was also processed in the cells to the mature enzyme, which contained an iron-sulfur in the molecule and interacted with P-450scc. Similarly, triple transfection of the COS1 cells with P-450scc, adrenodoxin and $P\text{-}450_{17\alpha}$ expression plasmids resulted in the construction of a concerted steroidogenic pathway from cholesterol to dehydroepiandrosterone in the COS7 cells (Zuber et al., 1988). Bovine adrenal mitochondrial $P\text{-}450_{11\beta}$ was also expressed in the COS1 cells, catalyzing 11-, 18- and 19-hydroxylation of deoxycorticosterone (Morohashi et al., 1990). These results clearly indicated that the COS1 cells are useful for the functional expression of both microsomal and mitochondrial P-450 monooxygenases.

Battula et al. (1987) utilized a vaccinia virus promoter for expression of mouse liver microsomal P_1-450 and P_3-450 cDNAs in several cultured cells. The murine and human cells infected with each of the constructed recombinant viruses efficiently produced the corresponding P-450 proteins, which were translocated into the microsomes and incorporated heme in the molecules to exhibit respective enzyme activities. Based on these studies, P_1-450 and P_3-450 cDNAs were found to encode aryl hydrocarbon hydroxylase and acetanilide p-hydroxylase, respectively (Battula et al., 1987). Also, P_3-450 expressed in human HeLa cells was found to catalyze 4-hydroxylation of the hepatocarcinogen aflatoxin B_1 leading to nonreactive aflatoxin M_1 (Faletto et al., 1988).

In contrast to the transient expression of membrane-bound P-450 species described above, Doehmer et al. (1988) developed a stable expression system for a P-450 cDNA in mammalian cells. The expression plasmid for rat liver microsomal P-450b was constructed by placing the cDNA under the control of SV40 early promoter and delivered to Chinese hamster ovary V79 cells with the selection marker neomycin phosphotransferase (resistant to G418). The establised cell lines produced the P-450b enzyme, which catalyzed the metabolic activation of aflatoxin B_1 to its mutagenic form by coupling with endogenous P-450 reductase.

These lines of studies suggested that the expression of P-450 monooxygenases in heterologous cells including bacterial, yeast and mammalian cells are useful for the identification, characterization and functional analysis of soluble and membrane-bound P-450 monooxygenases.

3. Functional analysis of P-450 monooxygenases

3.1. Site-directed mutagenesis of P-450 monooxygenases

The establishment of gene expression systems for P-450 monooxygenases by combination with the techniques for gene modification such as site-directed mutagenesis opened the way for the elucidation and analysing of the structure-function relationships of the enzymes. We focused on a specific amino acid residue the function of which was investigated. Sequence alignment of a number of P-450 molecules specified a cysteine residue commonly present in the carboxy-terminal regions of all P-450 species examined so far (GOTOH et al., 1983). X-ray analysis of the crystalized *P. putida* soluble P450cam indicated that the cysteine residue is the fifth ligand for heme binding (POULOS et al., 1985, 1986). This was also confirmed by the site-directed mutagenesis of P-450cam and rat liver microsomal P-450d. When Cys-357 of P-450cam was changed into His and Ser residues, the mutated P-450cam(His) produced in the *E. coli* cells showed a myoglobin-type spectrum and P-450cam(Ser) did not bind heme (UNGER and SLIGAR, 1985). Similar spectral changes were also observed with Cys-456 to His and Tyr mutants of P-450d when expressed in yeast (SHIMIZU et al., 1988). SHIMIZU et al. (1988) also introduced another amino acid substitution into P-450d. Measurement of spectra and catalytic activities of the mutant P-450d molecules suggested that some hydrophobic amino acid residues such as Phe-449, Leu-451, Gly-452, Ile-457 and Gly-458 are important for the protoheme to be held and to be incorporated at the active site of P-450d, and that Arg-454 interacts with the heme propionate. Recently, IMAI and NAKAMURA (1988) reported that Thr-301 of rabbit liver microsomal P-450(ω-1) and P-450(16α) participate in substrate binding, because the His-301 mutants showed no spectral changes in the presence of the substrates in spite of their typical P-450 absorption spectra.

The functional regions of NADPH-cytochrome P-450 reductase for FAD-, FMN- and NADPH-binding were estimated on the basis of sequence comparison of the reductase with other flavoproteins whose X-ray crystallographic structures have been revealed, together with the results of chemical modification of the reductases (PORTER and KASPER, 1986; KATAGIRI et al., 1986; YABUSAKI et al., 1988a). Site-directed mutagenesis of Tyr-140 and Tyr-178 in rat reductase to Phe and/or Asp revealed that both Tyr residues are major determinants for FMN-binding based on comparison of spectral properties and turnover numbers of the mutant enzymes with those of the wild-type (SHEN et al., 1988).

Analysis of naturally occurring mutant genes provided useful information on the structure-function relationships of the enzymes. Congenital adrenal hyperplasia due to a defect in any of the steroidogenic P-450s is a genetic disorder related to cortisol biosynthesis (WHITE et al., 1987). Analysis of the cloned mutant genes for P-450$_{C21}$ (originally identified in adrenal microsomes) from the patients revealed that the point mutations of Ser-269 and Asn-494, and Ile-172 in the P-450$_{C21}$ molecule resulted in a defect of the enzyme activity (RODRIGUES et al., 1987; AMOR et al., 1988). Further, sequencing analysis of four P-450$_{C21}$ mutant genes, coupled with expression of these genes in the COS1 cells, suggested that the changes in Ile-236 to Asn, Val-237 to Glu and/or Met-238 to Lys lead to inactivation of the P-450$_{C21}$ enzyme, while the change in Lys-102 to Arg and insertion of Leu after Leu-9 were not important for the activity (HIGASHI et al., 1988). This information was useful for us in understanding the physiological importance, as well as the structure-function relationships, of adrenal P-450$_{C21}$.

3.2. Chimeric P-450 monooxygenases

Construction of chimeric P-450 molecules among structurally related isozymes was a useful approach for functional analysis of the enzymes. Sequence analysis of rat P-450c (YABUSAKI et al., 1984a, 1984b) and P-450d cDNAs (KAWAJIRI et al., 1984) exhibited a totally 68% similarity in their amino acid sequences. Although both enzymes were immunochemically cross-reactive (RYAN et al., 1980), their enzymatic properties differed from each other; P-450c with a low-spin heme iron preferably metabolized benzo(a)pyrene and 7-ethoxycoumarin, while P-450d with a high-spin heme iron hydroxylated acetanilide as a good substrate. SAKAKI et al. (1987) constructed the expression plasmids for three chimeric P-450 molecules, P-450ccd, P-450cdc and P-450cdd, between P-450c and P-450d. The plasmids for the expression in yeast of these chimeric P-450s and P-450c contain a common nucleotide sequence from the ADH promoter to the coding region for the amino-terminal third of the P-450s. So, gene transcription and mRNA translation appeared to proceed similarly with these expression plasmids and microsomal insertion of the synthesized P-450 proteins also seemed to be similar among the chimeric P-450s and P-450c because these contained a common amino-terminal hydrophobic region. Chimeric P-450ccd, P-450cdc and P-450cdd were each produced in the yeast at the level of 4×10^5, 10^5 and 10^5 molecules per cell, respectively. The reduced CO-difference spectrum indicated that both P-450ccd and P-450cdd contained heme in their molecules, while P-450cdc did not. Monooxygenase activities in the transformed yeast cells and in the microsomal fractions prepared therefrom were measured against 7-ethoxycoumarin and acetanilide. P-450ccd showed high 7-ethoxycoumarin **O**-deethylation and low acetanilide **p**-hydroxylation activities as did P-450c, although both activities of P-450ccd were about 2.5 times higher than those of P-450c. On the other hand, P-450cdd showed a different substrate specificity from P-450c; high in acetanilide **p**-hydroxylation activity and low in 7-ethoxycoumarin **O**-deethylation activity. Thus, the substrate specificities of P-450ccd and P-450cdd appeared to be similar to those of P-450c and P-450d, respectively. These results suggested that the central third regions of P-450c and P-450d are responsible for substrate recognition and that the carboxy-terminal third regions are related to the interaction with yeast P-450 reductase.

KIMURA et al. (1987) constructed expression plasmids for chimeric P-450s between mouse liver microsomal TCDD-inducible P_1-450 and its noninducible mutant cDNAs. Determination of the enzyme activities of these chimeric P-450s expressed in the yeast revealed that the changes of Leu-118 and Arg-245 in P_1-450 caused loss of the catalytic activity. IMAI (1988) constructed three chimeric cDNAs for rabbit microsomal P-450s between pHP2-1 for laurate (ω-1)-hydroxylase and pHP3 for testosterone 16α-hydroxylase, and inserted each of them between the ADH promoter and terminator regions of pAAH5 to form the expression plasmids, pAH3P2, pAH3E2 and pAH3A2 (IMAI, 1988). pAH3P2 contained nearly the same coding sequence as that of pHP2-1, and both pAH3E2 and pAH3A2 were constructed by replacing the coding regions for the amino-terminal 210 and 262 amino acid residues of pAH3P2 with those of pHP3 respectively. Spectral changes were induced in both P-450(3P2) and P-450(3E2), expressed in the yeast by addition of the substrate, laurate or caprate, but not in P-450 (3A2). Fatty acid hydroxylation activity was detected only in P-450(3P2), but not in both P-450(3A2) and P-450(3E2). Thus, the region from amino acid residue 211 to 262 of the P450 seemed to be important for substrate binding, although more carboxy-terminal sequences were needed for the catalytic activity.

3.3. Function of the amino-terminal hydrophobic regions of P-450 monooxygenases

Microsomal P-450 and its reductase are synthesized in the rough endoplasmic reticulum and translocated into the microsomal membranes (FUJII-KURIYAMA et al., 1979; GONZALEZ and KASPER, 1980). The amino-terminal hydrophobic sequences found in all the microsomal P-450s resemble the signal sequences of the secreted proteins. These sequences are recognized by the signal recognition particle (SRP), but not cleaved during translocation into the microsomes. Yeast possesses SRP (SAKAGUCHI et al., 1984) which recognizes the recombinant mammalian P-450s synthesized in the yeast and translocated them into the yeast microsomal membranes. SAKAGUCHI et al. (1987) reported that the amino-terminal segment with less than 29 amino acid residues of a polycyclic aromatic hydrocarbon-inducible form of rabbit liver microsomal P-450s served as both insertion signal and stop-transfer sequence. MONIER et al. (1988) also demonstrated by using deletion variants and fusions of rat liver microsomal P-450b in similar experiments that the amino-terminal 20 residues of P-450b functioned as a combined insertion-halt-transfer signal.

YABUSAKI et al. (1988b) constructed the expression plasmid for the truncated rat liver microsomal P-450c lacking the amino-terminal hydrophobic 30 amino acid residues. Transformed yeast cells with the expression plasmid resulted in the production of the truncated P-450c protein, which contained a protoheme in the molecule and was associated with the yeast microsomes. The yeast cells carrying the truncated P-450c showed significant but reduced 7-ethoxycoumarin-O-deethylation activity. When the expression plasmid for the truncated bovine adrenal microsomal $P\text{-}450_{17\alpha}$ lacking the amino-terminal 17 amino acid residues was introduced into the COS1 cells, a significant portion of the truncated $P\text{-}450_{17\alpha}$ protein was associated with the microsomal membranes, although no $P\text{-}450_{17\alpha}$-dependent activity was detected (WATERMAN et al., 1989). Salt-washing of the microsomes did not completely abolish the membrane association of the truncated $P\text{-}450_{17\alpha}$ protein. These results suggested that the truncated P-450 proteins are probably associated with the microsomal membranes due to the presence of extra hydrophobic region(s) of the P-450 molecules, but not sufficient for the full enzyme activity.

NADPH-cytochrome P-450 reductase is anchored to microsomal membranes by its hydrophobic amino-terminal region. The reductase is easily released from the membranes by cleaving the hydrophobic region with the protease treatment to give a catalytically active hydrophilic domain (BLACK et al., 1979; KUBOTA et al., 1977). The released soluble domain did not interact with P-450s, but reduced cytochrome c. This was confirmed by the expression in yeast of the truncated rat soluble reductase lacking the amino-terminal 56 amino acid anchor region. The truncated reductase produced in the yeast was found in the soluble fraction free from the membranes and active in cytochrome c reduction (YABUSAKI et al., 1988b).

Mitochondrial P-450 enzymes are synthesized in free ribosomes as precursors with higher molecular weights and then transported across mitochondrial membranes into the inner membrane or matrix space (NABI and OMURA, 1980; NABI et al., 1980; DUBOIS et al., 1981; KRAMER et al., 1982). These precursors are processed by mitochondrial proteases to mature forms during translocation. Since the mature enzymes have no ability to move into the mitochondria in an in-vitro system, the extrapeptides are likely to contain all information for mitochondrial targeting (ONO and ITO, 1984; HORWICH et al., 1985). Sequence analysis of cDNAs coding for the respective precursors of bovine mitochondrial $P\text{-}450_{11\beta}$ and P-450scc, adrenodoxin, and adrenodoxin reductase revealed that the extrapeptides differed in length and amino acid sequence from each other (MOROHASHI et al., 1984, 1987; CHUA et al., 1987; OKAMURA et al., 1985; NONAKA et al., 1987; SAGARA et al., 1987). Modification of the cDNAs for the extrapeptides of bovine adrenal P-450scc and adrenodoxin precursors, together with the results of inhibition experiments of in vitro transport by synthetic oligopeptides for the extrapeptides (FURUYA et al., 1987), indicated that the amino-terminal 15 to 17 amino acid sequences including a couple of basic residues were essential for the import of these precursors into the mitochondria (KUMAMOTO et al., 1987; 1989).

3.4. P-450/reductase fused enzymes

The amino-terminal hydrophobic region of rat liver microsomal P-450c was important for the correct orientation of the newly synthesized peptide into microsomal membranes and for sufficient enzyme activity. On the other hand, rat soluble reductase lacking the amino-terminal anchor region was produced in the yeast in a large quantity and was active in cytochrome **c** reduction. Based on these results, MURAKAMI et al. (1987) constructed a P-450/reductase fused enzyme with whole P-450c and the soluble reductase moieties. A hybrid cDNA for the P-450/reductase fused enzyme was constructed by connecting the 3′-end of the P-450c cDNA with the 5′-end of the soluble reductase cDNA, and then expressed in yeast under the control of the ADH promoter and terminator. The amount of the fused enzyme produced in the transformed yeast cells was one sixth of that of P-450c as determined by the reduced CO-difference spectrum. The monooxygenase activity of the yeast cells carrying the fused enzyme on the basis of hemoprotein was about four times higher than that of the cells carrying P-450c alone, indicating that the fused enzyme is functional in the yeast cells. When the subcellular fractions prepared from the recombinant yeast cells were analysed for the fused enzyme, the enzyme with a molecular weight of 130 kilodaltons (kDa) was mainly detected in the microsomes by using both anti-P-450c and anti-reductase Igs. After treatment of the microsomes with trypsin, the P-450 moiety was found in the microsomes, while the reductase moiety was detected in the soluble fraction. The purified P-450/reductase fused enzyme showed spectral properties similar to those of both P-450c and rat reductase, and exhibited monooxygenase activity under in vitro emulsified conditions followed by first-order kinetics. These results suggested that the engineered fused enzyme is a novel self-catalytic P-450 monooxygenase with the P-450 moiety embedded in the microsomes and the reductase moiety located in the cytoplasm (MURAKAMI et al., 1987). Figure 2 illustrates the proposed conformation of the engineered P-450/reductase fused enzyme located in the microsomes.

NARHI and FULCO (1986) isolated a self-sufficient soluble P-450 monooxygenase, P-450_{BM-3}, with a molecular weight of 119 kDa from *B. megaterium*. P-450_{BM-3} contained one mole each of protoheme, FAD and FMN in the molecule. The P-450_{BM-3} molecule closely resembled the engineered fused enzyme described above in its construction; the P-450 moiety in the amino-terminal side and the reductase moiety in the carboxy-terminal side (NARHI and

FULCO, 1987), although the molecular weight of P-450_{BM-3} was somewhat smaller than that of the engineered fused enzyme. The difference in the molecular weight between P-450_{BM-3} and the engineered fused enzyme may be explained in part by the fact that P-450_{BM-3} is a soluble enzyme lacking the amino-terminal hydrophobic region, while the engineered fused enzyme con-

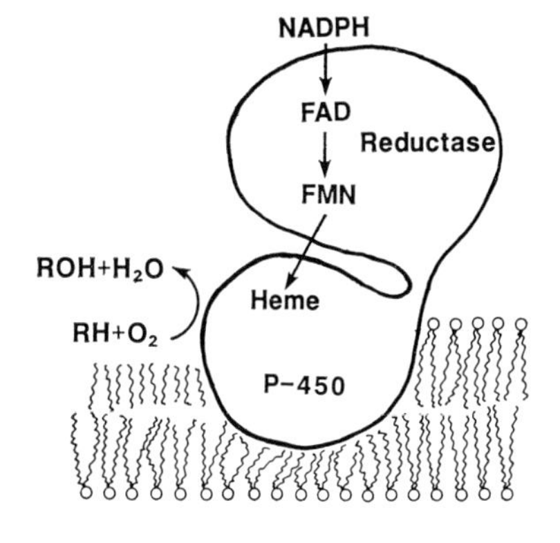

Fig. 2. The proposed illustration of the engineered P-450/reductase fused enzyme located in the microsomes.

tained it for microsomal localization. A DNA clone for P-450_{BM-3} was recently isolated (WEN and FULCO, 1987) and expressed in both *B. megaterium* and *E. coli* (NARHI et al., 1988). Determination of the primary structure of P-450_{BM-3} by sequencing the cloned DNA and its comparison with that of the engineered fused enzyme would give us interesting structural information on the P-450 monooxygenases.

4. Possible application of P-450 monooxygenases to bioconversion

P-450 monooxygenases have a great potential application for the practical bioconversion processes of a wide variety of lipophilic compounds including industrially important chemicals such as drugs and steroids. Recently developed genetic engineering technology has made it possible to produce P-450 monooxygenases in large quantities in *E. coli* and yeast, and to improve the enzymes in reactivity and stability for practical purposes. Taking an example, acetaminophene is widely used as an analgesic instead of aspirin because of the lack of some side effects and is manufactured annually in an amount of about 10 kilotons. SAKAKI et al. (1986) examined the possible application of a P-450 monooxygenase to a bioconversion process for the production of acetaminophene from acetanilide. The hydroxylation of acetanilide at 4-position leading to acetaminophene was catalyzed by rat liver microsomal P-450c,

P-450d and their engineered enzymes (Sakaki et al., 1987). After addition of acetanilide to the culture of the transformed yeast cells carrying P-450c, the culture supernatant was analysed for acetaminophene production by HPLC. The recombinant yeast cells were found to catalyze specifically p-hydroxylation of acetanilide to produce acetaminophene in the culture supernatant. No o- and m-hydroxyproducts were found.

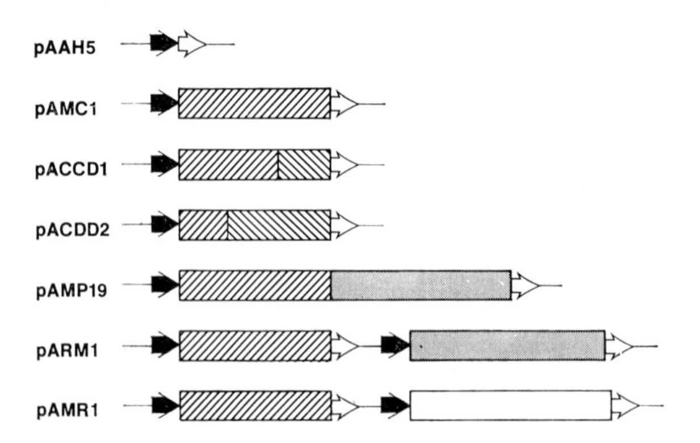

Fig 3. Expression plasmids for rat liver microsomal P-450c and the engineered P-450 monooxygenases.

P-450c; P-450d; rat reductase; yeast reductase; ADH promoter; ADH terminator.

To improve the monooxygenase activity in the transformed yeast cells, a number of genetically engineered P-450 monooxygenases were constructed and introduced into the yeast cells. Those included chimeric P-450ccd and P-450cdd (Sakaki et al., 1987), the P-450/reductase fused enzyme between P-450c and rat reductase (Murakami et al., 1987), and the simultaneous expression of P-450c and rat reductase (Murakami et al., 1986), or yeast reductase (Murakami et al., 1990), as shown in Figure 3. Table 4 summarizes the activity in the production of acetaminophene from acetanilide of a number of the recombinant yeast strains carrying each of the constructed enzymes. Chimeric P-450s and the P-450/reductase fused enzyme showed increased acetanilide p-hydroxylation activity on the basis of hemoprotein, but their activities per culture were little or slightly higher than that of the recombinant yeast cells carrying P-450c because of the decrease in the hemoprotein contents. On the other hand, the simultaneous expression of P-450c and the reductase resulted in an increase in the acetaminophene production per culture. Especially, the recombinant yeast cells carrying both P-450c and overproduced yeast reductase exibited the highest productivity per culture and on the basis of hemoprotein among the strains constructed. Therefore, the interaction of

Table 4. Bioconversion of acetanilide to acetaminophene by the recombinant yeast strains

Yeast strain	Produced enzyme	Acetaminophene formed	
		nmol/min × nmol P-450	nmol/ml culture
AH22/pAAH5	—	—	$<$ 1
AH22/pAMC1	P-450c	2	
AH22/pACCD1	P-450ccd	5	16
AH22/pACDD2	P-450cdd	12	9
AH22/pAMP19	P-450/reductase fused enzyme	7	4
AH22/pARM1	P-450c + rat reductase	4	12
AH22/pAMR1	P-450c + yeast reductase	41	96

$NHCOCH_3$ → $NHCOCH_3$
OH

P-450c with the reductase or the transfer of electrons from NADPH to P-450c through the reductase seemed to be one of the important steps for improvement of the P-450 monooxygenase activity in the yeast cells. These lines of studies implied the possibility of applying P-450 monooxygenases and/or the recombinant yeast cells carrying the engineered enzymes for practical bioconversion processes of industrially important chemicals, and the usefulness of the genetic engineering technology for improvement of the enzyme activity. Further studies are needed for evaluation of these enzymes and/or the recombinant strains for practical uses.

5. Concluding remarks

Genetic engineering technology rapidly improved the fundamental understanding on the biological, biochemical and physiological importance of the P-450 monooxygenases at the molecular level. Sequence analysis of a number of homologous P-450 genes revealed their primary structures but raised questions as to what molecular form is encoded in the predicted sequence and what is the function of a P-450 isozyme which were not purified and characterized at the protein level. The functional expression of a cloned cDNA in yeast and mammalian cells clarified the relationship between the predicted sequence and its function. Also, the gene expression techniques combined with the site-directed mutagenesis and chimera construction were useful for elucidation and analysis of the structure-function relationships of the P-450 monooxygenases as well as improvement of the enzymes in activity and stability. However, more precise information on the structure-function relationships is needed for protein engineering of the enzymes. Addition to bacterial P-450cam (POULOS et al., 1985, 1986), crystalization of P-450scc (IWAMOTO et al., 1987), adrenodoxin reductase (NONAKA et al., 1985), cytochrome b_5 (MATHEWS et al., 1972) and cytochrome b_5 reductase (MIKI et al., 1987) has been reported. X-ray analysis of the crystalized structures as well as measurement of physicochemical parameters by using NMR and ESR are also helpful for understanding fine the structures of the enzymes.

Genetic engineering technology also established the overproduction of the recombinant enzymes. This technology combined with the protein engineering of the enzymes will make it possible to improve the enzymes for practical purposes. An attempt was made to apply P-450c and its genetically engineered P-450 monooxygenases in a practical conversion process of acetanilide to acetaminophene. In addition, several P-450 species were found to be involved in the biosynthesis of physiologically important compounds such as steroid hormones, vitamine D_3 and prostaglandins. These enzymes have a possible potential for an application to bioconversion processes of the production of the pharmaceutically important compounds. For example, cortisol which is an important intermediate in the synthesis of steroid hormones in pharmaceutical industries, is synthesized in vivo from cholesterol through five enzyme steps, four of which were catalyzed by P-450 monooxygenases. Therefore, these steroidogenic P-450s may be applicable to practical conversion processes of the synthesis of steroid hormones when respective enzymes are overproduced and improved in activity and stability for practical uses by genetic and protein engineerings. Attempts have been made with the immobilization of microsomal P-450 enzymes in order to overcome instability of the enzymes (AZARI and WISEMAN, 1982; YAWETZ et al., 1984). Coupled with these lines of studies, P-450 monooxygenases may also be applied in future to several practical uses such as a bioreactor, biosensor and an artificial liver support for detoxification of toxicants.

Acknowledgement

We thank Dr. T. SAKAKI, Dr. H. MURAKAMI and Ms. M. SHIBATA for their contributions. Our work on P-450 monooxygenase was supported by the Research and Development Project of Basic Technologies for Future Industries from the Ministry of International Trade and Industry of Japan.

6. References

Adesnik, M. and M. Atchinson, (1986), CRC Crit. Rev. Biochem. **19**, 247–305.

Ammerer, G., (1983), Methods Enzymol. **101**, 192–201.

Amor, M., K. L. Parker, H. Globerman, M. I. New, and P. C. White, (1988), Proc. Natl. Acad. Sci. U. S. A. **85**, 1600–1604.

Aoyama, Y., Y. Yoshida, and R. Sato, (1984), J. Biol. Chem. **259**, 1661–1666.

Azari, M. R. and A. Wiseman, (1982), Enzyme Microb. Technol. **4**, 401–404.

Battula, N., J. Sagara, and H. V. Gelboin, (1987), Proc. Natl. Acad. Sci. U. S. A. **84**, 4073–4077.

Black, S. D., J. S. French, C. H. Williams Jr., and M. J. Coon, (1979), Biochem. Biophys. Res. Commun. **91**, 1528–1535.

Black, S. D. and M. J. Coon, (1986), in: Cytochrome P-450. Structure, Mechanism, and Biochemistry, (Ortiz de Montellano, P. R., ed.), Plenum Press, New York, pp. 161–216.

Black, S. D. and M. J. Coon, (1987), Adv. Enzymol. Relat. Area Mol. Biol. **60**, 35–87.

Chua, S. C., P. Szabo, A. Vitek, K.-H. Grzeschik, M. John, and P. C. White, (1987), Proc. Natl. Acad. Sci. U. S. A. **84**, 7193–7197.

Corbin, C. J., S. Graham-Lorence, M. McPhaul, J. I. Mason, C. R. Mendelson, and E. R. Simpson, (1988), Proc. Natl. Acad. Sci. U. S. A. **85**, 8948–8952.

Cullin, C. and D. Pompon, (1988), Gene **65**, 203–217.

Doehmer, J., S. Dogra, T. Friedberg, S. Monier, M. Adesnik, H. Glatt, and F. Oesch, (1988), Proc. Natl. Acad. Sci. U. S. A. **85**, 5769–5773.

DuBois, R. W., E. R. Simpson, J. Tuckey, J. D. Lambeth, and M. R. Waterman, (1981), Proc. Natl. Acad. Sci. U. S. A. **78**, 1028–1032.

Estabrook, R. W., J. I. Mason, C. Martin-Wixtrom, M. Zuber, and M. R. Waterman, (1988), Prog. Clin. Biol. Res. **274**, 525–540.

Faletto, M. B., P. L. Koser, N. Battula, G. K. Townsend, A. E. Maccubbin, H. V. Gelboin, and H. L. Gurtoo, (1988), J. Biol. Chem. **263**, 12187–12189.

Fujii-Kuriyama, Y., M. Negishi, R. Mikawa, and Y. Tashiro, (1979), J. Cell Biol. **81**, 501–519.

Furuya, S., M. Okada, A. Ito, H. Aoyagi, T. Kanmera, T. Kato, Y. Sagara, T. Horiuchi, and Omura, T., (1987), J. Biochem. **102**, 821–832.

Gluzman, Y., (1981), Cell **23**, 175–182.

Gonzalez, F. J. and C. B. Kasper, (1980), Biochemistry **19**, 1790–1796.

Gonzalez, F. J., B. J. Schmid, M. Umeno, O. W. Mcbride, J. P. Hardwick, U. A. Meyer, H. V. Gelboin, and J. R. Idle, (1988), DNA **7**, 79–86.

Gotoh, O., Y. Tagashira, T. Iizuka, and Y. Fujii-Kuriyama, (1983), J. Biochem. **93**, 807–817.

Gotoh, O. and Y. Fujii-Kuriyama, (1989), in: Frontiers in Biotransformation (Ruckpaul, K. and H. Rein, eds.), Akademie-Verlag, Berlin and Taylor and Francis, London, Vol. **1**, 195–243.

Hardwick, J. P., B.-J. Song, E. Huberman, and F. J. Gonzalez, (1987), J. Biol. Chem. **262**, 801–810.

Hayashi, S., T. Omura, T. Watanabe, and K. Okuda, (1988a), J. Biochem. **103**, 853–857.

Hayashi, S., K. Morohashi, H. Yoshioka, K. Okuda, and T. Omura, (1988b), J. Biochem. **103**, 858–862.

Higashi, Y., A. Tanae, H. Inoue, T. Hiromasa, and Y. Fujii-Kuriyama, (1988), Proc. Natl. Acad. Sci. U. S. A. **85**, 7486–7490.

Horwich, A. L., F. Kalousek, I. Mellman, and L. E. Rosenberg, (1985), EMBO J. **4**, 1129–1135.

Imai, Y., (1987), J. Biochem. **101**, 1129–1139.
Imai, Y., (1988), J. Biochem. **103**, 143–148.
Imai, Y. and M. Nakamura, (1988), FEBS Letters **234**, 313–315.
Ito, H., Y. Fukuda, K. Murata, and A. Kimura, (1983), J. Bacteriol **153**, 163–168.
Iwamoto, Y., M. Tsubaki, A. Hiwatashi, and Y. Ichikawa, (1987), Seikagaku (in Japanese) **59**, 713.
Kalb, V. F., C. W. Woods, T. G. Turi, C. R. Dey, T. R. Sutter, and J. C. Loper, (1987), DNA **6**, 529–537.
Katagiri, M., H. Murakami, Y. Yabusaki, T. Sugiyama, M. Okamoto, T. Yamano, and H. Ohkawa, (1986), J. Biochem. **100**, 945–954.
Kawajiri, K., O. Gotoh, K. Sogawa, Y. Tagashira, M. Muramatsu, and Y. Fujii-Kuriyama, (1984), Proc. Natl. Acad. Sci. U. S. A. **81**, 1649–1653.
Kimura, S., H. H. Smith, O. Hankinson, and D. W. Nebert, (1987), EMBO J. **6**, 1929–1933.
Kramer, R. E., R. N. Dubois, E. R. Simpson, C. M. Anderson, K. Kashiwagi, J. D. Lambeth, C. R. Jefcoate, and M. R. Waterman, (1982), Arch. Biochem. Biophys. **215**, 478–485.
Kubota, S., Y. Yoshida, H. Kumaoka, and A. Fumumichi, (1977), J. Biochem. **81**, 197–205.
Kumamoto, T., K. Morohashi, A. Ito, and T. Omura, (1987), J. Biochem. **102**, 833–838.
Kumamoto, T., A. Ito, and T. Omura, (1989), J. Biochem. **105**, 72–78.
Mathews, F. S., P. Argos, and M. Levine, (1972), Cold Spring Harbor Symp. Quant. Biol. **36**, 387–393.
McPhaul, M. J., J. F. Noble, E. R. Simpson, C. R. Mendelson, and J. D. Wilson, (1988), J. Biol. Chem. **263**, 16358–16363.
Miki, K., S. Kaida, N. Kasai, T. Iyanagi, K. Kobayashi, and K. Hayashi, (1987), J. Biol. Chem. **262**, 11801–11802.
Monier, S., P. VanLuc, G. Kreibich, D. D. Sabatini, and M. Adesnik, (1988), J. Cell. Biol. **107**, 457–470.
Morohashi, K., Y. Fujii-Kuriyama, Y. Okada, K. Sogawa, T. Hirose, S. Inayama, and T. Omura, (1984), Proc. Natl. Acad. Sci. U. S. A. **81**, 4647–4651.
Morohashi, K., H. Yoshioka, O. Gotoh, Y. Okada, K. Yamamoto, T. Miyata, K. Sogawa, Y. Fujii-Kuriyama, and T. Omura, (1987), J. Biochem. **102**, 559–568.
Morohashi, K., Y. Nonaka, S. Kirita, O. Hatano, A. Takakusu, M. Okamoto, and T. Omura, (1990), J. Biochem. **107**, 635–640.
Murakami, H., Y. Yabusaki, and H. Ohkawa, (1986), DNA **5**, 1–10.
Murakami, H., Y. Yabusaki, T. Sakaki, M. Shibata, and H. Ohkawa, (1987), DNA **6**, 189–197.
Murakami, H., Y. Yabusaki, T. Sakaki, M. Shibata, and H. Ohkawa, (1990), J. Biochem. **108**, 859–865.
Nabi, N. and T. Omura, (1980), Biochem. Biophys. Res. Commun. **97**, 680–686.
Nabi, N., S. Kominami, S. Takemori, and T. Omura, (1980), Biochem. Biophys. Res. Commun. **97**, 687–693.
Nagata, K., T. Matsunaga, J. Gillette, H. V. Gelboin, and F. J. Gonzalez, (1987), J. Biol. Chem. **262**, 2787–2793.
Nakamura, K., H. Murakami, K. Oeda, and H. Ohkawa, (1984), Seikagaku (in Japanese) **56**, 1093.
Namiki, M., M. Kitamura, E. Buczko, and M. L. Dufau, (1988), Biochem. Biophys. Res. Commun. **157**, 705–712.
Narhi, L. O. and A. J. Fulco, (1986), J. Biol. Chem. **261**, 7160–7169.
Narhi, L. O. and A. J. Fulco, (1987), J. Biol. Chem. **262**, 6683–6690.

NARHI, L. O., L.-P. WEN, and A. J. FULCO, (1988), Mol. Cell. Biochem. **79**, 63–71.
NEBERT, D. W. and F. J. GONZALEZ, (1987), Annu. Rev. Biochem. **56**, 945–993.
NONAKA, Y., S. AIBARA, T. SUGIYAMA, T. YAMANO, and Y. MORITA, (1985), J. Biochem. **98**, 257–260.
NONAKA, Y., H. MURAKAMI, Y. YABUSAKI, S. KURAMITSU, H. KAGAMIYAMA, T. YAMANO, and M. OKAMOTO, (1987), Biochem. Biophys. Res. Commun. **145**, 1239–1247.
OEDA, K., T. SAKAKI, and H. OHKAWA, (1985), DNA **4**, 203–210.
OKAMURA, T., M. E. JOHN, M. X. ZUBER, E. R. SIMPSON, and M. R. WATERMAN, (1985), Proc. Natl. Acad. Sci. U. S. A. **82**, 5705–5709.
OKAYAMA, H. and P. BERG, (1983), Mol. Cell. Biol. **3**, 280–289.
ONO, H. and A. ITO, (1884), J. Biochem. **95**, 345–352.
POMPON, D., (1988), Eur. J. Biochem. **177**, 285–293.
PORTER, T. D. and C. B. KASPER, (1986), Biochemistry **25**, 1682–1687.
PORTER, T. D., T. E. WILSON, and C. B. KASPER, (1987), Arch. Biochem. Biophys. **254**, 353–367.
POULOS, T. L., B. C. FINZEL, I. C. GUNSALUS, G. C. WAGNER, and J. KRAUT, (1985), J. Biol. Chem. **260**, 16122–16130.
POULOS, T. L., B. C. FINZEL, and A. J. HOWARD, (1986), Biochemistry **25**, 5314–5322.
RODRIGUES, N., I. DUNHAM, C. Y. YU, M. C. CARROLL, R. R. PORTER, and R. D. CAMPBELL, (1987), EMBO J. **6**, 1653–1661.
RYAN, D. E., P. E. THOMAS, and W. LEVIN, (1980), J. Biol. Chem. **255**, 7941–7955.
SAGARA, Y., Y. TAKATA, T. MIYATA, T. HARA, and T. HORIUCHI, (1987), J. Biochem. **102**, 1333–1336.
SAKAGUCHI, M., K. MIHARA, and R. SATO, (1984), Proc. Natl. Acad. Sci. U. S. A. **81**, 3361–3364.
SAKAGUCHI, M., K. MIHARA, and R. SATO, (1987), EMBO J. **6**, 2425–2431.
SAKAKI, T., K. OEDA, M. MIYOSHI, and H. OHKAWA, (1985), J. Biochem. **98**, 167 to 175.
SAKAKI, T., K. OEDA, Y. YABUSAKI, and H. OHKAWA, (1986), J. Biochem. **99**, 741 to 749.
SAKAKI, T., M. SHIBATA, Y. YABUSAKI, and H. OHKAWA, (1987), DNA **6**, 31–39.
SAKAKI, T., M. SHIBATA, Y. YABUSAKI, H. MURAKAMI, and H. OHKAWA, (1989), DNA **8**, 409–418.
SAKAKI, T., M. SHIBATA, Y. YABUSAKI, H. MURAKAMI, and H. OHKAWA, (1990), DNA Cell Biol. **9**, 603–614.
SHEN, A., T. PORTER, T. WILSON, and C. B. KASPER, (1988), FASEB J. **2**, A355.
SHIMIZU, T., K. SOGAWA, Y. FUJII-KURIYAMA, M. TAKAHASHI, Y. OGOMA, and M. HATANO, (1986), FEBS Letters **207**, 217–221.
SHIMIZU, T., K. HIRANO, M. TAKAHASHI, M. HATANO, and Y. FUJII-KURIYAMA, (1988), Biochemistry **27**, 4138–4141.
UMENO, M., O. W. MCBRIDE, C. S. YANG, H. V. GELBOIN, and F. J. GONZALEZ, (1988), Biochemistry **27**, 9006–9013.
UNGER, B. P. and S. G. SLIGAR, (1985), International Workshop on P-450 Genes and Their Regulation Airlie, U. S. A.
UNGER, B. P., I. C. GUNSALUS, and S. G. SLIGAR, (1986), J. Biol. Chem. **261**, 1158 to 1163.
WATERMAN, M. R., J. I. MASON, M. X. ZUBER, M. C. LORENCE, B. J. CLARK, J. M. TRANT, H. J. BARNES, E. R. SIMPSON, and R. W. ESTABROOK, (1989), Arch. Toxicol., Suppl. **13**, 155–163.
WEN, L.-P. and A. J. FULCO, (1987), J. Biol. Chem. **262**, 6676–6682.
WENZEL, E., J. FRIEBERTSHAUSER, D. HAUSTEIN, K. J. NETTER, and G. F. FUHRMANN, (1988), J. Biotechnol. **7**, 147–160.

WHITE, P. C., M. I. NEW, and B. DUPONT, (1987), New Engl. J. Med. **316**, 1519–1524, 1580–1586.
WISEMAN, A. and L. F. J. WOODS, (1979), J. Chem. Tech. Biotechnol. **29**, 320–324.
YABUSAKI, Y., M. SHIMIZU, H. MURAKAMI, K. OEDA, K. NAKAMURA, and H. OHKAWA, (1984a), Nucleic Acids Res. **12**, 2929–2938.
YABUSAKI, Y., H. MURAKAMI, K. NAKAMURA, N. NOMURA, M. SHIMIZU, K. OEDA, and H. OHKAWA, (1984b), J. Biochem. **96**, 793–804.
YABUSAKI, Y., H. MURAKAMI, and H. OHKAWA, (1988a), J. Biochem. **103**, 1004–1010.
YABUSAKI, Y., H. MURAKAMI, T. SAKAKI, M. SHIBATA, and H. OHKAWA, (1988b), DNA **7**, 701–711.
YASUMORI, T., N. MURAYAMA, Y. YAMAZOE, A. ABE, Y. NOGI, T. FUKAZAWA, and R. KATO, (1989), Mol. Pharmacol. **35**, 443–449.
YOSHIDA, Y. and Y. AOYAMA, (1984), J. Biol. Chem. **259**, 1655–1660.
YAWETZ, A., A. S. PERRY, A. FREEMAN, and E. KATCHALSKI-KATZIR, (1984), Biochim. Biophys. Acta **798**, 204–209.
ZUBER, M. X., E. R. SIMPSON, and M. R. WATERMAN, (1986), Science **234**, 1258–1261.
ZUBER, M. X., J. I. MASON, E. R. SIMPSON, and M. R. WATERMAN, (1988), Proc. Natl. Acad. Sci. U. S. A. **85**, 699–703.
ZURBRIGGEN, B., E. BÖHLEN, D. SANGLARD, O. KÄPPELI, and A. FIECHTER, (1989), J. Biotechnol. **9**, 255–272.

Chapter 7

Biochemistry and Physiology of Plant Cytochrome P-450

F. Durst

1. Introduction

The first indices of cytochrome P-450 in plants appeared more than a decade after its discovery in rat liver, and progress was slow at first. Besides the obvious lack of the incentive of medical research, progress was hampered by the difficulties confronting enzyme studies in plants: i) high mechanical forces that must be exerted to disrupt the plant cell wall, ii) the ensuing mixing of phenolics stocked in the vacuole with peroxidases and phenoloxidases from other cellular compartments leading to quinones formation and redox cycling, and iii) the fact that plants contain elevated amounts of proteases and lipases. This situation is aggravated by the hydrophobicity of the cytochrome P-450 system and by its very low titer in most plants.

Although the great majority of cytochrome P-450 enzymes known today are hydroxylases acting upon physiological substrates, the first solid evidence for a cytochrome P-450 activity in plants concerned the N-demethylation of monuron, a substituted phenylurea herbicide, by cotton microsomes (FREAR et al., 1969). A series of studies by WEST and his colleagues (MURPHY and WEST, 1969; LEW and WEST, 1971; COOLBAUGH et al., 1978) suggested the involvement of cytochrome P-450 in a sequence of four oxidative steps leading from kaurene, a precursor of the gibberellin family of plant hormones, to 7β-hydroxykaurenoic acid. Meanwhile, spectroscopic evidence for cytochrome P-450 in microsomal fractions from several plant species was obtained (MARKHAM et al., 1972; YAHIEL et al., 1974). In 1974, the involvement of cytochrome P-450 in the 4-hydroxylation of *trans*-cinnamic acid, a precursor of lignins and numerous secondary plant products, was established on the basis of photochemical action spectra (POTTS et al., 1974) and type I substrate-binding difference spectra (BENVENISTE and DURST, 1974).

During the last few years, the study of plant cytochrome P-450 has expanded rapidly and new cytochrome P-450 dependent reactions have been described. Several reviews are already available (WEST, 1980; HIGASHI, 1985; RIVIÈRE and CABANNE, 1987; HIGASHI, 1988), the latter concerned with ecotoxicological aspects of plant cytochrome P-450. This review is focused on physiological aspects of cytochrome P-450 in higher plants. Because of lack of space, fungi and yeasts are not covered here (for a review see KÄPPELI, 1986).

2. Occurrence and tissue distribution

Several lists of plants containing cytochrome P-450 have been published (RICH and BENDALL, 1975; HIGASHI, 1985; HENDRY, 1986). However, it is becoming obvious that this cytochrome is present in all higher plants. At least, all those assayed in our laboratory contained some cytochrome P-450.

2.1. Distribution in plant organs

The occurrence and content of cytochrome P-450 in different organs is not easy to evaluate for several reasons. First, cytochrome P-450 is much more difficult to measure in some plant organs (leaves, flowers, fruits) that are more richly pigmented than others, like radicles, shoots and cotyledons of etiolated seedlings or in storage tissues. Second, in a same organ, cytochrome P-450 content will vary according to age. Furthermore, its concentration may be enhanced by various factors such as damage, light, infection by fungi, exposure to xenobiotics. Even when cytochrome P-450 cannot be determined by spectrophotometry, its presence may be indirectly demonstrated using a sensitive assay for a well documented and ubiquitous cytochrome P-450 dependent activity like the cinnamic acid 4-hydroxylase (CA4H).

Up to now, cytochrome P-450 has been detected by spectrophotometry in endosperm of germinating seeds (Young and Beevers, 1976; Lew and West, 1971), and in the different parts of etiolated seedlings: radicle (Rich and Bendall, 1975), shoot (Rich and Bendall, 1975; Soliday and Kolattukudy, 1977), hypocotyl (Markham et al., 1972; Rich and Bendall, 1975), cotyledons (Hasson and West, 1976; Rich and Lamb, 1977), apical buds (Benveniste et al., 1978). It is also present in dormant bulbs like tulip (Rich and Bendall, 1975), *Lillium* (Salaün, unpublished) and *Amaryllis* (Fonné et al., 1988) and storage organs such as Jerusalem artichoke tubers (Benveniste and Durst, 1974), potato (Rich and Lamb, 1977) and sweet potato tubers (Fujita et al., 1982) and *Iris* rhizomes (Rich and Bendall, 1975). There is also cytochrome P-450 in the avocado fruit (Hendry, 1986), in the spadix of *Arum maculatum* (Yahiel et al., 1974) and in cauliflower inflorescences (Rich and Bendall, 1975).

2.2. Cell cultures

Use of liquid cell suspension cultures for physiological and biochemical studies is developing rapidly. Cytochrome P-450 dependent activities have been reported in cell cultures from *Catharanthus roseus* (Spitzberg et al., 1981), *Acer* and Jerusalem artichoke (Simon, 1982), bean (Bolwell and Dixon, 1986), soybean (Hagmann et al. 1984; Kochs and Grisebach, 1986), parsley (Hagmann et al., 1983; Heller and Kühnl, 1985), *Digitalis lanata* (Petersen and Seitz, 1985). Both activity and cytochrome P-450 were measured in *Acer* and Jerusalem artichoke cell cultures (Salaün et al., 1982; Simon, 1982), in parsley cell cultures (Hagmann et al., 1983) and in *Digitalis lanata* cells (Petersen, 1986), in *Ammi majus* cells (Hamerski and Matern, 1988a, 1988b), and wheat cells (Canivenc et al., 1989).

2.3. Specific content

The specific content of cytochrome P-450 per mg microsomal protein is generally quite low as compared to most animal tissues. Comparisons are difficult to make and may be very misleading because the figures found in the literature were obtained with various plant organs of different ages, grown under different light and temperature regimes and possibly exposed to various inducing chemicals or to injury. Since few works were performed under strictly sterile conditions, infection by pathogens may also be a factor of variation. Typical figures in non-induced tissues, e. g. dark grown seedlings or undamaged storage organs are 0.005 to 0.050 nmol · mg protein^{-1}. However, some organs in certain species show a much higher content: 0.2 nmol · mg protein^{-1} in dormant bulbs of *Tulipa gesneraria* and 0.4 nmol · mg protein^{-1} in the mesocarp of avocado pear (Rich and Lamb, 1977). Yahiel et al. (1974) found 0.070 nmol · mg protein^{-1} in the spadix of *Arum maculatum*. However, a survey made in our laboratory (Werck-Reichhart, unpublished) showed that, at the peak of the thermogenic crisis, much higher values may be observed: less than 0.05 nmol · mg protein^{-1} in stem and leaves, 0.1 in the flower and up to 0.602 nmol · mg protein^{-1} in the spadix. The function of this cytochrome P-450 remains unknown. NADPH-cytochrome c activity was comparatively low in these microsomes and only trace amounts of CA4H were measured.

We found 0.029 and 0.070 nmol cytochrome P-450 per mg protein^{-1} in cell suspension cultures from *Acer* and Jerusalem artichoke respectively (Salaün et al., 1982), and 0.110 to 0.140 nmol · mg protein^{-1} were recorded in *Digitalis lanata* cells (Petersen, 1986).

3. Subcellular localization

In all plants studied so far, cytochrome P-450 and cytochrome P-450 dependent activities were found to be localized in the microsomal fraction, defined as those cellular membranes sedimenting between 10,000 and 100,000 g. However, microsomes constitute a poorly defined and composite mixture of different fragments of the endomembrane system: smooth and rough endoplasmic reticulum, peroxisomes, tonoplaste, plasmalemma. Therefore, assessing a precise subcellular localization requires more than differential centrifugation. A few thorough studies, associating gradient centrifugation and biochemical characterization, yield a contrasting picture.

3.1. Localization on the endoplasmic reticulum

Most studies showed cytochrome P-450 and oxygenase activities associated with the ER. In germinating castor bean, CA4H and *p*-chloro-N-methylaniline N-demethylase were assigned to ER, based on gradient and marker studies

(YOUNG and BEEVERS, 1976). However, substantial demethylase activity was associated with mitochondrial and glyoxysomal fractions as well. Using similar techniques, cytochrome P-450 and CA4H were found associated with ER ($d = 1.12$ g/cm^3) in Jerusalem artichoke tuber (BENVENISTE et al., 1977) and in pea seedlings (BENVENISTE et al., 1978). In the latter study, velocity gradient centrifugation was used to separate ER from plastids to exclude a possible plastidial CA4H, as claimed by CZICHI and KINDL (1977). CA4H was also localized on the ER ($d = 1.12$ g/cm^3) in *Sorghum* seedlings (SAUNDERS et al., 1977), and both CA4H and cytochrome P-450 were associated with ER ($d = 1.11-1.14$ g/cm^3) in potato tubers (RICH and LAMB, 1977). Similarly, in *Vicia faba,* palmitic acid ω-hydroxylase and hydroxypalmitate in-chain hydroxylase were both assigned to the ER ($d = 1.12$ g/cm^3) (SOLIDAY and KOLATTUKUDY, 1978).

3.2. Localization on other membrane fractions

In *Catharanthus roseus* seedlings, cytochrome P-450 and geraniol 10-hydroxylase were associated with a provacuolar fraction (MADYASTHA et al., 1977), isolated by linear and discontinuous sucrose gradient centrifugation ($d = 1.09$ to 1.13 g/cm^3, depending on the technique used). The fraction had high alkaline phosphatase activity and showed a dense vacuolar population on electron micrographs. Cytochrome P-450 (essentially P-420) was found associated with phase partition-enriched plasma membranes from cauliflower inflorescences (KJELLBOM et al., 1985). In sweet potato roots infected with *Ceratocystis fimbriata*, the distribution of two cytochrome P-450 enzymes, CA4H and ipomeamarone 15-hydroxylase over a density gradient was different (FUJITA and ASAHI, 1985a). The 15-hydroxylase banded as a single peak ($d = 1.12-1.14$ g/cm^3) while CA4H appeared bimodal with a minor peak coinciding with the 15-hydroxylase, and a large one at $d = 1.17-1.19$ g/cm^3. A similar high density was also reported for CA4H in wound-induced sweet potato root tissue (TANAKA et al., 1974).

To conclude, cytochrome P-450 appears essentially, but not exclusively localized on the ER. The present endomembrane concept, as a developmental and sometimes physical continuum (HARRIS, 1986) may help us to conceive that certain cytochromes P-450, dedicated to particular catalytical tasks, may be located at some stage on other components of the endomembrane system.

4. Components of the plant cytochrome P-450 system

Plant cytochromes P-450 resemble the enzymes from mammalian liver microsomes. There is no evidence for mitochondrial cytochrome P-450 nor for the involvement of an adrenodoxin-like iron-sulfur protein in electron transfer to the oxygenase.

4.1. Cytochrome P-450

4.1.1. Spectral studies

Spectral studies show that, unlike mammalian microsomes, plant microsomes contain more cytochrome b_5 than P-450. In fact, what is generally referred to as cytochrome b_5 (measured after reduction with dithionite), is a mixture

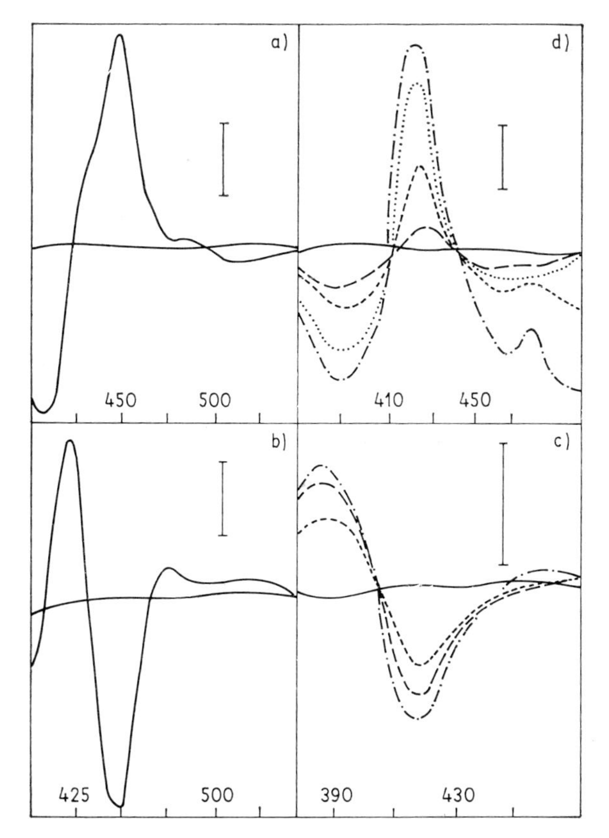

Fig. 1. Spectral interactions in Jerusalem artichoke (*Helianthus tuberosus* L.) microsomes. Different microsomal preparations were used (2—3 mg protein per assay).

a: Carbon monoxide difference spectrum of dithionite reduced Jerusalem artichoke microsomes.

b: ClO_4Na difference spectrum of dithionite reduced and CO bubbled Jerusalem artichoke microsomes.

c: Cinnamic acid induced type I difference spectrum of Jerusalem artichoke microsomes. ——— 5μM — — — 15 μM ·—·— 40 μM cinnamic acid.

d: Tetcyclacis induced type II difference spectrum of Jerusalem artichoke microsomes. — — 10 μM — — — 20 μM - - - - - 30 μM ·—·— 40 μM.
Bars represent 0.01 A scalers.

of NAD(P)H-reducible cytochrome b_5 and several other b-type cytochromes (RICH and BENDALL, 1975; HENDRY et al., 1981) among which an ascorbate-reducible cytochrome, with a Soret band at 429 nm (DURST et al., unpublished) of unknown function.

In most cases, reduced CO-difference spectra of plant microsomes show a major peak at 450 nm (448–452) and a shoulder of varying importance at 420–422 nm (Fig. 1a). Peaks at wavelengths higher than 452 nm (FUJITA et al., 1982; LARSON and BUSSARD, 1986) could be due to contaminating cytochrome aa_3 from mitochondrial debris. Cytochrome P-450 is readily converted to the P-420 form upon treatment of microsomes with heat or chaotropic agents (Fig. 1b). Cytochrome P-420 content is often high in plant microsomes, with P-420/P-450 ratios up to 6/1 in mung beans (HENDRY et al., 1981). In the latter case, in vivo spectrophotometry showed a similar P-420/P-450 ratio already present in the intact tissue. It is however the experience of our laboratory that cytochrome P-420 content can be reduced to very low level using speed and care when preparing microsomes. EPR signals typical of both high-spin and low-spin forms of cytochrome P-450 have been detected in plant microsomes (RICH et al., 1975).

Typical type I binding spectra (Fig. 1c) were first observed with *tr*-cinnamic acid and Jerusalem artichoke microsomes (BENVENISTE and DURST, 1974) and potato microsomes (RICH and LAMB, 1977). Type II binding spectra were obtained with aniline and cauliflower microsomes (RICH and BENDALL, 1975), and Jerusalem artichoke microsomes and aniline and various azoles in our laboratory (Fig. 1d) and others (KOCHS and GRISEBACH, 1986; TATON et al., 1988).

4.1.2. Solubilization and reconstitution

Cytochrome P-450 may be solubilized from plant microsomes by 1–2% (v/v) detergent solutions. Non-ionic (Emulgen 911, Renex 690) or zwitterionic (Chaps) detergents yield more stable preparations than cholate or deoxycholate. Purification of plant cytochrome P-450 has been the goal of many laboratories, with little success until now. A pure cytochrome P-450 preparation was obtained from Emulgen 911 solubilized tulip bulb microsomes (HIGASHI et al., 1983; HIGASHI 1985). The purified protein, with molecular weight 52,500, had an absorption peak in the reduced-CO difference spectrum at 448 nm (450 nm in the microsomes). The absolute spectrum showed maxima at 392, 525 and 645 nm in the oxidized form. The Soret band shifted to 412 nm upon reduction. Unfortunately, the function of this hemoprotein remains unknown since no catalytic activity was reported. In our laboratory, a highly enriched preparation of *tr*-cinnamic acid 4-hydroxylase (CA4H), has been obtained (GABRIAC et al., 1985). Upon SDS-gel electrophoresis, the major band, presumably the cytochrome, had a molecular weight of 56,000 and the preparation showed a maximum at 450 nm in the reduced-CO difference spectrum. When reconstituted with NADPH-cytochrome P-450 reductase, this preparation catalyzed cinnamate hydroxylation.

Taken together, all available evidence supports the view that plant cytochrome P-450 are ferriprotoporphyrin IX-containing hemoproteins with a thiolate fifth ligand, similar to mammalian cytochrome P-450.

4.2. NADPH-cytochrome P-450 reductase

NADPH-cytochrome P-450 reductase has been purified from microsomes from several plant species. A partial purification of the enzyme from *Catharanthus roseus* was first reported by Madyastha and Coscia (1979). More recently, Fujita and Asahi (1985b) reported the purification of a sweet potato reductase and we purified and characterized the enzyme from Jerusalem artichoke (Benveniste et al., 1986). That the purified enzymes were not just NADPH-cytochrome *c* reductases but effectively cytochrome P-450 reductase was demonstrated by reconstitution of geraniol 10-hydroxylase with partially purified *Catharanthus* cytochrome P-450 (Madyastha and Coscia, 1979) and, in the case of the Jerusalem artichoke enzyme, by reconstitution of CA4H activity. Furthermore, a polyclonal antibody raised against the reductase inhibited completely cinnamate hydroxylation by Jerusalem artichoke microsomes (Benveniste et al., 1986). This antibody has since been shown to inhibit also other cytochrome P-450 dependent monooxygenases in microsomes from different plant origins (Benveniste et al., 1989). Immunochemical studies

Table 1. Characteristics of three purified NADPH-cytochrome c (P450) reductases

Enzyme source	*Catharanthus roseus*	Sweet potato	Jerusalem artichoke
M_r	78,000	81,000	82,000
M_r (proteolytic form)	63,000	72,000	77,000
Specific activity (μM Cytc/min/mg)	17 μM	29 μM	33 μM
Active in reconstituted system	yes (G10H)	n. d.	yes (CA4H)
Km NADPH	5.7 μM	7.7 μM	20 μM
Km NADH	n. d.	n. d.	46 μM—54 mM
Mechanism	n. d.	n. d.	sequential
Flavins	FMN and FAD	n. d.	FMN and FAD

using polyclonal and monoclonal antibodies to the artichoke reductase, showed that there is no cross reaction between the plant enzyme and that from insect or mammalian microsomes (BENVENISTE et al., 1989).

Some characteristics of the three reductases are summarized in Table 1.

4.3. Other microsomal electron transfer components

Reconstitution experiments indicate that the minimal hydroxylating system is constituted from cytochrome P-450, the reductase and a lipid phase. This does not preclude the participation of other electron transfer components when cytochrome P-450 is acting in microsomes. In view of the ability (albeit limited) of NADH to sustain most monooxygenase reactions and several reports of NADPH/NADH synergism (see below), the role of cytochrome b_5 and NADH-cytochrome b_5 reductase must be considered.

4.3.1. Cytochrome b_5

Cytochrome b_5 has been purified from *Catharanthus roseus* microsomes (MADYASTHA and KRISHNAMACHARY, 1986), cytochrome b_5 and NADH-cytochrome b_5 reductase have been purified from potato microsomes (BONNEROT et al., 1985; GALLE et al., 1984; GALLE and KADER, 1986) and from pea microsomes (JOLLIE et al., 1987). Cytochrome b_5 from all plant sources had very similar M_r: 16,700; 16,400; and 16,500 in potato, pea and *Catharanthus*, respectively. In our laboratory, a cytochrome b_5 with higher M_r (22,000 in native state and 17,000 after trypsine treatment) has been purified from Jerusalem artichoke tuber microsomes (GABRIAC et al., unpublished). This hemoprotein reconstituted cytochrome *c* reductase activity with purified NADH-cytochrome b_5 reductase prepared from the same plant material. Interestingly, the pea cytochrome b_5 cross reacts with antibody raised to rat liver cytochrome b_5 (JOLLIE et al., 1987).

4.3.2. NADH-cytochrome b_5 reductase

Reductases with different molecular weights have been reported: 44,000 for the potato and 34,500 for the pea enzyme. The pea reductase has extremely high affinity for cytochrome b_5: apparent K_m is 7 nM, which is almost three orders of magnitude lower than that of the mammalian NADH-cytochrome b_5 reductase (SPATZ and STRITTMATTER, 1973). Although the flavin in NADH-cytochrome b_5 reductase has not yet been characterized, taken together, these reports indicate that the components of the fatty acid desaturase electron transfer route are very similar in plant and mammalian microsomes.

5. Reaction mechanism and catalytic cycle

Several lines of evidence indicate that the type of reactions, reaction mechanism and catalytic cycle of plant cytochrome P-450 enzymes are similar to those of the mammalian microsomal systems (see for example REICHHART et al., 1980). Oxidations of endogenous substrates catalyzed by cytochrome P-450 in plants (described in more detail in section 5) comprise aliphatic and aromatic hydroxylation, epoxidation, demethylation, and intramolecular rearrangements. With xenobiotic substrates, dealkylation, but not oxidation, of nitrogen has also been reported (FREAR et al., 1969; YOUNG and BEEVERS, 1976; FONNÉ, 1985; FONNÉ et al., 1988).

Irreversible inactivation of several plant oxygenases by mechanism-based inhibitors of cytochrome P-450 like terminal olefins and terminal acetylenes (SALAÜN et al., 1984, 1988), or compounds with N-N functions (REICHHART et al., 1982; WERCK-REICHHART, 1985) indicate that identical or closely related reactive intermediates are generated during catalysis by plant and animal cytochromes. Furthermore, nitrogen-containing compounds like pyrimidine and azole derivatives (COOLBAUGH et al., 1978; KOCHS and GRISEBACH, 1986, 1987; WENDORF and MATERN, 1986; VANDEN BOSSCHE et al., 1988; HAMERSKI and MATERN, 1988a and b; TATON et al., 1988), elicit type II binding spectra and inhibit plant cytochrome P-450, probably through bond formation between their hetero atomic lone pair and prosthetic heme, as demonstrated for numerous pyrimidine and imidazole derivatives with animal microsomes.

We have measured a 1:1:1 NADPH/O_2/substrate stoichiometry during cinnamate hydroxylation in Jerusalem artichoke microsomes (unpublished). However, in pea microsomes that contain high NADPH-oxidase activity, more NADPH and O_2 than substrate were consumed. NADPH/NADH synergism has been reported for many but not all plant cytochrome P-450 reactions (see section 6). Based on inhibitor studies, it was proposed that, in the presence of NADPH and NADH, the first electron would be transferred from NADPH via NADPH-cytochrome P-450 reductase but that the second electron would be preferentially donated by NADH, via the NADH/cytochrome b_5 pathway (BENVENISTE et al., 1982b). More recently, polyclonal and monoclonal antibodies to the Jerusalem artichoke NADPH-cytochrome P-450 reductase were used to probe the pyridine nucleotide dependency of plant cytochrome P-450 enzymes (BENVENISTE et al., 1989). NADPH and NADH dependent oxidation of cinnamic acid, lauric acid and the herbicide chlortoluron were almost completely inhibited by the antibodies. This could imply that with NADH as the electron source, the first electron is donated via NADPH-cytochrome P-450 reductase since NADH-cytochrome *c* reductase was not inhibited.

Mammalian cytochrome P-450 may react with different organic hydroperoxides to directly yield the ferryl-oxo complex required for hydroxylation. Studies in our laboratory (unpublished) showed that cumene and linoleyl hydroperoxides had only limited capacities to sustain cinnamic acid and lauric acid hydroxylation in Jerusalem artichoke microsomes. This was probably due to rapid inactivation of the enzymes.

6. Oxidation of endogenous substrates

In this section, I will discuss the different cytochrome P-450 dependent reactions demonstrated so far (November 1988), starting with CA4H, the archetypal and most studied plant monooxygenase. Other reactions will be briefly described, with emphasis on specific features and those characteristics that deviate from the 'CA4H model'. The classification used here is for the sake of clarity and is rather arbitrary.

6.1. Oxidation of C6—C3 compounds

6.1.1. *tr*-Cinnamic acid 4-hydroxylase (CA4H)

Deamination of phenylalanine by phenylalanine ammonia-lyase (PAL) produces cinnamic acid, thus forming the C6—C3 phenylpropanoid building block of lignins, flavonoids, and cinnamic esters. The 4-hydroxylation of this central secondary metabolite, common to all major phenylpropanoid pathways, seems ubiquitous among plant species.

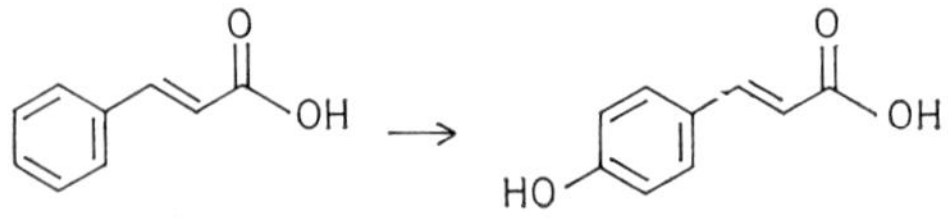

O OH → O OH HO

Fig. 2. Conversion of *tr*-cinnamic acid to 4-hydroxycinnamic (*p*-coumaric) acid catalyzed by cinnamic acid 4-hydroxylase.

CA4H (Fig. 2) was first accurately described in pea microsomes (RUSSELL, 1971). Although some properties of the enzyme have not been confirmed later, the monooxygenase nature of the enzyme was clearly recognized, and the involvement of cytochrome P-450 suggested, based on requirement for NADPH and light-reversible CO-inhibition. Further evidence for the cytochrome P-450 nature of the hydroxylase was obtained from type I binding spectra recorded in Jerusalem artichoke microsomes (BENVENISTE and DURST, 1974) and from the action spectrum of light reversal of CO-inhibition in Sorghum microsomes (POTTS et al., 1974). The importance of membrane lipids was shown by delipidation experiments with parsley microsomes (BÜCHE and SANDERMANN, 1973). Since then, numerous reports have confirmed and completed our understanding of this enzyme.

The requirement for NADPH-cytochrome P-450 reductase is clearly demonstrated by the use of antibody raised to purified Jerusalem artichoke reductase. NADPH dependent cinnamate hydroxylation was 90–99% inhibited by this antibody in artichoke microsomes (Benveniste et al., 1986), and in microsomes from 15 other plant species tested (Benveniste et al., 1989).

6.1.2. *p*-Coumaroylshikimate 3′-hydroxylase

In the plant, *p*-coumaric acid is further hydroxylated at C-3 to caffeic acid. *In vitro* hydroxylation of free *p*-coumaric acid by soluble phenolases has been reported (Stafford and Dressler, 1972; Bolwell and Butt, 1983). Recently, it was found that microsomes from elicitor-induced parsley cells catalyzed the hydroxylation of the shikimate ester of *p*-coumaric acid (Heller and Kühnl, 1985), thus forming *tr*-5-O-caffeoylshikimate from *tr*-5-O-(4-coumaroyl)shikimate (Fig. 3). The reaction required O_2, NADPH (K_m NADPH = 19 μM), and showed NADPH/NADH synergism. It was inhibited by CO and this inhibition was partially reversed by 450 nm light. This hydroxylase showed high substrate specificity: the *cis* isomer of the substrate was not oxidized, nor were the quinate or CoA esters of *p*-coumaric acid. Microsomes from carrot cell cultures hydroxylate both shikimate (K_m = 12 μM) and quinate (K_m = 4.5 μM) esters. Inhibitor studies suggest that these reactions are catalyzed by the same isoform. The physiological importance of cytochrome P-450 versus phenolases remains to be clarified.

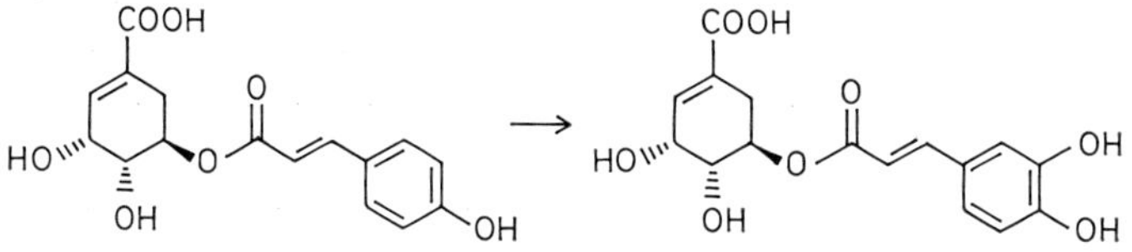

Fig. 3. Conversion of *tr*-5-O-(4-coumaroyl)shikimic acid to *tr*-5-O-caffeoylshikimate by *p*-coumaroylshikimic acid 3′-hydroxylase.

6.1.3. Ferulic acid 5-hydroxylase

Ferulic acid, formed by O-methylation of caffeic acid, is a precursor of sinapic acid, one of the lignin monomers. From *in vivo* labelling experiments, involvement of a ferulate 5-hydroxylase was postulated in this process. Recently, the 5-hydroxylation of ferulic acid by microsomes from poplar seedlings (Fig. 4) has been reported (Grand, 1984). Specific activity (per mg microsomal protein) was higher in sclerenchyma than in xylem. Cytochrome P-450 involvement was demonstrated by O_2 and NADPH requirement, NADPH/NADH synergism, and light reversible CO-inhibition. The K_m for ferulate was 63 μM, and the reaction was inhibited by cinnamic acid and, to a lesser extent, by 5-hydroxy-ferulic acid. *p*-Coumaric acid had no effect on the hydroxylase activity.

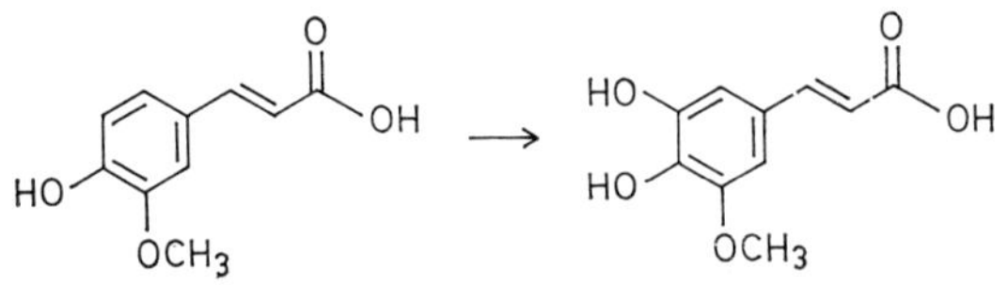

Fig. 4. Conversion of ferulic acid (3-methoxy. 4-hydroxycinnamic acid) to 5-hydroxy-ferulic acid catalyzed by ferulic acid 5-hydroxylase.

6.2. Oxidation of flavonoids

Flavonoids form a complex and widely distributed group of secondary metabolites which display a C6–C3–C6 skeleton. Since the diversity of flavonoids is substantially based on their oxidation state, numerous oxidation reactions are involved in this biosynthetic pathway. Although other oxidases (e. g. 2-oxo-glutarate-dependent dioxygenases) also participate in flavonoid synthesis, in recent years an increasing number of cytochromes P-450 involved in this biosynthetic pathway has been characterized, to a great part by Grisebach and his colleagues.

6.2.1. Flavonoid 3′-hydroxylase

Although the hydroxylation of the B ring can also be catalyzed by phenolases, these enzymes are perhaps not specifically involved in flavonoid synthesis. Recently, Grisebach and co-workers characterized a flavonoid 3′-hydroxylase (Fig. 5) which converted (S)-naringenin to eriodictyol in microsomes from light-induced parsley cell cultures (Hagmann et al., 1984). Based on O_2 and NADPH requirement, inhibition by CO and reversion by light, this activity was assigned to cytochrome P-450. Dihydrokaempferol, kaempferol and apigenin were also 3′-hydroxylated to dihydroquercetin, quercetin and luteolin, respectively (Fig. 5). Apparent K_m were in the 0.15–0.7 μM range, and pH 7.5 was optimum. (S)-Naringenin and apigenin, but not kaempferol, produced substrate inhibition at concentrations higher than 2.5 μM. The (R)-isomer of naringenin was not recognized. Surprisingly, 1 mM N-ethylmaleimide stimulated this activity instead of the expected inhibition due to cytochrome P-450 reductase inhibition.

A flavonoid 3′-hydroxylase was also demonstrated in microsomes from maize seedlings (Larson and Bussard, 1986). The activity was highest in shoots but also present in roots. This maize enzyme much resembles the parsley hydroxylase excepted that dihydrokaempferol was not substrate and pH optimum was 8.5. K_m values for kaempferol and NADPH were 7.15 and 5.8 μM respectively.

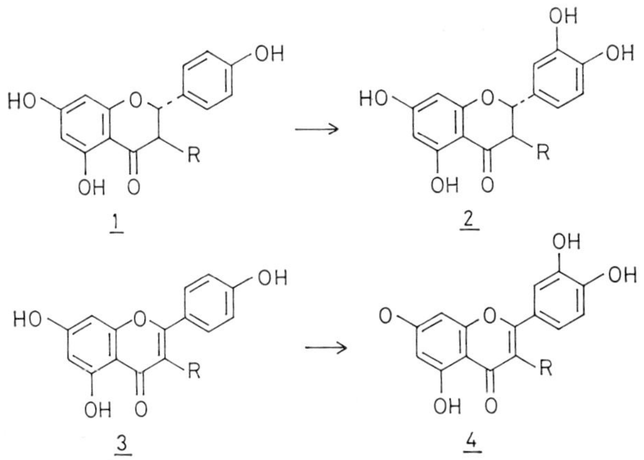

Fig. 5. Four reactions catalyzed by flavonoid 3′-hydroxylase:
a) R = H: conversion of naringenin (1) to eriodictyol (2) and of apigenin (3) to luteolin (4).
b) R = OH: conversion of dihydrokeampferol (1) to dihydroquercetin (2) and of kaempferol (3) to quercetin (4).

6.2.2. Isoflavone synthase

During biosynthesis of isoflavonoids, an important subgroup of flavonoids, the B ring of flavonoids undergoes a 2,3-aryl migration. The intramolecular rearrangement of a flavanone, (2S)-naringenin, to the isoflavone genistein (Fig. 6), is catalyzed by microsomes from elicitor-challenged soybean cell cultures (see section 7.3.). The reaction was oxygen and NADPH dependent and was not supported by ascorbate, FMN, FAD or pteridin (HAGMANN and GRISEBACH, 1984). Low concentrations of oxidized cytochrome *c* were inhibitory. It was suggested that the reaction proceeds via cytochrome P-450 cataly-

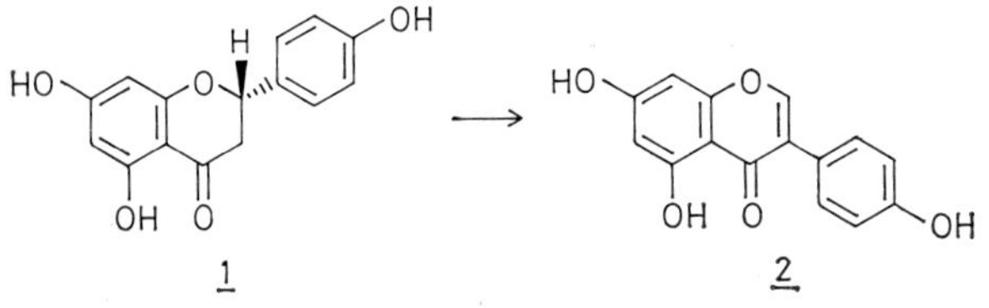

Fig. 6. Conversion of (2S)-naringenin (1) to genistein (2) by isoflavone synthase.

zed 3,4-epoxidation of naringenin. Involvement of cytochrome P-450 in this reaction was later confirmed (KOCHS and GRISEBACH, 1986). Evidence was provided by CO-inhibition, light reversion of this effect, and inhibition by SKF 525A and different fungicides and growth retardants known to interfere with cytochrome P-450 (see section 8.2.). NADH did not sustain the reaction but NADPH/NADH synergism was observed. Optimum pH for this isoflavone synthase was higher than usual, between 8.0 and 8.6. From detailed investigations of the reaction mechanism, it was concluded that conversion of naringenin to genistein proceeds in two distinct steps. First, naringenin is converted to 2,5,7,4'-tetrahydroxyisoflavanone, which is then dehydrated. While the first step is clearly microsomal, the second seems catalyzed by a soluble system (160,000 g supernatant) (KOCHS and GRISEBACH, 1986).

6.2.3. Flavone synthase II

The conversion of (S)-naringenin to apigenin (Fig. 7) is catalyzed by microsomes from osmotically stressed soybean cells (KOCHS and GRISEBACH, 1987). Cytochrome P-450 dependence is evidenced by O_2 and NADPH dependence, light reversible CO-inhibition and inhibition by SKF 525A, metyrapone and by a

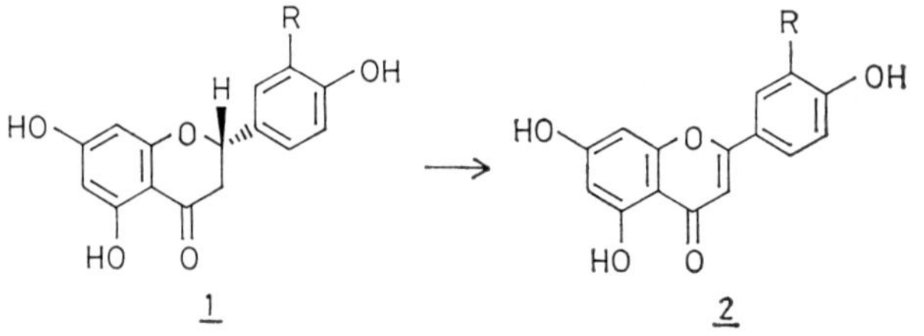

Fig. 7. Two reactions catalyzed by flavone synthase II.
a) R = H: conversion of (2S)-naringenin (1) to apigenin (2).
b) R = OH: conversion of eriodictyol to luteolin.

series of azole derivatives. Apparent K_m values were 11.4 μM for (S)-naringenin and 39 μM for NADPH. The enzyme was termed flavone synthase II because the same reaction is also catalyzed in irradiated parsley cells, by a previously characterized flavone synthase (synthase I) which is a soluble 2-oxo-glutarate dependent dioxygenase (BRITSCH et al., 1981).

6.2.4. Pterocarpan 6α-hydroxylase

Soybeans infected by the fungus *Phytophthora megasperma* accumulate the isoflavonoid glyceollin, a pterocarpanoid phytoalexin. Microsomal preparations from elicitor-challenged soybean cell cultures catalyze the stereospecific hydroxylation at position 6α of 3,9-dihydroxypterocarpan to 3,6α,9-trihydroxypterocarpan (Fig. 8) which is a precursor for glyceollin (HAGMANN et al., 1984). The enzyme activity was very unstable, with a half-life of 9 min in glycerol-containing phosphate buffer. Activity showed a broad pH optimum between 7.2 and 8.2, required O_2, was completely NADPH-dependent and inhibited by cytochrome *c*. Taken together, these findings suggested that this 6α-hydroxylase is cytochrome P-450 dependent.

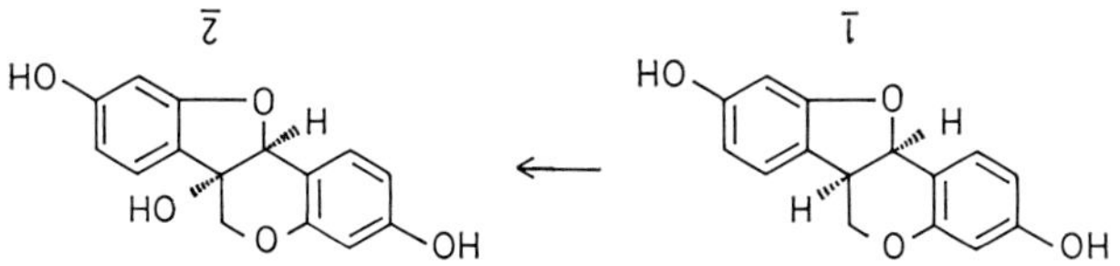

Fig. 8. Conversion of 3,9-dihydroxypterocarpan (1) to 3,6α,9-trihydroxypterocarpan (2) by pterocarpan 6α-hydroxylase.

Pisatin, another pterocarpan phytoalexin, is produced by pea plants challenged by fungi. Pisatin synthesis in pea, and its metabolism by the fungus *Nectria haematococca* has been studied in detail by VANETTEN and coworkers. In the fungus, which is tolerant to pisatin, pisatin is O-demethylated at the 3-methoxy group by a cytochrome P-450 that has been solubilized and reconstituted (DESJARDINS and VANETTEN, 1984). In pea, labelling experiments suggest that pisatin is synthesized from the precursor (+)-maakiain via hydroxylation at the 6α position (Fig. 9). Some isolates of *Nectria haematococca* detoxify (−)-maakiain by 6α-hydroxylation. The two hydroxylase reactions, in the plant and in the fungus, have been compared by $^{18}O_2$ labelling experiments followed by mass spectrometry analysis of reaction products (MATTHEWS et al., 1987). Surprisingly, molecular oxygen was incorporated at the 6α position during maakiain hydroxylation by the fungus but not during the synthesis of

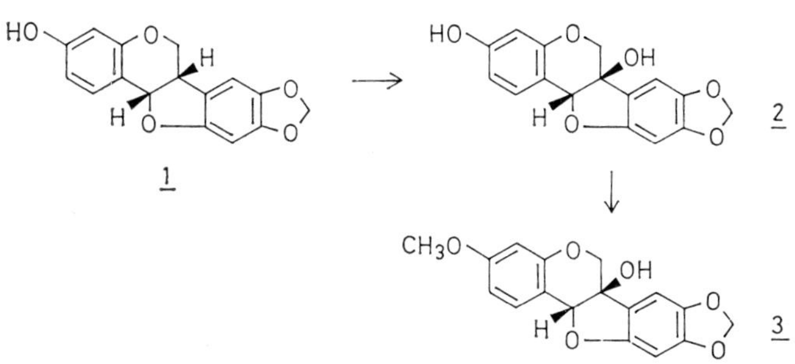

Fig. 9. Conversion of (+)-maakiain (1) to (+)-6α-hydroxymaakiain (2) by pterocarpan 6α-hydroxylase. (2) is further methylated to pisatin (3).

pisatin by pea. This is in contrast to the results by HAGMANN et al. (1984) showing that 6α-hydroxylation during glyceollin synthesis in soybean is O_2 and NADPH dependent. Clearly, the mechanism for oxygen incorporation at the 6α position of pterocarpans warrants further investigation.

6.2.5. Marmesin synthase

It was recently reported that demethylsuberosin conversion to (+)-marmesin (Fig. 10) is effectively catalyzed by microsomes isolated from elicitor-induced cell cultures of *Ammi majus* (HAMERSKI and MATERN, 1988b). No activity was found in non-induced cells. Marmesin synthase activity was maximal at pH 7.5 and 20 °C, and remained stable for 10 min only at this temperature. K_m values

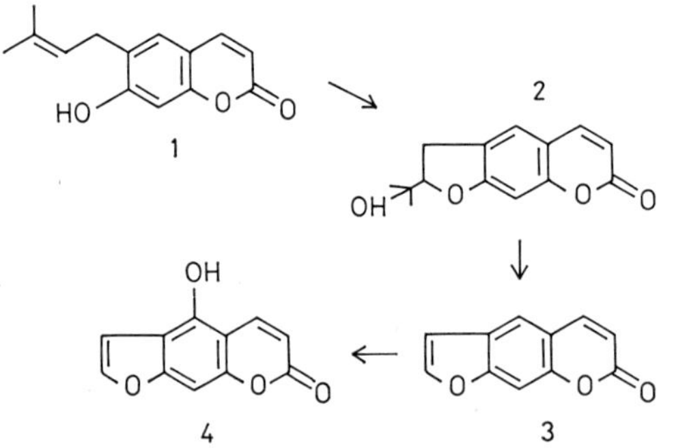

Fig. 10. Conversion of demethylsuberosin (1) to marmesin (2) by marmesin synthase of (2) to psoralen (3) by psoralen synthase and of (3) to bergaptol (4) by psoralen 5-monooxygenase.

were 10.3 μM for demethylsuberosin and 19.6 μM for NADPH. Supporting evidence for cytochrome P-450 involvement in marmesin synthase is similar to that for psoralen synthase: O_2 and NADPH requirement, blue light-reversible CO inhibition and action of several cytochrome P-450 inhibitors. Marmesin synthase, psoralen synthase and prenyltransferase were all localized on the ER. This, and *in vivo* labelling experiments showing that exogeneously supplied umbelliferone was much better incorporated into psoralen than marmesin, may suggest that umbelliferone enters the ER lumen where the whole reaction sequence to psoralen would take place (HAMERSKI and MATERN, 1988b).

6.2.6. Psoralen synthase

Dark-grown parsley cells accumulate linear furanocoumarins (psoralens) in response to elicitor challenge. Microsomes prepared from elicitor-induced parsley cells catalyze the conversion of marmesin to psoralen (Fig. 10) (WEN-

DORF and MATERN, 1986). This activity was not detectable in microsomes from non-induced cells. Half-life of the enzyme was 50 min at 25 °C and 20 min at 37 °C. The optimal pH was 7–7.2 and the reaction rate was linear for 45 min at 20 °C. The reaction was dependent upon O_2 and NADPH, NADH was not effective but NADPH/NADH synergism was observed. K_m value for NADPH was 52 μM and not determined for marmesin. The activity was inhibited by $NADP^+$ and cytochrome *c*. The cytochrome P-450 nature of this psoralen synthase was further evidenced by blue light-reversible inhibition by CO, as well as inhibition by several fungicides and growth retardants acting on cytochrome P-450. A mechanism for marmesin to psoralen conversion was proposed. It involves oxidation at the 3′ position *cis* to the isopropyl side chain, followed by water and acetone loss and 2′–3′ double bond formation.

6.2.7. Psoralen 5-monooxygenase

Psoralen is further oxidized to bergaptol (5-hydroxypsoralen) (Fig. 10) by microsomes from elicitor treated *Ammi majus* cell cultures (HAMERSKI and MATERN, 1988a). K_m values of 12 μM for psoralen and 200 μM for NADPH were determined. Activity was maximal after 20–24 h induction. The activity is inhibited by tetcyclacis (80% at 5 μM) and blue light-reversible inhibition by CO was observed. Gradient centrifugation showed that psoralen monooxygenase is localized on the E. R.

6.3. Ipomeamarone 15-hydroxylase

Sweet potato roots infected by the fungus *Ceratocystis fimbriata* produce and accumulate various antifungal furano-sesquiterpenes, among which ipomeamarone. This phytoalexin is further metabolized to ipomeamaronol (Fig. 11)

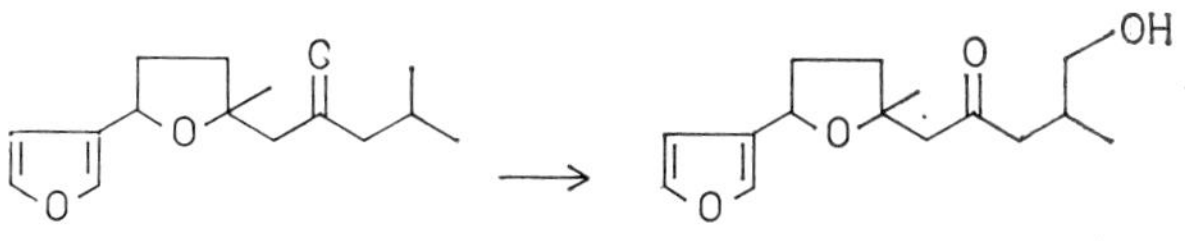

Fig. 11. Conversion of ipomeamarone (1) to ipomeamaronol (2) by ipomeamarone 15-hydroxylase.

in the plant by a microsomal 15-hydroxylase (FUJITA et al., 1981). The enzyme required O_2 and NADPH. NADH did not sustain the reaction but NADPH/NADH synergism was observed (FUJITA et al., 1982). Activity was inhibited by CO and this inhibition was partially prevented by light. K_m values of 60 mM for ipomeamarone and 2 μM for NADPH were determined. Competition experiments showed that this enzyme is distinct from CA4H which is also present in sweet potato microsomes.

6.4. Oxidation of terpenoids

Numerous secondary plant metabolites are terpene derivatives which may display biological (hormones, phytoalexins) or pharmacological (alkaloids) activity, and/or are constituents of membranes (sterols). There is evidence that cytochrome P-450 is implicated in the biosynthesis of several terpenoids.

6.4.1. Geraniol 10-hydroxylase

Geraniol and nerol (*cis* and *trans* isomers, respectively) are precursors of indole alkaloids in *Catharanthus roseus*. Both monoterpenols are hydroxylated at C-10 (Fig. 12) by microsomal fractions from *C. roseus* seedlings (MEEHAM and

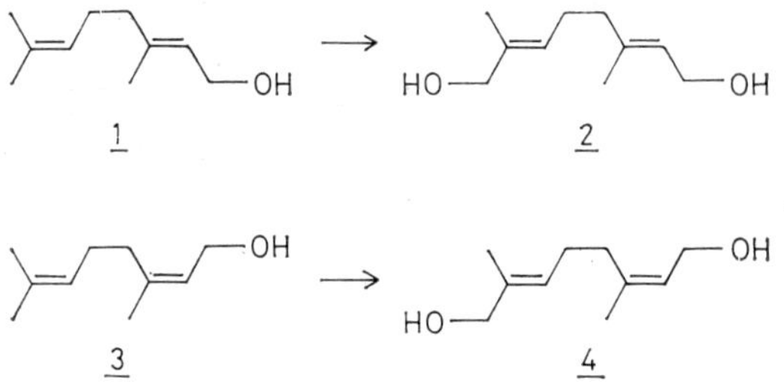

Fig. 12. Hydroxylation of geraniol (1) and nerol (3) to the corresponding diols (2) and (4) by geraniol 10-hydroxylase.

COSCIA, 1973; MADYASTHA et al., 1977). The K_m (5.4 μM) and V_{max} values are similar for both substrates, suggesting that the same isoenzyme catalyzes these oxidations. The enzyme is NADPH- and O_2-dependent, showed light reversible CO-inhibition and was 60% inhibited by 1 mM SKF 525 A and by 1 mM NaN3 (MEEHAN and COSCIA, 1973). The reaction was also inhibited by several indole alkaloid end products: vinblastine, vindoline and catharanthine. The latter compound acted as a non-competitive inhibitor with $K_i = 1$ mM suggesting a feedback control mechanism (McFARLANE et al., 1975).

6.4.2. Abscisic acid 6′-hydroxylase

Metabolism of the plant hormone abscisic acid (ABA) to phaseic acid (Fig. 13) proceeds via 6′-hydroxymethyl-ABA formation. This hydroxylation is probably catalyzed by cytochrome P-450. The reaction, studied in microsomes prepared from the liquid endosperm of *Echinocystis lobata* seeds, required O_2 and NADPH (GILLARD and WALTON, 1976). It is inhibited by CO, but light reversal was not reported. (±)-ABA-methylester and the 1′,4′-diol of (±)-ABA were not oxidized, and the enzyme appears highly specific for the naturally occuring (+)-ABA.

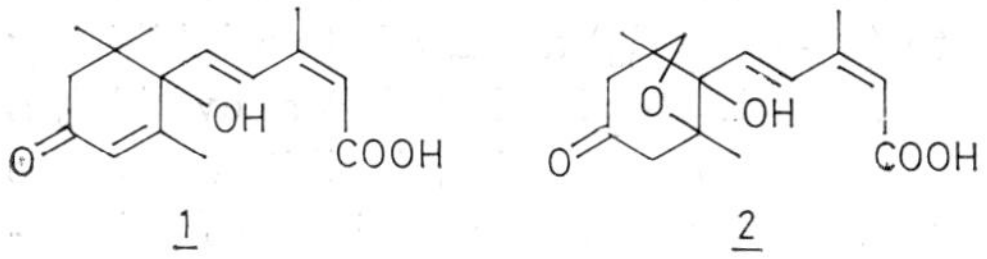

OH COOH O 1 O OH COOH O 2

Fig. 13. Conversion of abscisic acid (1) to phaseic acid (2) by abscisic acid 6′-hydroxylase.

6.4.3. Kaurene oxidase

The biosynthesis of gibberellins, a large family of plant hormones, involves several cytochrome P-450-catalyzed steps in the common pathway from *ent*-kaurene to gibberellin A12-aldehyde (for a recent review, see GRAEBBE, 1987). There is very good evidence that a sequence of 4 consecutive oxidative steps converting kaurene (*ent*-kaur-16-ene) to 7β-hydroxykaurenoic acid (Fig. 14), are catalyzed by cytochrome P-450 (MURPHY and WEST, 1969; LEW and WEST, 1971; COOLBAUGH and MOORE, 1971, COOLBAUGH et al., 1978). The study by

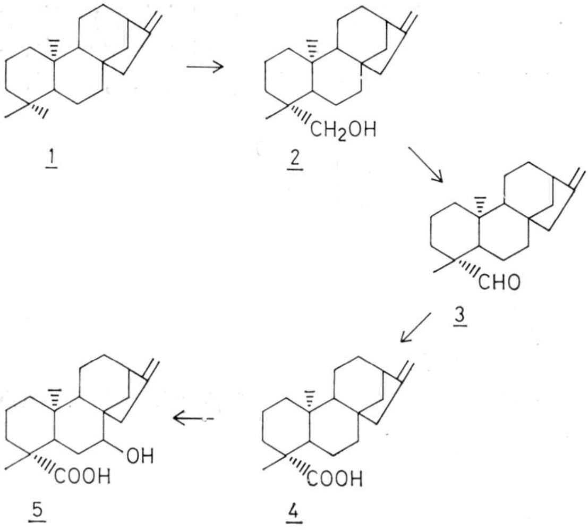

1 CH2OH 2 CHO 3 OH COOH 5 COOH 4

Fig. 14. A sequence of 4 consecutive oxidative steps converting kaurene (*ent*-kaur-16-ene) (1) to 7β-hydroxykaurenoic acid (5), via kaurenol (2), kaurenal (3) and kaurenoic acid (4).

Coolbaugh and Moore was with microsomes from pea seeds and the others with the more favorable endosperm of *Marah macrocarpus* seeds. Evidence for cytochrome P-450 involvement is based on NADPH and O_2 requirement, light reversible CO-inhibition, and incorporation of ^{18}O after incubation with $^{18}O_2$. K_m values for the different substrates (kaurene, kaurenal and kaurenol) are about 0.5 μM (West, 1980). Different isoenzymes may be involved because high concentrations of one substrate (10—100 K_m) are needed to inhibit by 50% the oxidation of another (West, 1980). The finding that ancymidol, a pyrimidine derivative with growth retardant properties, inhibited kaurene oxidation (Coolbaugh and Hamilton, 1976) provided a rationale for interpreting the effect of such compounds (see section 8.2.).

6.4.4. Digitalin 12-hydroxylase

First evidence for cytochrome P-450 dependent oxidation of a steroid-like-molecule in higher plants was the conversion of the cardiac glycoside digitoxin to digoxin (Fig. 15) by a 12β-hydroxylase from *Digitalis lanata* cell cultures (Petersen and Seitz, 1985). The enzyme is NADPH-dependent, no NADPH/NADH synergism was observed, pH optimum was 7.5. K_m value for β-methyldigitoxin, the substrate used in this study, was 7.1 μM, but other cardenolides like acetyldigitoxin, digitoxin and digitoxigenin were also hydroxylated by *Digitalis* microsomes. The stability of this enzyme is remarkable. While most plant cytochrome P-450 reactions are linear only over short periods of time (5—20 min), digitoxin 12-hydroxylation was linear for 4 h and the enzyme retained activity after 20 h (Petersen, 1986). A reconstituted system containing partially purified cytochrome P-450, NADPH-cytochrome P-450 reductase and a microsomal lipid extract catalyzed digitoxin 12-hydroxylation (Petersen, 1986).

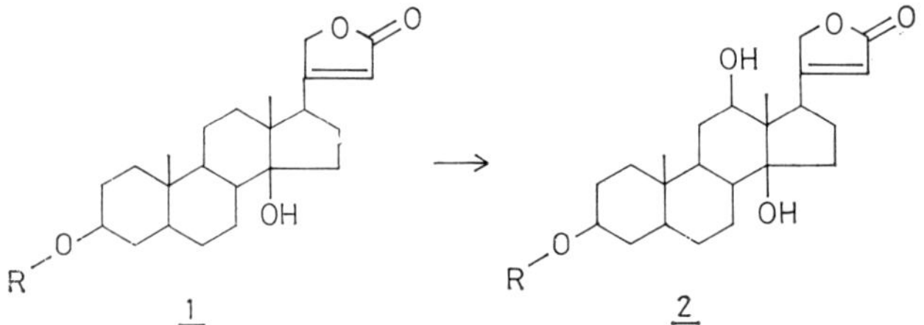

Fig. 15. Hydroxylation of cardiac glycosides by digitoxin 12β-hydroxylase. With R = H (1) is digitoxigenin and (2) is digoxigenin.

6.4.5. Obtusifoliol 14α-demethylase

The loss of the 14α-methyl group is a key step in sterol biosynthesis. This reaction is catalyzed by cytochrome P-450 in mammals and also in yeasts (Aoyama et al., 1984). Recently, a cytochrome P-450 dependent obtusifoliol

14α-demethylase (Fig. 16) has been demonstrated in corn embryos microsomes (RAHIER and TATON, 1986). Demethylation at C-14 creates a C-14, C-15 double bond which is then reduced by a distinct C-14 reductase. The demethylase is

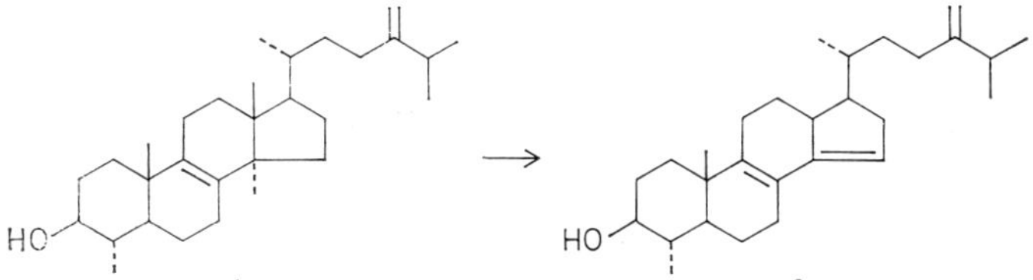

Fig. 16. Conversion of obtusifoliol (1) to 4α-methyl-5α-ergosta,8,14-24-trien-3β-ol (2) by obtusifoliol 14α-demethylase.

strictly NADPH-dependent, shows light-dependent CO inhibition and is highly sensitive to triazoles like propiconazole and triadimeton which inhibit lanosterol 14a-demethylation in yeasts (GADHER et al., 1983). A common property with digitoxin 12-hydroxylase is the stability of the demethylase, the reaction proceeding linearly for at least 5 h at 30 °C.

6.5. Fatty acid oxidation

6.5.1. Oxidation of long chain fatty acids

Polyesters derived from hydroxylated and epoxidated fatty acids are main constituents of cutins and suberins (KOLATTUKUDY, 1977). The enzymes involved in oxidation of C16 and C18 acids have been extensively studied and partially characterized by KOLATTUKUDY's group.

$$HOOC-(CH_2)_n-CH_3 \longrightarrow HOOC-(CH_2)_n-CH_2OH$$

1 2

Fig. 17. Oxidation of fatty acids at the methyl terminus by fatty acid ω-hydroxylase. (1) is capric acid with n = 8, lauric acid with n = 10, myristic acid with n = 12 and palmitic acid with n = 14.

A microsomal preparation from *Vicia faba* catalyses the ω-hydroxylation (Fig. 17) of palmitic acid. Hydroxylation is O_2 and NADPH-dependent, inhibited by CO, but no light reversal could be observed (SOLIDAY and KOLATTUKUDY, 1977). This ω-hydroxyacid is further hydroxylated at C-10 by a 27,000 g supernatant from the epidermis of *Vicia* leaves in presence of O_2, NADPH, ATP and CoA, suggesting that activation of the carboxyl group is required

(WALTON and KOLATTUKUDY, 1972). Later it was found that microsomes from *Vicia* shoots catalyze C-9 or C-10 hydroxylation of 16-hydroxypalmitic acid (K_m 50 μM) without requiring CoA or ATP. Photoreversible inhibition of this latter reaction by CO was demonstrated (SOLIDAY and KOLATTUKUDY, 1978).

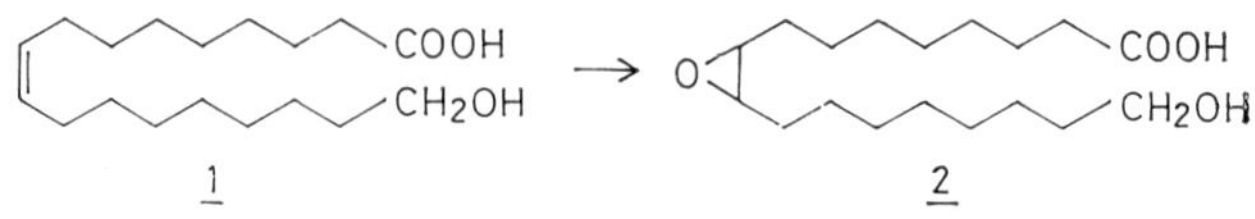

Fig. 18. 18-hydroxyoleic acid is epoxidated by a 9,10-epoxidase.

Young spinach leaves catalyze the 9–10 epoxidation of 18-hydroxyoleic acid (Fig. 18) (CROTEAU and KOLATTUKUDY, 1975). The activity, located in a 3,000 g particulate fraction, required O_2, NADPH, ATP and CoA. Apparent K_m was 75 μM. Light reversible inhibition by CO suggests the involvement of cytochrome P-450, despite the unusual subcellular localization of this epoxidase and its high pH optimum near 9.0.

6.5.2. Lauric acid hydroxylases

In 1978, SALAÜN et al. showed that aerobic incubation of Jerusalem artichoke microsomes with lauric acid (dodecanoic acid) and NADPH, produced a mixture of 8-, 9- or 10-hydroxylaurate (10, 65 and 15% respectively). This in-chain hydroxylase (IC-LAH) (Fig. 19) is a typical cytochrome P-450 dependent oxidase with an apparent K_m of 1.0 μM. Subsequently, we discovered a second laurate monooxygenase, strictly ω-hydroxylating (Fig. 17) (ω-LAH), in pea microsomes (BENVENISTE et al., 1982a). The apparent K_m was 20 μM and,

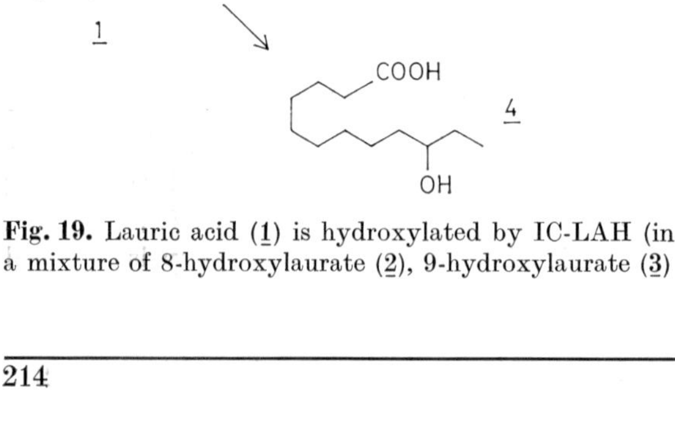

Fig. 19. Lauric acid (1) is hydroxylated by IC-LAH (in chain laurate hydroxylase) to a mixture of 8-hydroxylaurate (2), 9-hydroxylaurate (3) and 10-hydroxylaurate (4).

contrary to the in-chain hydroxylase, this enzyme showed no NADPH/NADH synergism. Since then the ω-hydroxylase has been found in several other species, all of which are leguminosae, while the in-chain hydroxylase was found in all other plants (SALAÜN et al., 1982). Although the survey is quite limited (circa 20 species), it is striking that both activities were never found in the same plant.

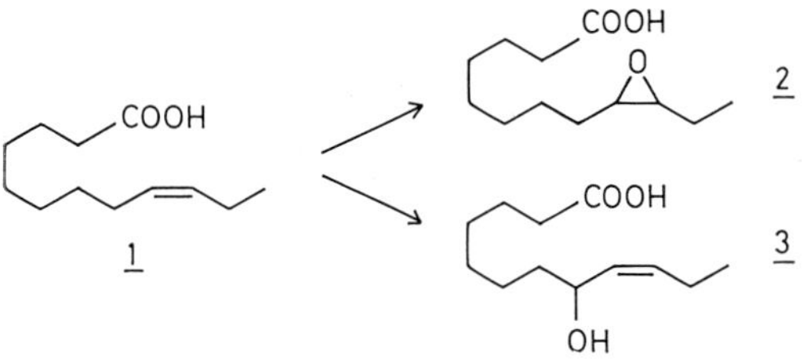

Fig. 20. 9-Dodecenoic acid (1) is oxidized by IC-LAH to a mixture of 9,10-epoxylaurate (2) and 8-hydroxy-9-dodecenoate (3).

The two hydroxylases appeared rather specific for lauric acid since incubations with radiolabeled laurate + unlabelled fatty acids with 12 or 14 carbons, lauric acid methylester, 1-dodecanol or dodecanedioic acid showed little inhibition of hydroxylaurate formation (SALAÜN et al., 1981; Benveniste et al., 1982a). Only decanoate did compete to some extent with laurate.

In an effort to explore the regio- and stereo-selectivity of these enzymes, we have synthesized radiolabeled *cis* and *trans* dodecenoic acids unsaturated at positions 7–8, 8–9, 9–10, 10–11 or 11–12 (SALAÜN et al., submitted). Preliminary results show that the *Vicia* enzyme ω-hydroxylates all these substrates, with the already described suicide effect of 11-dodecenoic acid (SALAÜN et al., 1988). IC-LAH on the other hand, catalyzed the epoxidation of sp2 carbons (Fig. 20), concurrently with hydroxylation at saturated positions (SALAÜN et al., unpublished). When carbon C-9 is in an allylic position, its hydroxylation is greatly favored over epoxidation of the double bond. Surprisingly, apparent affinities for the *cis* and *trans* isomers appear very similar and epoxidation takes place with retention of configuration.

6.5.3. Caprate and myristate hydroxylases

More recently we have studied the oxidation of radiolabeled capric (C10) and myristic (C14) acids in *Vicia* microsomes (SIMON, 1987). Both fatty acids were actively ω-hydroxylated (Fig. 17). Cytochrome P-450 involvement was demonstrated by O_2 and NADPH requirement, light reversible CO-inhibition and inhibition by anti NADPH-cytochrome P-450 reductase antibody. Furthermore, induction and inhibition studies showed that caprate, laurate and myristate are substrates for the same cytochrome P-450 isozyme, indicating that in *Vicia*, ω-LAH is a medium chain fatty acid hydroxylase (SIMON, 1987). Capric acid is substantially more polar than laurate and myristate, but the apparent K_m values fall within the same range (20—30 μM) if corrected for substrate partition between the lipidic and aqueous phases of the incubation medium (SIMON et al., unpublished). Preliminary results show that Jerusalem artichoke microsomes, containing IC-LAH, also catalyze the cytochrome P-450 dependent in-chain oxidation of caprate and myristate.

The physiological role of these widely distributed enzymes remains unclear. It is worth noting that the positions hydroxylated by IC-LAH (8, 9 or 10) are also those hydroxylated by the palmitate mid-chain hydroxylase described by SOLIDAY and KOLATTUKUDY (1978) in *Vicia* microsomes. Similarly, the epoxidase from spinach leaves oxidizes at position 9—10 of hydroxyoleate (CROTEAU and KOLATTUKUDY, 1975). However, palmitic acid is apparently not a substrate for the LAHs and medium chain hydroxyacids are at best minor components of plant cutins and suberins. Work is now being undertaken to unravel the mechanisms of unsaturated long chain fatty acid hydroxylation and epoxidation in our plant materials.

6.6. Reactions that might be catalyzed by cytochrome P-450

The enzymes that catalyze the countless oxygenation steps of biosynthesis or catabolism of secondary metabolites remain largely unknown. In some cases, the involvement of cytochrome P-450, although not demonstrated, is suggested.

Ethylene, a plant hormone, is oxidized to the *cis*-epoxide by *Vicia faba* microsomes incubated with O_2 and NADPH (SMITH et al., 1985). K_m for ethylene was 0.02 μM, pH optimum was 7.9 and the reaction was inhibited by CO. However, CO seems to compete with ethylene rather than with O_2 in this system.

Solanidin, a steroidal alkaloid from *Solanum chacoense*, is metabolized by microsomes from the plant tuber (OSMAN et al., 1987). The reaction product, 23β-hydroxysolanidine, is neither formed in the absence of NADPH nor with the glycosylated substrate.

Several reports suggest that cytochrome P-450 may be implicated in the metabolism of cytokinins, an important group of plant hormones. N^6-(isopentenyl)adenine and N^6-(isopentenyl)adenosine are hydroxylated at position 4 on the isopentenyl side-chain in the presence of O_2, NADPH and cauliflower microsomes (CHEN and LEISNER, 1984). The activity was totally inhibited by 10 mM metyrapone and by 100% CO, but this may be an effect of anaerobiosis alone. Similarly, the 30—40% inhibition produced by flushing the reaction mixture for 10 minutes with ethylene might be due to suicidal inactivation of a cytochrome P-450 (see section 8.1) or simply to deprivation of oxygen. *In vivo* metabolism of N^6-benzylaminopurine, another cytokinin, is inhibited by metyrapone in tomato pericarp (LONG and CHISM, 1987).

Several other claims for the participation of cytochrome P-450 in diverse metabolic pathways have been made. They are based on weaker, thus more indirect, evidence. It has been suggested that cytochrome P-450 is involved in the biosynthesis of valepotriates, the pharmacologically active compounds found in valerian roots (VIOLON and VERCRUYSSE, 1985). But this proposal is sustained only by the linear relation found by the authors between valepotriate contents and cytochrome P-450 titer in *Centranthus macrosiphon* tissue cultures. The inhibition of betacyanin synthesis in *Amaranthus* seedlings treated with SKF 525A was taken as evidence for the implication of cytochrome P-450 in light-dependent and independent betacyanin biosynthesis (OBRENOVIC, 1986). Metyrapone inhibited growth, chlorophyll, carotenoid and anthocyanin synthesis in mung bean seedlings. This might be due to the inhibition of cytochromes P-450 involved in the biosynthesis of growth promoting corticosteroids (GEUNS, 1977). Such findings are difficult to interpret since inhibitory effects of metyrapone and SKF 525A are often weak in the *in vitro* assays of well defined plant cytochrome P-450 activities.

7. Induction and regulation

The finding that plant cytochrome P-450 enzymes may be selectively enhanced in response to various factors (light, injury, pathogens, xenobiotics) offered a potent tool for studies on multiplicity, regulation and the characterization of cytochromes P-450 which take minor, sometimes undetectable, forms in the non-induced state.

7.1. Induction by light

Phytochrome-mediated induction of an enzyme of the phenylpropanoid pathway was first demonstrated for PAL (phenylalanine ammonia-lyase) (DURST and MOHR, 1966). Since then this photoreceptor has been implicated in the regulation of numerous other plant enzymes. A 5-fold increase of CA4H following a short white light irradiation of etiolated pea seedlings was already reported in the early work of RUSSELL (1971). Involvement of phytochrome in this light effect was subsequently demonstrated in pea seedlings submitted to either short pulses of red and far-red light, or to continuous far-red or blue irradiation (BENVENISTE et al., 1978). The lag-phase for CA4H and NADPH-cytochrome P-450 reductase induction was very short, about 40 minutes. Total cytochrome P-450 was only slightly stimulated, presumably because other cytochrome P-450 enzymes were not phytochrome dependent. This was confirmed by the finding that lauric acid ω-hydroxylase was induced in the same plants only after a 24 h lag-phase, with a time course very different from that of CA4H (BENVENISTE et al., 1982a). Flavonoid 3′-hydroxylase was found in microsomes from light-grown but not from dark-grown parsley seedlings (HAGMANN et al., 1983). Similarly, the enzyme was not detectable in dark-grown parsley cell cultures, but was induced by white light and reached a maximum after 24 h irradiation, synchronous with PAL and CA4H. This coordinate induction of PAL and CA4H by light has been reported by several groups (HAHLBROCK et al., 1976; BILLET and SMITH, 1980). In contrast, it was found that cytochrome P-450 content decreased rapidly when 3 day-old dark-grown mung bean seedlings were transferred to light (HENDRY et al., 1981). When the seedlings were returned to darkness, cytochrome P-450 accumulated once more. It was suggested that this light-mediated decrease might be due to heme breakdown. An alternative explanation would be that, among the numerous enzymes induced by light, only a few are cytochrome P-450 isoenzymes. Hence, bulk cytochrome P-450 would be stimulated less than bulk protein, resulting in apparent decrease of cytochrome P-450 specific content. This could also explain that when ω-LAH in irradiated pea was related to apical bud it seemed induced (BENVENISTE et al., 1982a), but when expressed per mg microsomal protein the hydroxylase was depressed (BENVENISTE et al., 1980).

7.2. Induction by injury

Injury and "ageing", i. e. slicing of plant tissues and incubation of the slices in moist atmosphere or in well aerated solutions, result in activation of numerous metabolic processes, including enzyme synthesis. CA4H is induced by ageing in Jerusalem artichoke tuber disks (BENVENISTE and DURST, 1974; BENVENISTE et al., 1977). This induction coordinates with that of PAL (DURST, 1976) and occurs within 40 minutes after slicing. It is abolished by slicing in a nitrogen atmosphere or in the presence of cycloheximide. Activity, which is undetectable in the intact tuber, goes through a maximum 24 h after excission and declines thereafter. Cytochrome P-450 and NADPH-cytochrome P-450 reductase are also induced. It was subsequently found that IC-LAH is also induced in ageing tuber disks (SALAÜN et al., 1978). In this system, the induction of CA4H, but not of IC-LAH was strongly repressed by exogenously supplied 4-hydroxycinnamate (SALAÜN et al., 1981). Comparable results were reported in ageing sweet potato (TANAKA et al., 1974) and in potato slices (RICH and LAMB, 1977). In the potato system however, activity and cytochrome P-450 were detected in the intact tuber and the lag-phase for induction was about 8 h.

Osmotic stress (i. e. transfer to a medium containing 0.4 M glucose) applied to soybean cells induced flavone synthase II from 0 to 270 nkat/kg but repressed isoflavone synthase (Kochs and Grisebach, 1987).

7.3. Induction by pathogens and elicitors

Phytoalexins are low-M_r compounds produced in plants exposed to pathogens or to elicitors, i. e. various fractions derived from pathogenic and nonpathogenic fungi. It is not surprising that several cytochromes P-450 involved in the synthesis of flavovoid and terpenoid phytoalexins are strongly and selectively induced in elicitor-challenged plants or cell cultures. In several cases the enzyme is detectable only in elicited cells or plants.

CA4H is the first cytochrome P-450 in the general phenylpropanoid pathway. This enzyme, and NADPH-cytochrome P-450 reductase, are induced in parsley cells challenged with elicitor from *Phytophthora megasperma*. In contrast, flavonoid 3′-hydroxylase was not affected (Hagmann et al., 1983). This *P. megasperma* elicitor induced also the 5-O-(4-coumaroyl)shikimate 3′-hydroxylase from parsley cell cultures (Heller and Kühnl, 1985). Activity was not detectable in non-induced cells.

In soybean cell cultures, isoflavone synthase was induced 2-fold after 19 h of treatment by elicitor from *Alternaria carthami* and 3-fold after exposure to *Phytophthora megasperma* elicitor (Kochs and Grisebach, 1986). It is remarkable that the flavone synthase II from the same soybean cells was not induced by these elicitors.

The *P. megasperma* elicitor induced also the pterocarpan 6α-hydroxylase involved in glyceollin synthesis in soybean cell cultures (Hagmann et al., 1984). After 50 h treatment, the hydroxylase, which was low but detectable in control cells, was stimulated 6-fold. Treatment of soybean seedlings with the elicitor produced an 8.6-fold increase in hydroxylase activity. A 6α-hydroxylase involved in pisatin synthesis is also induced in elicitor-challenged pea, but the nature of this enzyme remains unclear (Matthews et al., 1987).

Psoralen synthase, the enzyme forming psoralen from marmesin, was characterized in parsley cells challenged with elicitors from *P. megasperma* and *A. carthami* (Wendorf and Matern, 1986). Both fungi are non-pathogenic to parsley. This activity was not measurable in non-induced cells. Similarly, marmesin synthase could be identified and studied in *Alternaria*- and *Phytophthora*-induced microsomes from *Ammi majus* cells but not in microsomes from non-induced cell cultures (Hamerski and Matern, 1988a). Similar results were reported for psoralen 5-monooxygenase (Hamerski and Matern, 1988b).

Synthesis of furano-sesquiterpenes and activity of ipomeamarone 15-hydroxylase are induced in sweet potato tissue either by injury or by infection with

Ceratocystis fimbriata (FUJITA et al., 1981). The hydroxylase was not found in intact roots and was 3 to 5-fold higher in sickened than in injured tissues. It is noteworthy that CA4H was induced more than the 15-hydroxylase in damaged tissues whilst the reverse was observed in *C. fimbriata* infected tissues (FUJITA et al., 1982).

Treatment of bean cell cultures with an elicitor from *Colletotrichum lindemuthianum* produced a 3-fold increase of CA4H activity (BOLWELL and DIXON, 1986). A monoclonal antibody raised to a rat liver cytochrome P-450 inhibited cinnamate hydroxylation to 40% and recognized a 48 kD peptide in Western blots. The 48 kD protein comigrates with a hemeprotein on SDS-PAGE analysis. However, it appears that the antibody will also recognize some other protein unrelated to P-450 (BOLWELL, personal communication), and the identity of the 48 kD protein needs confirmation.

7.4. Induction by endo- and exochemicals

Since several years, evidence has accumulated that cytochrome P-450 is induced in plants exposed to xenobiotics.

7.4.1. Induction by metals

Divalent manganese ions produce a coordinate increase of PAL and CA4H activities in ageing Jerusalem artichoke tuber tissues (DURST, 1976). Time-course studies (REICHHART et al., 1979, 1980) showed that CA4H was stimulated 3 to 4-fold in tuber slices aged in 25 mM $MnCl_2$ and that the activity peak was time-shifted from 24 h in control tissues aged in water to 96 h. Cytochrome P-450 was induced with a similar time-course whilst NADPH-cytochrome P-450 reductase was not increased but shifted from 24 to 50 h. In some experiments, CA4H and cytochrome P-450 were further stimulated by complementing the medium with 10 μM $FeCl_3$ (REICHHART et al., 1979). Subsequently it was found that IC-LAH, which is very low in noninduced slices, was stimulated 15-fold in slices aged on manganese (SALAÜN et al., 1981). Mn^{2+} is very specific since other divalent cations Cu^{2+}, Co^{2+} or Ni^{2+} could not mimic its effect, but its mode of action remains unknown. Manganese also induced CA4H in potato tuber slices but was inhibitory in pea seedlings (unpublished).

In sweet potato root discs exposed for one day to 3.68 mM $CdSO_4$ or $HgCl_2$ and further incubated for one day, cytochrome P-450 was increased by 50% and 100% respectively (FUJITA, 1985). These inductions were accompanied by an increase of sequiterpenes synthesis.

7.4.2. Induction by organic xenobiotics

Cytochrome P-450 and CA4H were induced, with differing time-courses and intensities, in Jerusalem artichoke tuber slices exposed to various compounds: 300 mM ethanol, 4 mM phenobarbital and 200 μM of several herbicides such as monuron, dichlobenil and chloro-IPC (Reichhart et al., 1979, 1980). Phenobarbital showed very different effects on CA4H and IC-LAH: whilst CA4H was stimulated 1.5-fold, a 26-fold increase of IC-LAH was observed (Salaün et al., 1981). Similarly, 2,4-dichlorophenoxyacetic acid (2,4-D), a synthetic auxine with herbicide properties, induced CA4H and IC-LAH with different time-courses and intensities (Adelé et al., 1981).

The effect of phenobarbital was studied in several plant systems and seems of general importance (Salaün et al., 1982; Fonné et al., 1988). Consistently, the laurate hydroxylases were much more induced than CA4H by phenobarbital. An example of selective isozyme induction by xenobiotics is shown in Table 2. Clofibrate, a hypolipidemic drug which induces specifically ω-LAH in mammalian systems, caused a 28-fold increase of ω-LAH in *Vicia* but only a two-fold stimulation of IC-LAH in artichoke tissues (Salaün et al., 1986). A recent study by Simon (1987) in our laboratory showed that several mono- and dichloro-phenoxypropanoic and phenoxybutanoic analogs of clofibrate are also potent inducers of cytochrome P-450 and ω-LAH in *Vicia faba* seedlings. In some cases, the effect of phenobarbital seems restricted to a precise developmental stage. For example, we have recently observed much higher inductions of the enzyme hydroxylating the herbicide dichlofop in 2–3 days wheat seedlings than in older ones (Zimmerlin et al., to be published).

Finally, one must be aware that stimulation depends on many factors besides the nature and concentration of the inducer: i) the basal level of activity, which

Table 2. Comparative effects of clofibrate on cytochrome P-450, lauric acid ω- and in-chain hydroxylase and CA4H activities

	H. tuberosus		***V. sativa***	
	Water	**Clofibrate**	**Water**	**Clofibrate**
Cytochrome P-450	53	119	58	89
ω-LAH	0	0	15	429
IC-LAH	7	16	0	0
CA4H	580	3510	172	281

Microsomes were extracted from *Vicia sativa* seedlings or Jerusalem artichoke tuber slices incubated in water or 2 mM clofibrate solution. Enzyme activities are in pmol $\times$ min^{-1} $\times$ mg^{-1} protein. Cytochrome P-450 content is in pmol $\times$ mg^{-1} protein.

in turn is drastically influenced by the handling of the tissues (cutinjury, light or dark growth, composition of aging or culture medium, etc. ...), ii) the timing of observations since it appears that inductions take the form of waves reaching a maximum within hours or even minutes (elicitor effects on cell suspension cultures) or several days (manganese induction in tuber slices), iii) the plant species, the organ under study and its stage of development.

7.4.3. Induction by substrates

Cytochrome P-450 was also induced by feeding substrates to mung bean seedlings (Hendry and Jones, 1984). Exposure of 80 h, seedlings for 16 h, to 10 mM cinnamate or geraniol increased cytochrome P-450 by 1.8 and 2.7-fold, respectively. These increases were suppressed in the presence of either cycloheximide or laevulinate indicating that protein and heme synthesis were required for the induction to take place.

7.5. Regulation

Expression of cytochrome P-450 activities may be regulated at the level of synthesis, stability and activity of the individual isoenzymes, but also by the rates of heme synthesis and of electron transfer from NADPH-cytochrome P-450 reductase.

7.5.1. Heme synthesis

Unlike animals, yeasts or bacteria, higher plants synthesize heme for cytochrome P-450 induction through a glutamate-dependent pathway, and not via mitochondrial 5-ALA-synthase (Werck-Reichhart et al., 1988). Involvement of this C5 pathway for heme synthesis was demonstrated by the use of gabaculine and other inhibitors of pyridoxal-linked transaminases, which strongly repressed the induction of cytochrome P-450 and monoxygenase activities in Jerusalem artichoke tuber slices. Not only cytochrome P-450, but also cytochrome b_5, peroxidase and chlorophyll synthesis were blocked by these compounds. NADPH-cytochrome *c* reductase was only slightly affected. The turnover of cytochrome b_5 appeared much shorter than that of cytochrome P-450.

7.5.2. Electron transfer

There is no evidence available to show that electron transfer may limit oxygenase activity *in vivo*. *In vitro*, when microsomes are incubated under V_{max} conditions, NADPH-cytochrome P-450 reductase never seems to be a limiting factor: 60% inhibition of cytochrome *c* reduction by $NADP^+$ results in only 30% inhibition of CA4H or LAH in artichoke microsomes. The possible participation of NADH-cytochrome c reductase in plant cytochrome P-450 reactions is supported by numerous reports of NADPH/NADH synergism.

Very indirect evidence for cytochrome b_5 participation stems from stopped-flow experiments showing that cytochrome b_5 is transiently reoxidized in artichoke tuber microsomes reduced by low NADPH (5 μM + regenerating system) upon addition of cinnamic acid (Durst, unpublished), suggesting that electrons are drawn from cytochrome b_5 when CA4H is turning over.

7.5.3. Protein synthesis

There are several reports showing that induction of cytochrome P-450 is inhibited by cycloheximide. Until antibody and cDNA probes become available it will not be possible to study in more detail the molecular mechanisms nnderlying induction of plant cytochrome P-450 isoenzymes. In the case of (NADPH-cytochrome P-450 reductase, where specific antibody was obtained Benveniste et al., 1986), ^{35}S-methionine *in vivo* labeling experiments followed by immunoprecipitation showed good agreement between induction of reductase activity and enzyme synthesis in artichoke tuber slices (Benveniste et al., unpublished). Also, Western blot analysis of translation products of polyadenylated RNA extracted from 0 and 16 h tuber slices, showed that the messenger RNA for the reductase is virtually absent in dormant tubers, and induced upon slicing (Lesot et al., unpublished).

8. In vitro and in vivo inhibition

Like other enzymes, cytochromes P-450 may be inhibited by substrate analogs and, in rare cases, by reaction products or end-products (Russel, 1969; Grand, 1984; McFarlane et al., 1975). Also, many redox compounds will compete with the cytochrome by accepting electrons from cytochrome P-450 reductase. It is surprising to note that the classical broad-spectrum inhibitors of animal cytochrome P-450 enzymes like metyrapone, SKF 525A or piperonyl butoxide, when used in the 10^{-5} to 10^{-6} M concentration range, are only weak, if at all, inhibitors of the plant enzymes. Some notable exceptions are inhibition of isoflavone synthase and flavone synthase II by SKF 525A (Kochs and Grisebach, 1986, 1987) and inhibition of geraniol hydroxylase by the same compound (Meehan and Coscia, 1973). Also, some *in vivo* studies suggest that they may inhibit herbicide detoxication (Gaillardon et al., 1985), oxidation of cytokinins (Chen and Leisner, 1984), synthesis of betacyanins (Obrenovic, 1986) and of corticosteroids (Geuns, 1977).

The following discussion will be concerned with only two types of inhibitors interacting specifically with plant cytochrome P-450: mechanism-based inactivators (suicide-substrates) and nitrogen containing heterocycles binding to ferric protoheme iron. An excellent discussion of cytochrome P-450 inhibition is given in a recent review by ORTIZ DE MONTELLANO (1986).

8.1. Mechanism-based inhibitors

Suicide inhibitors are substrates that are catalytically activated by the enzyme to reactive species that bind, irreversibly in most cases, to the active center. The overall efficiency of such inhibitors is primarily governed by the partition of the transition state between the formation of an alkylating species and the 'normal' course of reaction (for example forming an epoxide from an olefinic substrate). In a collaborative effort with the laboratory of ORTIZ DE MONTELLANO, we have studied the susceptibility of plant P-450 to several such compounds. A summary of results appears in Table 3.

When oxidized by cytochrome P-450, ABT (1-aminobenzotriazole) yields benzyne which reacts immediately with two heme nitrogens (ORTIZ DE MON-

Table 3. Loss of cytochrome P-450 and inhibition of oxygenase activities in plant microsomes treated with different inhibitors

Compound	% P-450	%CA4H	% IC-LAH	% ω-LAH
AIA	5 (J. a.)	2 (J. a.)	0 (J. a.)	0 (Pea)
ABT	32 (J. a.)	97 (J. a.)	0[a] (J. a.)	0[a] (Pea)
DCBT	19 (J. a.)	0 (J. a.)	n. m.	n. m.
PP	12 (J. a.)	95 (J. a.)	0 (J. a.)	n. m.
$PPCl_2$	18 (J. a.)	94 (J. a.)	19 (J. a.)	n. m.
9-DNA	25 (J. a.)	5 (J. a.)	45	0 (V. f.)
11-DDNA	43 (J. a.)	67 (J. a.)	80 (J. a.)	
11-DDYA	15 (J. a.)	0 (J. a.)	12 (J. a.)	75 (V. f.)
Phenelzine	75 (J. a.)	73 (J. a.)	n. m.	n. m.

[a]: 20% at very high (2 mM) ABT concentration.

Microsomes were preincubated at 25 °C with NADPH and the compounds (20—100 μM) prior to substrate addition. Preincubation time was 10 or 20 min for activity and 30 min for P-450 measurements. Results are expressed as % inhibition as compared to controls preincubated with NADPH alone.
AIA: 2-isopropyl, 4-pentenamide; ABT: 1-aminobenzotriazole; DCBT: 5,6-dichloro 1,2,3-benzothiadiazole; PP: 3-phenoxy,1-propyne; $PPCl_2$: 3(2,4-dichlorophenoxy) 1-propyne; 9-DNA: 9-decenoic acid; 11-DDNA: 11-dodecenoic acid; 11-DDYA: 11-dodecynoic acid. Phenelzine: 2-phenylethylhydrazine; J. a.: Jerusalem artichoke; V. f.: *Vicia faba*; n. m.: not measured.

TELLANO, 1986). It was first shown that ABT determines the irreversible loss of CA4H in Jerusalem artichoke microsomes (REICHHART et al., 1982).

Inactivation followed pseudo-first order kinetics and the half-life of the enzyme was under one min at 25 °C and saturating ABT. Lauric acid hydroxylase and bulk cytochrome P-450 were less affected. ABT was effective not only *in vitro* but also was able to destroy about 90% of CA4H activity *in vivo* in artichoke tuber tissues (REICHHART, 1982). This observation has been extended to other plant materials and cytochrome P-450 enzymes. Several groups have used the *in vivo* inhibition of metabolism of xenobiotics by ABT as an indication of cytochrome P-450 involvement in this process (GAILLARDON et al., 1985; CABANNE et al., 1987; COLE and OWEN, 1987; GONNEAU et al., 1987; KEMP and CASELEY, 1987; CANIVENC et al., 1989).

Other N-N functions, such as phenylhydrazones and substituted hydrazines, are also activated to reactive intermediates by cytochrome P-450 (ORTIZ DE MONTELLANO, 1986). Phenelzine (2-phenylethylhydrazine) produced a very rapid loss of about 75% of cytochrome P-450 and CA4H activity in Jerusalem artichoke microsomes (WERCK-REICHHART, 1985).

In many cases, cytochrome P-450 catalyzed epoxidation of terminal olefins is accompanied by inactivation of the enzyme due to N-alkylation of the prosthetic heme group (ORTIZ DE MONTELLANO, 1986). Inactivation of plant fatty acid hydroxylases by several fatty acid analogs with terminal double and triple bonds has been extensively studied in our laboratory. The in-chain lauric acid hydroxylase is effectively inactivated by 9-decenoic acid with a half-life at saturating inhibitor of 10 min (SALAÜN et al., 1984) whereas the ω-hydroxylase is specifically inhibited by 11-dodecynoic acid (11-DDYA), the half-life at saturating inhibitor being 2.4 min (SALAÜN et al., 1988). Recently, it was found that 11-DDYA inactivated both capric acid and myristic acid hydroxylases from *Vicia sativa* microsomes (SIMON, 1988). On the other hand, 11-dodecenoate appeared much less specific and, at higher concentrations, markedly reduced the microsomal cytochrome P-450 content and CA4H activity (SALAÜN et al., 1984, 1988).

8.2. Nitrogen containing heterocyclic inhibitors

Several fungicides and plant growth retardants inhibit cytochrome P-450 reactions. Most of these compounds contain a heterocyclic nitrogen atom, and some are able to produce type II binding spectra with plant microsomes. This suggests that they bind to the cytochrome through the nitrogen lone pair excluding oxygen from the reaction center.

First evidence was provided when Coolbaugh and colleagues showed that ancymidol, a pyrimidine derivative which is a potent growth regulator in higher plants, inhibited the oxidative steps from kaurene to kaurenoic acid (see section 5.4.3.) in *Marah oreganus* and *M. macrocarpus* microsomes with K_i values in the order of 10^{-9} (COOLBAUGH and HAMILTON, 1976; COOLBAUGH et al., 1978). In contrast, 7-hydroxylation of kaurenoic acid and cinnamate 4-hydroxylation were not inhibited. The inhibition of gibberellin biosynthesis provides a rationale for the biological activity of this and other growth regulators. Cytokinins, which have structural similarities to ancymidol also inhibited kaurene oxidation in wild cucumber microsomes (COOLBAUGH, 1984). Since then, several other growth retardants like tetcyclacis and various triazoles (BAS 110, BAS 111 ...) which share a sp^2-hybridized nitrogen atom at the periphery of the molecule were shown to inhibit the kaurene oxidases with I_{50} values in the range 10^{-5} to 10^{-8} M (RADEMACHER et al., 1988). It is interesting that tetcyclacis and the triazoles tested in this study did not inhibit pterocarpan 6α-hydroxylase and only very weakly or not at all (depending on plant material) CA4H. Inhibition of sterol 14α-demethylase with I_{50} values around 10^{-5} M and type II binding spectra were observed in maize microsomes with tetcyclacis, BAS 110 and BAS 111 and triazole fungicides (TATON et al., 1988).

The interaction of antifungal azoles with microsomes from yeast, plants (Jerusalem artichoke, pea, tulip and maize) and mammals has been studied using the decrease of CO difference spectrum produced by azole binding as a measure of sensitive cytochrome P-450 isozymes content (VANDEN BOSSCHE et al., 1988). Measured in this way, the I_{50} values for azaconazole, penconazole, propiconazole and imazalil appeared 200 to 500-fold higher for plant than for yeast microsomes. Of course, the implicit assumption that the proportion of sensitive isozymes is similar in plants and yeast may not be true. Indeed, when enzyme activity is measured instead, the plant 14α-demethylase was readily inhibited by propiconazole (RAHIER and TATON, 1986).

9. Physiological significance of cytochrome P-450 in plants

I will distinguish two groups of plant cytochromes P-450, those found, at least at one developmental stage, in all plants, and those peculiar to families or species.

The first group is at present small and may remain so in the future. The archetype of these enzymes is the cinnamic acid 4-hydroxylase. Due to its strategic position in the biosynthesis of the ubiquitous phenylpropanoids, this enzyme is probably present in all higher plants and may be the most abundant cytochrome P-450 on earth. Surprisingly, one of the two forms of the lauric acid hydroxylases (LAHs), whose roles remain obscure (this is also true for the animal LAHs), have been found in each of the tens of plants assayed in our laboratory. Tentatively, one may also include in this group the longer chain fatty hydroxylases and epoxidases which contribute to the biosynthesis of the plant cuticles cutin and suberin and the cytochromes P-450 involved in plant hormone synthesis and catabolism.

Except perhaps for the laurate hydroxylases, all these enzymes catalyze reactions vital to plants, which warrants their widespread distribution. It should be stressed that functional similarities do not imply identity at the sequence level. Molecular cloning studies are required to determine the degree of homology of functionally similar enzymes in different plant species.

There are over 10,000 known secondary metabolites in plants but there may be as many as 400,000 (SWAIN, 1977). A number of these secondary metabolites are phytoalexins produced by plants challenged by pathogens or chemical deterrents to phytophagous animals. A glance at the structures of such compounds evokes the conviction that oxygenation, in the form of hydroxylation, oxidative dealkylation, ring closure, isomerisation..., takes place during their biosynthesis. In view of the limited number of plant oxygenase classes, it seems inevitable that cytochromes P-450 catalyze a substantial number of these reactions. This implies that hundreds of cytochromes P-450, some of them unique to a family or even a species, exist in plants and await characterization. The cytochromes P-450 already described and which are involved in the synthesis and metabolism of phytoalexins belong to this rapidly growing group.

The cytochromes P-450 that metabolize foreign compounds are probably also members of this group. It is believed that the relative unspecificity of many animal monooxygenases stems from evolutionary pressure imposed by allochemicals, mainly of plant origin, present in animal diets (BOXENBAUM, 1983). Such evolutionary pressure did not exist for the plant cytochrome P-450 system and there are no compelling reasons why cytochromes P-450 with broad substrate specificity should have evolved in plants. More likely, xenobiotics enter by chance in the site(s) of one or several of the cytochromes P-450 that are present in a particular plant (or organ). This would explain the wide interspecific differences in detoxification capacity that form the basis for metabolism dependent selectivity of herbicides.

It has recently been suggested (HENDRY, 1966) that cytochrome P-450 is mainly restricted to young and achlorophyllous tissues and that its principal role is detoxification of endogenous compounds. In mature tissues, most cytochromes P-450 would be turned off and defence chemicals would accumulate. Attractive as it is, the proposal does not stand to facts. Indeed, almost all published work is concerned with either etiolated seedlings or storage tissues devoid of chlorophyll. The difficulties of measuring cytochrome P-450 and monooxygenase activities outlined above are severely aggravated in green mature tissues. Chlorophylls are not only an obvious nuisance for spectrophotometric determinations, but mature cells have large vacuoles with elevated contents of inhibitory compounds. Nevertheless, we have measured 'normal' amounts of cytochrome P-450, CA4H and LAH in such materials, even in the green parts of the marine plant *Posidonia* (GABRIAC and SALAÜN, unpublished). Moreover, CA4H and LAH remained inducible by xenobiotics in these tissues. Furthermore, there is growing evidence that cytochrome P-450 is at least as important for the synthesis of phytoalexins and chemical deterrents as for their detoxification.

Another role of this enzyme system could be to mediate some of the morphogenetic blue light effects in plants. These effects are most probably mediated by flavines and the flavine-mediated photoreduction of *b* and *c* type cytochromes in plant membranes has been reported (Schmidt and Butler, 1976). The finding that blue light stimulated CA4H in pea seedlings (Benveniste et al., 1977) led us to propose that blue light sensitization of NADPH-cytochrome P-450 reductase could enhance monooxygenase activity. However, at least in the case of CA4H, this hypothesis could not be substantiated by experimental evidence (Benveniste et al., 1980). Recently , a similar role has been proposed for a plasma membrane cytochrome P-450 from cauliflower (Kjellbom et al., 1985).

10. Concluding remarks

The study of plant cytochrome P-450 appears more complex than anticipated ten years ago, when it appeared that the redox chains in plant and mammalian microsomes bore striking similarities. It is clear now that the overall organization as a membrane-bound flavoprotein/hemoprotein complex, and its function, i. e. production of a reactive ferryl-oxo group, are shared by the plant and the animal systems. However, with the exception of fatty acids, the substrates are different. Hence, it is not surprising that inhibition of plant monooxygenase by antibodies to animal cytochromes P-450 has not been observed, with the exception of the report by Bolwell and Dixon (1986) (see section 7.3.). Even the reductases, albeit of similar M_r and diplaying the same set of flavins, have different antigenic properties in plants and animals since no cross-reactivity was observed between plant reductase and antibodies against the animal enzyme, and vice versa. Furthermore, attempts to 'fish' genes coding for plant cytochrome P-450 with cDNA probes for animal enzymes have remained unsuccessful until now.

This situation is likely to change in the near future. Indeed, the incentive to study plant cytochrome P-450 at the molecular level is much higher now that it is recognized that genes coding for these enzymes are factors of resistance to pesticides (herbicide detoxification), to pathogenes (synthesis of phytoalexins) and have an important potential for biotechnology (synthesis of pharmacodynamic compounds). For example, it has recently been shown (Petersen et al., 1987) that microsomes from *Digitalis lanata* cells may be immobilized within alginate beads and retain digitoxin 12β-hydroxylase activity.

Acknowledgement

I thank numerous colleagues for allowing me to use unpublished data. I am most grateful to my co-workers for their help and patience.

This work was supported by grants from C.N.R.S. and I.N.R.A.

11. References

ADELE, P., D. REICHHART, J.-P. SALAÜN, I. BENVENISTE, and F. DURST, (1981), Plant Sci. Lett. **22**, 39–46.

AOYAMA, Y., Y. YOSHIDA, and R. SATO, (1984), J. Biol. Chem. **259**, 1661–1666.

BENVENISTE, I. and F. DURST, (1974), C. R. Hebd. Seances Acad. Sci. **278**, 1487–1490.

BENVENISTE, I., J.-P. SALAÜN, and F. DURST, (1977), Phytochemistry **16**, 69–73.

BENVENISTE, I., J.-P. SALAÜN, and F. DURST, (1978), Phytochemistry **17**, 359–363.

BENVENISTE, I., (1978), Thèse de Doctorat d'Etat, Université Louis Pasteur Strasbourg.

BENVENISTE, I., J.-P. SALAÜN, D. REICHHART, and F. DURST, (1980), in: Photoreceptors and Plant Development Antwerpen, 1979, Antwerpen University Press, pp. 297–304.

BENVENISTE, I., J.-P. SALAÜN, A. SIMON, D. REICHHART, and F. DURST, (1982a), Plant Physiol. **70**, 122–126.

BENVENISTE, I., B. GABRIAC, R. FONNÉ, D. REICHHART, J.-P. SALAÜN, A. SIMON, and F. DURST, (1982b), in: Cytochrome P-450, Biochemistry, Biophysics and Environnemental Implications, Kuopio, 1982, Elsevier Biomedical Press, pp. 201–208.

BENVENISTE, I., B. GABRIAC, and F. DURST, (1986), Biochem. J. **235**, 365–373.

BENVENISTE, I., A. LESOT, M.-P. HASENFRATZ, and F. DURST, (1989), Biochem. J. **259**, 847–853.

BILLET, E. E. and H. SMITH, (1980), Phytochemistry **19**, 1035–1041.

BOLWELL, G. P. and V. S. BUTT, (1983), Phytochemistry **22**, 37–45.

BOLWELL, G. P. and R. A. DIXON, (1986), Eur. J. Biochem. **159**, 163–169.

BONNEROT, C., A. GALLE, A. JOLLIOT, and J. C. KADER, (1985), Biochem. J. **226**, 331–334.

BOXENBAUM, H., (1983), Drug Metab. Rev. **14**, 1057–1097.

BRITSCH, L., W. HELLER, and H. GRISEBACH, (1981), Z. Naturforsh. **36**c, 742–750.

BÜCHE, T. and H. SANDERMANN, (1973), Arch. Biochem. Biophys. **158**, 445–447.

CABANNE, F., D. HUBY, P. GAILLARDON, R. SCALLA, and F. DURST, (1987), Pestic. Biochem. Physiol. **28**, 371–380.

CANIVENC, M.-C., B. CAGNAC, F. CABANNE, and R. SCALLA, (1989), Plant Physiology and Biochemistry **27**, 193–196.

CHEN, C.-M. and S. M. LEISNER, (1984), Plant Physiol. **75**, 442–446.

COLE, D. J. and W. J. OWEN, (1987), Plant Science **50**, 13–20.

COOLBAUGH, R. C. and T. C. MOORE, (1971), Phytochemistry **10**, 2401–2412.

COOLBAUGH, R. C. and R. HAMILTON, (1976), Plant Physiol. **57**, 243–248.

COOLBAUGH, R. C., S. S. HIRANO, and C. A. WEST, (1978), Plant Physiol. **62**, 571–576.

COOLBAUGH, R. C., (1984), J. Plant Growth Regul. **3**, 97–109.

CROTEAU, R. and P. E. KOLATTUKUTY, (1975), Arch. Biochem. Biophys. **170**, 61–72.

CZICHI, U. and H. KINDL, (1977), Planta **134**, 133–143.

DESJARDINS, A. E., D. E. MATTHEWS, and H. D. VANETTEN, (1984), Plant Physiol. **75**, 611–616.

DURST, F. and H. MOHR, (1966), Naturwiss. **53**, 531.

DURST, F., (1976), Planta **132**, 221–227.
FONNÉ, R., (1985), Thèse de Doctorat Université Louis Pasteur, Strasbourg, France.
FONNÉ, R., A. SIMON, J.-P. SALAÜN, and F. DURST, (1988), Plant Sci. **55**, 9–20.
FREAR, D. S., H. R. SWANSON, and F. S. TANAKA, (1969), Phytochemistry **8**, 2157 to 2169.
FUJITA, M., K. OBA, and I. URITANI, (1981), Agric. Biol. Chem. **45** 1911–1913.
FUJITA, M., K. OBA, and I. URITANI, (1982), Plant Physiol. **70**, 573–578.
FUJITA, M., (1985), Agric. Biol. Chem. **49** 3045–3047.
FUJITA, M. and T. ASAHI, (1985a), Plant Cell Physiol. **26**, 389–395.
FUJITA, M. and T. ASAHI, (1985b), Plant Cell Physiol. **26**, 397–405.
GABRIAC, B., I. BENVENISTE, and F. DURST, (1985), C. R. Hebd. Séances Acad. Sci. **301**, 753–758.
GADHER, P., E. I. MERCER, B. C. BALDWIN, and T. E. WIGGINS, (1983), Pestic. Biochem. Physiol. **19**, 1–10.
GAILLARDON, P., F. CABANNE, R. SCALLA, and F. DURST, (1985), Weed Res. **25**, 397 to 402.
GALLE, A., C. BONNEROT, A. JOLLIOT, and J. C. KADER, (1984), Biochem. Biophys. Res. Commun. **122**, 1201–1205.
GALLE, A. and J. C. KADER, (1986), J. Chromatogr. **366**, 422–426.
GEUNS, J. M. C., (1977), Biochem. Physiol. Planzen **171**, 435–447.
GILLARD, D. F. and D. C. WALTON, (1976), Plant Physiol. **58**, 790–795.
GONNEAU, M., B. PASQUETTE, F. CABANNE, R. SCALLA, and B. G. LOUGHMAN, (1987), Proc. Br. Crop Prot. Conf. **1**, 329–336.
GRAEBBE, J. E., (1987), Ann. Rev. Plant. Physlol. **38**, 419–465.
GRAND, C., (1984), FEBS, Lett. **169**, 7–11.
HAGMANN, M.-L., W. HELLER, and H. GRISEBACH, (1983), Eur. J. Biochem. **134**, 547 to 554.
HAGMANN, M.-L. and H. GRISEBACH, (1984), FEBS Lett. **175**, 199–202.
HAGMANN, M.-L., W. HELLER, and H. GRISEBACH (1984) Eur. J. Biochem. **142**, 127–131.
HAHLBROCK, K., K. H. KNOBLOCH, F. KREUZALER, J. R. M. POTTS and E. WELLMANN, (1976), Eur. J. Biochem. **61**, 199–206.
HAMERSKI, D. and U. MATERN, (1988a), Eur. J. Biochem. **171**, 369–375.
HAMERSKI, D. and U. MATERN, (1988b), FEBS Letters, **239**, 263–265.
HARRIS, N., (1986), Ann. Rev. Plant Physiol. **37**, 73–92.
HASSON, E. P. and C. A. WEST, (1976), Plant Physiol. **58**, 473–484.
HELLER, W. and T. KÜHNL, (1985), Arch. Biochem. Biophys. **241**, 453–460.
HENDRY, G., J. D. HOUGHTON and O. T. G. JONES, (1981), Biochem. J. **194**, 743–751.
HENDRY, G. and O. T. G. JONES, (1984), New Phytol. **96**, 153–159.
HENDRY, G., (1986), New Phytol. **102**, 239–247.
HIGHASHI, K., K. IKEUCHI, Y. KARASAKI, and M. OBARA, (1983), Biochem. Biophys. Res. Commun. **115**, 46–52.
HIGASHI, K., K. IKEUCHI, M. OBARA, Y. KARASAKI, H. HIRANO, S. GOTOH, and Y. KOGA, (1985), Agric. Biol. Chem. **49**, 2399–2405.
HIGASHI, K., (1985) GANN Monograph on Cancer Research **30**, 49–66.
HIGASHI, K., (1988), Mut. Res. (in press).
JOLLIE, D. R., S. G. SLIGAR, and M. SCHULER, (1987), Plant Physiol. **85**, 457–462.
KÄPPELI, O., (1986), Microbiol. Rev. **50**, 244–258.
KEMP, M. S. and J. C. CASELEY, (1987), Proc. Br. Crop Prot. Conf. **3**, 895–899.
KJELLBOM, P., C. LARSSON, P. ASKERLUND, C. SCHELIN, and S. WIDELL, (1985), Photochem. Photobiol. **42**, 779–783.
KOCHS, G. and H. GRISEBACH, (1986), Eur. J. Biochem. **155**, 311–318.

Kochs, G., and H. Grisebach, (1987), Z. Naturforsch. **42c**, 343–348.
Kolattukudy, P. E., (1977), in: Lipids and lipid polymers in higher plants, Springer-Verlag Berlin, Heidelberg, New-York (Tevini, M. and H. K. Lichtenthaler, eds.) pp. 271–292.
Larson, R. L. and J. B. Bussard, (1986), Plant Physiol. **80**, 483–486.
Lew, F. T. and C. A. West, (1971), Phytochemistry **10**, 2065–2076.
Long, A. R. and G. W. Chism, (1987), Biochem. Biophys. Res. Commun. **144**, 109–114.
Madyastha, K. M., J. E. Ridgway, J. G. Dwyer, and C. J. Coscia, (1977), J. Cell Biol. **72**, 302–313.
Madyastha, K. M. and C. J. Coscia, (1979), J. Biol. Chem. **254**, 2419–2427.
Madyastha, K. M., and N. Krishnamachary, (1986), Biochem. Biophys. Res. Commun. **136**, 570–576.
Markham, A., G. C. Hartman, and D. V. Parke, (1972), Biochem. J. **130**, 90P.
Matthews, D. E., E. J. Weiner, P. S. Matthews, and H. D. VanEtten, (1987), Plant Physiol. **83**, 365–370.
McFarlane, J., Madyastha, K. M., and C. J. Coscia, (1975), Biochem. Biophys. Res. Commun. **66**, 1263–1269.
Meehan, T. D. and C. J. Coscia, (1973), Biochem. Biophys. res. Commun. **53**, 1043 to 1048.
Murphy, P. J. and C. A. West, (1969), Arch. Biochem. Biophys. **133**, 395–407.
Obrenovic, S., (1986), Physiol. Plant. **67**, 626–629.
Ortiz de Montellano, P. R., (1986), in: Cytochrome P-450. Structure, Mechanism and Biochemistry. Plenum Press, New York (Ortiz de Montellano, ed.) pp. 273–314.
Osman, S., S. L. Sinden, K. Deahl, and R. Moreau, (1987), Phytochemistry **26**, 3163–3165.
Petersen, M. and H. U. Seitz, (1985), FEBS Lett. **188**, 11–14.
Petersen, M., (1986), Thesis, Eberhard-Karls-Universität, Tubingen (GFR).
Petersen, M., A. W. Alfermann, E. Reinhard, and H. U. Seitz, (1987), Plant Cell Rep. **6**, 200–203.
Potts, J. R. M., R. Weklych, and E. E. Conn, (1974), J. Biol. Chem. **25**, 5019 to 5026.
Rademacher, W., H. Fritsch, J. E. Graebe, H. Sauter, and J. Jung, (1988), Pestic. Sci. **21**, 241–252.
Rahier, A. and M. Taton, (1986), Biochem. Biophys. Res. Commun. **140**, 1064–1072.
Reichhart, D., J.-P. Salaün, I. Benveniste, and F. Durst, (1979), Arch. Biochem. Biophys. **196**, 301–303.
Reichhart, D., J.-P. Salaün, I. Benveniste, and F. Durst, (1980), Plant Physiol. **66**, 600–604.
Reichhart, D., A. Simon, F. Durst, J. M. Mathews, and P. R. Ortiz de Montellano, (1982), Arch. Biochem. Biophys. **216**, 522–529.
Rich, P. R. and D. S. Bendall, (1975), Eur. J. Biochem. **55**, 33–341.
Rich, P. R., R. Cammack and D. S. Bendall, (1975), Eur. J. Biochem. **59**, 281–286.
Rich, P. R. and C. J. Lamb, (1977), Eur. J. Biochem. **72**, 353–360.
Rivière, J.-L. and F. Cabanne, (1987), Biochimie **69**, 743–752.
Russell, D. W., (1971), J. Biol. Chem. **246**, 3870–3878.
Salaün, J.-P., I. Benveniste, D. Reichhart, and F. Durst, (1978), Eur. J. Biochem. **90**, 155–159.
Salaün, J.-P., I. Benveniste, D. Reichhart, and F. Durst, (1981), Eur. J. Biochem. **119**, 651–655.
Salaün, J.-P., I. Benveniste, R. Fonné, B. Gabriac, D. Reichhart, A. Simon, and F. Durst, (1982), Physiol. Vég. **20**, 613–621.

SALAÜN, J.-P., D. REICHHART, A. SIMON, F. DURST, N. O. REICH, and P. R. ORTIZ DE MONTELLANO, (1984), Arch. Biochem. Biophys. **232**, 1–7.

SALAÜN, J.-P., A. SIMON, and F. DURST, (1986), Lipids **21**, 776–779.

SALAÜN, J.-P., A. SIMON, F. DURST, N. O. REICH, and P. R. ORTIZ DE MONTELLANO, (1988), Arch. Biochem. Biophys. **260**, 540–545.

SAUNDERS, J. A., E. E. CONN, C. L. LIN, and M. SHIMADA, (1977), Plant Physiol. **60**, 629–634.

SCHMIDT, W. and W. L. BUTLER, (1976), Photochem. Photobiol. **24**, 77–80.

SIMON, A., (1982), Thèse de Doctorat de Troisième Cycle. Université Louis Pasteur, Strasbourg, France.

SIMON, A., D. REICHHART, F. DURST, and P. R. ORTIZ DE MONTELLANO, (1984), Xenobiotica **14s**, 83–84.

SIMON, A., (1987), Thèse de Doctorat d'Etat, Université Louis Pasteur, Strasbourg, France.

SMITH, P. G., M. A. VENIS, and M. A. HALL, (1985), Planta **163**, 97–104.

SOLIDAY, C. L. and P. E. KOLATTUKUTY, (1977), Plant Physiol. **59**, 1116–1121.

SOLIDAY, C. L. and P. E. KOLATTUKUTY, (1978), Arch. Biochem. Biophys. **188**, 338–347.

SPATZ, L. and P. STRITTMATTER, (1974), J. Biol. Chem. **248**, 793–799.

SPITZBERG, V., C. J. COSCIA, and R. KRUEGER, (1981), Plant Cell Rep. **1**, 43–48.

STAFFORD, H. A. and S. DRESSLER, (1972), Plant Physiol. **49**, 590–595.

SWAIN, T., (1977), Ann. Rev. Plant Physiol. **28**, 479–501.

TANAKA, Y., M. KOJIMA, and I. URITANI, (1974), Plant Cell Physiol. **15**, 843–854.

TATON, M., P. ULLMANN, P. BENVENISTE, and A. RAHIER, (1988), Pestic. Biochem. Physiol. **30**, 178–189.

VANDEN BOSSCHE, H., P. MARICHAL, J. GORRENS, D. BELLENS, H. VERHOEVEN, M.-C. COENE, W. LAUWERS, and P. A. J. JANSSEN, (1988), Pestic. Sci. **21**, 289–306.

VIOLON, C. J. I. and A. A. VERCRUYSSE, (1985), Phytochemistry **24**, 2205–2209.

WALTON, T. J. and P. E. KOLATTUKUDY, (1972), Biochem. Biophys. Res. Commun. **46**, 16–21.

WENDORF, H. and U. MATERN, (1986), Eur. J. Biochem. **161**, 391–398.

WERCK-REICHHART, D., (1985), Thèse de Doctorat d'Etat, Université Louis Pasteur, Strasbourg, France.

WERCK-REICHHART, D., O. T. G. JONES, and F. DURST, (1988), Biochem. J. **249**, 473–480.

WEST, C. A., (1980), in: The Biochemistry of Plants, Vol. **2**, (STUMPF and CONN, eds.) Academic Press Inc. New-York, pp. 317–363.

YAHIEL, V., M. G. COTTE-MARTINON, and G. DUCET, (1974), Phytochemistry **13**, 1649–1651.

YOUNG, O. and H. BEEVERS, (1976), Phytochemistry **15**, 379–385.

List of Authors

Y. Aoyama
Mukogawa Women's University
Faculty of Pharmaceutical Sciences
Nishinomiya

Hyogo 663
Japan

O. Asberger
Karl-Marx-University
Division of Biosciences
Talstraße 33

7010 Leipzig
FRG

K. Breskvar
Medical Faculty
Institute of Biochemistry
Vrazov Trg 2

61000 Ljubljana
Yugoslavia

F. Durst
CNRS — Institut de Biologie Moléculaire des Plantes
Departement d'Enzymologie Cellulaire et Moléculaire
Institut de Botanique
28 Rue Goethe

67083 Strasbourg Cedex
France

I. C. Gunsalus
University of Illinois
Department of Biochemistry
415 Roger Adams Laboratory
1209 West California Street

Urbana Illinois 61801
USA

T. Hudnik-Plevnik
Medical Faculty
Institute of Biochemistry
Vrazov Trg 2

61000 Ljubljana
Yugoslavia

E. Kärgel
Central Institute of
Molecular Biology
Robert-Rössle-Straße 10

1115 Berlin
FRG

P. Kleber
Karl-Marx-University
Division of Biosciences
Talstraße 33

7010 Leipzig
FRG

S. A. Martinis
University of Illinois
Department of Biochemistry
415 Roger Adams Laboratory
1209 West California Street

Urbana Illinois 61801
USA

H.-G. Müller
Central Institute of
Molecular Biology
Robert-Rössle-Straße 10

1115 Berlin
FRG

H. Ohkawa
Takarazuka Research Center
Biotechnology Laboratory
2-1, 4-Chome, Takatsukasa
Takarazuka-Shi

Hyogo-Ken 665
Japan

J. D. Ropp
University of Illinois
Department of Biochemistry
415 Roger Adams Laboratory
1209 West California Street

Urbana Illinois 61801
USA

W.-H. Schunck
Central Institute of
Molecular Biology
Robert-Rössle-Straße 10

1115 Berlin
FRG

S. G. Sligar
University of Illinois
Department of Biochemistry
415 Roger Adams Laboratory
1209 West California Street

Urbana Illinois 61801
USA

Y. Yabusaki
Takarazuka Research Center
Biotechnology Laboratory
2-1, 4-Chome, Takatsukasa
Takarazuka-Shi

Hyogo-Ken 665
Japan

Y. Yoshida
Mukogawa Women's University
Faculty of Pharmaceutical Sciences
Nishinomiya

Hyogo 663
Japan

Subject Index